Proceedings of the ASME

INTERNATIONAL MECHANICAL ENGINEERING CONGRESS AND EXPOSITION

– 2010 –

VOLUME 1

ADVANCES IN AEROSPACE TECHNOLOGY

presented at

ASME 2010 INTERNATIONAL MECHANICAL ENGINEERING CONGRESS AND EXPOSITION
NOVEMBER 12-18, 2010
VANCOUVER, BC, CANADA

A S M E

Three Park Avenue * New York, N.Y. 10016

ISBN No. 978-0-7918-4425-0

WELCOME FROM THE CHAIRS

Welcome to the 2010 ASME International Mechanical Engineering Congress and Exhibition (IMECE) at the Vancouver Convention Center. The IMECE represents the one major conference where all of the broad disciplines within Mechanical Engineering come together under one roof. This provides a unique opportunity for colleagues to meet and discuss a range of challenges - from technical problems requiring multi- and inter-disciplinary expertise to how we should be educating our future mechanical engineers. Representative of this type of interactive discussion, we are very pleased to have the participation of the National Science Foundation (NSF) Engineering Directorate, represented by John McGrath and Steve McKnight, the Division Directors for Chemical, Biological, Environmental, and Transport Systems (CBET) and Civil, Mechanical, Manufacturing, and Innovation (CMMI), respectively, and several of their program directors. The NSF highlight sessions will include opportunities for participants to provide input into research funding directions and learn about current efforts to increase academia-industry-government partnerships across multiple agencies.

Overall, this year's IMECE includes 18 multi-disciplinary tracks with over 2500 slide and poster presentations. Posters will be available for viewing when the exhibit hall opens each morning, through lunch, and with the presenters available for discussion during the 2:45-4:45pm poster session. There is an increased emphasis on new emerging technologies in all tracks. The Micro & Nanotechnology Society-wide Forum will bring together ASME members and others from different disciplines/divisions with a focus on micro and nanotechnology. The Energy Systems Track will take a look at how emerging technologies and research can provide solutions to today's global energy problems. A particular highlight of this year's Congress is the Energy Grand Challenge series. The goal of these three sessions is to bring renowned speakers from industry and government agencies to provide clear and comprehensive insight on the key technical challenges that need solutions from our community.

There will also be many events hosted by the ASME divisions and committees. Your participation in ASME divisions and committees is the pathway to helping us continue to improve the Society and future conferences such as IMECE. Enjoy the tours of local industries and sites, including several that show both the behind-the-scenes engineering involved in the 2010 Winter Olympics and the beauty of Vancouver. We hope that you will find the 2010 Congress to be a rewarding experience, both professionally and socially.

An event of this magnitude requires the hard work and dedication of many people, most working behind the scenes and unfortunately too numerous to mention individually here. While in Vancouver, please take the time to thank the staff and volunteers who make IMECE run. In brief, I want to thank the organizational support of our steering committee, our operations group including our local University of British Columbia volunteers, and everyone involved at ASME. A special thank you goes out to our track chairs, topic and symposium organizers, and session chairs and reviewers, all who have ensured that this conference has papers and presentations of integrity, substance and quality. Finally, we recognize the contributions of all the authors, plenary and keynote speakers who have helped to make this conference a success.

I look forward to meeting many of you in the halls of the convention center, at the special events, presentations and in technical sessions. Have a productive and enjoyable Congress and we hope to see you in Denver next year.

Sincerely,
Julie Chen
2010 Congress Technical Program Chair

2010 Congress Chairs

2010 Congress Steering Committee Chair

Urmila Ghia
University of Cincinnati
Email: Urmila.Ghia@uc.edu

2010 Congress Steering Committee Vice Chair

Greg Kowalski
Northeastern University
Email: gkowal@coe.neu.edu

2010 General Conference Chair

Pradeep Lall
Auburn University
Email: lall@eng.auburn.edu

2010 Technical Program Chair

Julie Chen
University of Massachusetts Lowell
Email: julie_chen@uml.edu

Track Organizers

Track 1 Advances in Aerospace Technology
Track Organizer: M. Erol Ulucakli, *Lafayette College*
Email: ulucakli@lafayette.edu

Track 2 Biomedical and Biotechnology Engineering
Track Organizer: A. M. Al-Jumaily, *Institute of Biomedical Technologies*
Email: ahmed.aljumaily@aut.ac.nz
Track Co-Organizer: Sara Wilson, *University of Kansas*
Email: sewilson@ku.edu
Track Co-Organizer: M. Erol Ulucakli, *Lafayette College*
Email: ulucakli@lafayette.edu

Track 4 Design and Manufacturing
Track Organizer: Rama P. Koganti, *Ford Motor Company*
Email: ramakoganti@yahoo.com
Track Co-Organizer: Endel Iarve, *University of Dayton Research Institute*
Email: Endel.Iarve@WPAFB.AF.MIL
Track Co-Organizer: Sanjeev K. Khanna, *University of Missouri - Columbia*
Email: khannas@missouri.edu
Track Co-Organizer: Kimberly Cook-Chennault, *Rutgers, The State University of New Jersey*
Email: kacook@jove.rutgers.edu
Track Co-Organizer: Shuichi Fukuda, *Stanford University*
Email: shufukuda@cdr.stanford.edu
Track Co-Organizer: M. S. Fofana, *Worcester Polytechnic Institute*
Email: msfofana@wpi.edu

Track 5 Electronics and Photonics
Track Organizer: Kaustubh Nagarkar, *GE Global Research Center*
Email: nagarkar@research.ge.com
Track Co-Organizer: Cemal Basaran, *State University of New York at Buffalo*
Email: cjb@buffalo.edu

Track 6 NanoEngineering for Energy
Track Organizer: Vikas Prakash, *Case Western Reserve University*
Email: vikas.prakash@case.edu
Track Co-Organizer: Hornsen (HS) Tzou, *Zhejiang University*
Email: hstzou@engr.uky.edu
Track Co-Organizer: H. Eliot Fang, *Sandia National Laboratories*
Email: hefang@sandia.gov
Track Co-Organizer: Thomas George Thundat
Track Co-Organizer: Ajit Roy, *AFRL*
Email: ajit.roy@wpafb.af.mil
Track Co-Organizer: Bruce LaMattina
Track Co-Organizer: Xi Chen, *Columbia University*
Email: xichen@columbia.edu

Track Co-Organizer: Ronggui Yang
Track Co-Organizer: Tony Huang, *The Pennsylvania State University*
Email: junhuang@psu.edu
Track Co-Organizer: Vikas Tomar, *Purdue University*
Email: tomar@purdue.edu
Track Co-Organizer: Hanchen Huang, *University of Connecticut*
Email: hanchen@uconn.edu
Track Co-Organizer: Liming Dai

Track 7 Energy Systems: Analysis, Thermodynamics and Sustainability
Track Organizer: Wilson K. S. Chiu, *University of Connecticut*
Email: wchiu@engr.uconn.edu
Track Co-Organizer: Vittorio Verda, *Politecnico di Torino*
Email: vittorio.verda@polito.it
Track Co-Organizer: Christian Sattler, *German Aerospace Center - DLR*
Email: Christian.Sattler@dlr.de
Track Co-Organizer: Martina Neises, *German Aerospace Center DLR E.V.*
Email: martina.neises@dlr.de

Track 8 Engineering Education and Professional Development
Track Organizer: Ahmad Fakheri, *Bradley University*
Email: ahmad@bradley.edu
Track Co-Organizer: M. S. Fofana, *Worcester Polytechnic Institute*
Email: msfofana@wpi.edu
Track Co-Organizer: Sanjeev K. Khanna, *University of Missouri - Columbia*
Email: khannas@missouri.edu
Track Co-Organizer: Mohammad Naraghi, *Manhattan College*
Email: mohammad.naraghi@manhattan.edu

Track 9 Engineering to Address Climate Change
Track Organizer: Andy Miller, *US EPA*
Email: miller.andy@epa.gov

Track 10 Fluid Flow, Heat Transfer and Thermal Systems
Track Organizer: David Halt, *New PAX, Inc*
Email: dwhalt@gmail.com
Track Co-Organizer: S. A. Sherif, *University of Florida*
Email: sasherif@ufl.edu

Track 11 Dynamic Systems and Control
Track Organizer: Albert Luo, *Southern Illinois University Edwardsville*
Email: aluo@siue.edu
Track Co-Organizer: Liming Dai, *University of Regina*
Email: Liming.Dai@uregina.ca

Track 12 Mechanics of Solids, Structures and Fluids
Track Organizer: Kenneth Liechti, *University of Texas at Austin*
Email: kml@mail.utexas.edu
Track Co-Organizer: Larry Bergman

Track Co-Organizer: Javid Bayandor, *Virginia Tech*
Email: Bayandor@vt.edu
Track Co-Organizer: Tayfun Tezduyar, *Rice University*
Email: tezduyar@gmail.com

Track 13 Micro and Nano Systems
Track Organizer: EH Yang, *Stevens Institute of Technology*
Email: eyang@stevens.edu
Track Co-Organizer: Tony Huang, *The Pennsylvania State University*
Email: junhuang@psu.edu

Track 14 New Developments in Simulation Methods and Software for Engineering Applications
Track Organizer: Jayathi Murthy, *Purdue*
Email: jmurthy@ecn.purdue.edu
Track Co-Organizer: Partha Mukherjee, *Oak Ridge National Laboratory*
Email: mukherjeepp@ornl.gov
Track Co-Organizer: Xiulin Ruan, *Purdue University*
Email: ruan@purdue.edu

Track 15 Processing and Engineering Applications of Novel Materials
Track Organizer: Vikas Prakash, *Case Western Reserve University*
Email: vikas.prakash@case.edu

Track Co-Organizer: Karl Jacob

Track 17 Safety Engineering, Risk Analysis and Reliability Methods
Track Organizer: W.R. (Bill) Byrd, *RCP Inc.*
Email: wrb@rcp.com
Track Co-Organizer: John Wiechel, *SEA, Limited*
Email: jwiechel@sealimited.com
Track Co-Organizer: Clarence Ross

Track 19 Transportation Systems
Track Organizer: Khaled Zaazaa, *Ensco, Inc.*
Email: Zaazaa.Khaled@ensco.com

Track 20 Sound, Vibration and Design
Track Organizer: Robert (Bob) Tomko, *Bechtel Marine Propulsion Corporation*
Email: tomkorp@yahoo.com
Track Co-Organizer: Sue Sung, *General Motors Research & Development Ctr.*
Email: shsung@aol.com
Track Co-Organizer: Brent Paul, *Alion Science and Technology*
Email: bpaul@alionscience.com
Track Co-Organizer: Stephen Hambric, *ARL/Penn State*
Email: sah19@arl.psu.edu
Track Co-Organizer: Adam Smith, *NIOSH*
Email: ASmith9@cdc.gov

Track Co-Organizer: Liang-Wu Cai, *Kansas State University*
Email: cai@ksu.edu
Track Co-Organizer: Jeffrey S Vipperman, *University of Pittsburgh*
Email: jsv@pitt.edu
Track Co-Organizer: Lonny Thompson, *Clemson University*
Email: lonny@clemson.edu

Track 21 Posters

Track Organizer: Assimina Pelegri, *Rutgers - The State University of New Jersey*
Email: pelegri@jove.rutgers.edu

CONTENTS

ADVANCES IN AEROSPACE TECHNOLOGY

ADVANCED TOPICS IN AEROSPACE MATERIALS AND STRUCTURES

ADVANCES IN AEROSPACE TECHNOLOGY: GENERAL PAPERS

Proceedings of the ASME 2010 International Mechanical Engineering Congress & Exposition
IMECE2010
November 12-18, 2010, Vancouver, British Columbia, Canada

IMECE2010-37653

STRUCTURAL HEALTH MONITORING USING PZT TRANSDUCER NETWORK AND LAMB WAVES

J. Q. Mou, L. Martua, Y. Q. Yu, Z. M. He, C. L. Du, J. L. Zhang, E. H. Ong

Data Storage Institute, 5 Engineering Drive 1, Singapore 117608
Mou_jianqiang@dsi.a-star.edu.sg

ABSTRACT

Structure health monitoring (SHM) using ultrasonic waves is an emerging technology that can be applied to real-time detect, locate and quantify the structural damages in aircraft structures. In this paper, the monitoring of crack growth at rivet holes in an aluminum test plate using a PZT transducer network and Lamb waves is investigated. The thin disc PZT transducers surface mounted at the test plate are used as actuators to transmit the windowed sinewave bursts and sensors to receive the ultrasonic Lamb waves. The symmetrical S_0 mode and anti-symmetrical A_0 mode of the Lamb waves in the structures are studied with correlated theoretical, experimental and numerical analysis. The optimal excitation frequency is determined for the test plate. Finite element method (FEM) numerical models for simulations of the wave propagations and interactions with the holes and cracks in the plate are developed and verified with the experimental results. The wave responses modes and characteristics for detection of the cracks at the rivet holes are analyzed. The Lamb wave signals in the PZT transducer network are processed with the short time Fourier transform (*STFT*). It is demonstrated that the time of flight and the energy transmission ratio of the S_0 wave are sensitive to the cracks in the structure.

Keyword: SHM, PZT, ultrasonic, Lamb wave, network, FEM

INTRODUCTION

In recent years, structure health monitoring (SHM) using ultrasonic Lamb waves excited and detected by PZT transducer network has been extensively studied because of its sensitivity to detect defects inside the thin plate-like aircraft structures. S. Grondle, C. Delebarre, J. Assaad *et al* [1] investigated the feasibility of using the Lamb waves and surface mounted PZT transducers for the fatigue crack monitoring of riveted aluminum joints by experimental studies. J.B. Ihn and F. K. Chang [2, 3] studied the ultrasonic guided Lamb waves signal generation, signal processing and damage interpretation for monitoring of fatigue crack growth using a built-in piezoelectric transducer network, and applied the technique to monitor fatigue crack growth at riveted fuselage joints and a cracked metallic plate repaired with a bonded composite patch. S. Park, C. B. Yun and Y. Roh [4] reported a surface mounted PZT network system for identification of cracks at a welded zone of a steel truss member with the wavelet coefficients corresponding to the anti-symmetric Lamb wave mode. D. Wang, L. Ye *et al* [5] evaluated a diagnostic algorithm based on the probability of the presence of the damage in the form of a through-thickness hole in a composite panel using an active PZT sensor network and transmission of the symmetric Lamb wave mode.

In this paper, the monitoring of crack growth at rivet holes in an aluminum test plate using a PZT transducer network and Lamb waves is further investigated by correlated theoretical, experimental and numerical analysis. The Lamb wave modes and the dispersion of the velocities of the waves are studied with the theoretical analysis. Experimental system is set up and tests are carried out for the test plates with the PZT transducer network and Lamb waves. The explicit LS-DYNA FEM numerical models for the test plates are developed, and verified with the experimental results. The interaction of Lamb waves with the rivet holes and the cracks are analyzed. Finally, signal processing and damage index for the crack growth monitoring using the PZT transducer network and Lamb waves are discussed.

LAMB WAVE PROPAGATION

Lamb wave is a type of ultrasonic wave that propagates along thin structure bounded by two free surfaces. The Lamb wave was named after H. Lamb who firstly introduced the theory in 1917 [6]. Lamb waves are able to propagate over a relatively long distance and are particularly useful for detecting damages such as cracks and delamination in plate and shell structures. Lamb waves are dispersive, meaning that the speeds of the waves vary with frequency. The Lamb wave modes can be categorized into symmetric modes S_i and anti-symmetric modes A_i, based on whether the out-of-plane displacements on either side of the plate are in or out of phase. For isotropic materials, the dispersion equation for symmetric modes can be written as [6]:

$$\frac{\tan(0.5qh)}{\tan(0.5ph)} = -\frac{4k^2qp}{(q^2-k^2)^2} \qquad (1)$$

and for anti-symmetric modes formulated as [6]:

$$\frac{\tan(0.5qh)}{\tan(0.5ph)} = -\frac{(q^2-k^2)^2}{4k^2qp} \qquad (2)$$

$$p^2 = -\frac{\omega^2}{c_L^2} - k^2, \quad q^2 = -\frac{\omega^2}{c_T^2} - k^2,$$

and $k = \dfrac{\omega}{c_p}$

where h is the plate thickness and k is the wave number, and c_L, c_T, c_P, ω are the bulk longitudinal and bulk transverse wave speeds, phase velocity and wave circular frequency respectively.

When the wavelengths are large in relation to the thickness of the plate, the fundamental symmetric S_0 mode and the fundamental anti-symmetric A_0 mode are equivalent to the extensional and flexural waves respectively. The fundamental S_0 mode and A_0 mode have been widely used for structural health monitoring due to their sensitivity to damages in plate and shell structures [3, 7].

The above equations are transcendental and have to be solved in an iterative way to obtain the phase velocity. Numerical codes have been developed with software Matlab to solve the Lamb wave dispersion equations. For a given material property and plate thickness, the phase velocity of Lamb wave is solved, which is a function of frequency. The group velocity is then calculated and the group velocity dispersion curves are produced, which is critical for selection of the specific excitation frequency in damage detection using Lamb waves. The group velocity dispersion curves for an ideal aluminum plate with thickness 1.6 mm that is equal to the thickness of the test specimen are shown in Figure 1. It is found that the fundamental modes are the symmetric S_0 mode and anti-symmetric A_0 mode, and the S_0 mode has a substantially faster group velocity and hence would arrive at the sensors before A_0 mode. To isolate the S_0 mode and A_0 mode from other higher order modes, the excitation frequency should be kept below 1 MHz. It is also noticed that the S_0 modes are less dispersive in the frequency range from 200 kHz to 600 kHz.

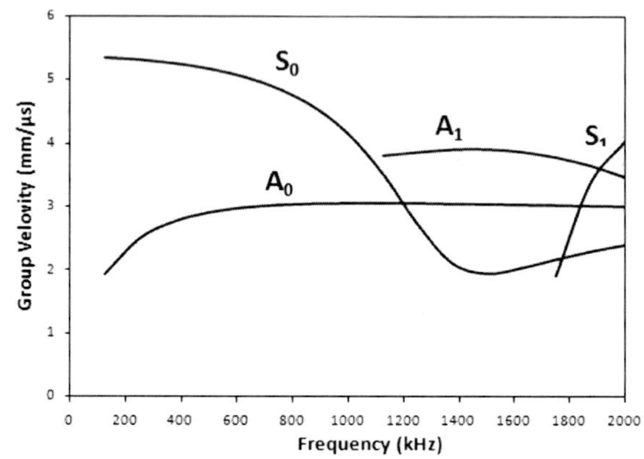

Fig. 1 Group velocity dispersion curves of 1.6 mm thick aluminum plate.

EXPERIMENTAL SETUP AND TESTING

An aluminum plate measuring 500 mm long by 500 mm wide by 1.6 mm thick as shown in Figure 2 was used as the test specimen.

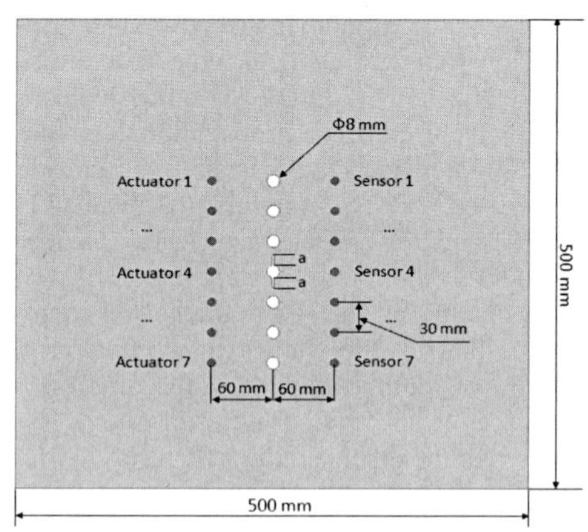

Fig. 2 The test specimen and PZT transducer network.

A linear array of seven rivet holes (diameter 8mm) spaced 30 mm apart were fabricated along the center line of the plate. Through-thickness 0.5 mm wide cuts starting from the holes were created in the plate using an electrical discharge machining (EDM) machine to simulate the edge cracks at the holes. The transducer network was configured by fourteen piezoelectric transducers surface mounted using epoxy adhesives at the plate in two parallel linear arrays corresponding to the holes. The piezoelectric transducers used in the study are piezoelectric lead zirconate titanate (PZT) ceramic discs with a diameter of 6.35 mm and a thickness of 0.254 mm, which are thickness poled and operated in a radial extension mode. The array of seven PZT transducers to transmit the windowed sinewave bursts are worked as actuators and the other array of seven PZT transducers to receive the ultrasonic Lamb waves are functioned as sensors. The actuators are activated in sequence to scan the network enclosed area to form totally forty nine cross sensing paths from actuator i to sensor j denoted as Pi-j.

The five-cycle Hanning-windowed sinewave bursts are used as excitation signals. The experimental setup consists of a Tektronix Arbitrary Wave Generator (AWG) 2041, amplifiers, junction connectors, and a programmable GAGE Compuscope 82G card connected with a personal computer (PC) running Labwindows/CVI. The data acquisition subsystem includes a PCI interface controlled with the PC running Labwindows/CVI. The actuation and response signals are amplified by using high bandwidth amplifiers. Matlab software is installed in the PC for experimental data processing. The schematic diagram of the experimental setup is shown in Figure 3.

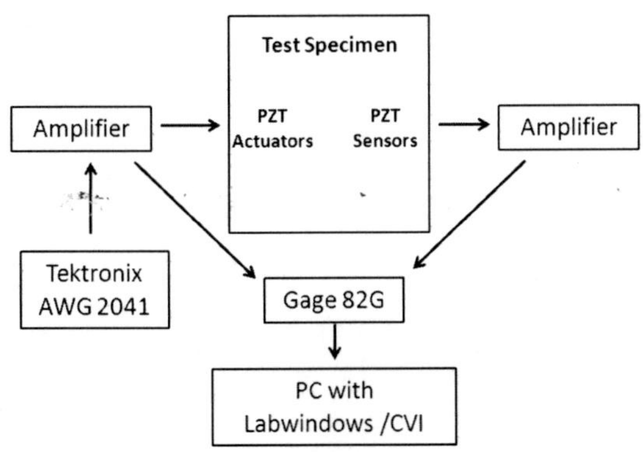

Fig. 3 The schematic of experimental setup.

To determine the optimal excitation frequency and placement of the transducers, the intact test specimen with the linear array of holes but no cracks is tested. The five-cycle Hanning-windowed sinewave bursts with the central frequency varied from 200 kHz to 600 kHz are applied to the actuators, and the lamb waves received by the sensors are recorded. The sampling rate is set at 20 MHz. The 60 mm horizontal spacing from the holes to actuators and sensors respectively are determined to separate the S_0 waves from the A_0 waves and eliminate the influence of the electromagnetic interference (EMI). The signals recorded in the tests for the path P4-4 are shown in Figure 4. The dot lines represent the input signals of the actuator 4 and solid lines represent the signals received by sensor 4. Distinct pulse bursts are visible in the signals received by the sensor 4. With reference to the group velocity dispersion curves for an ideal 1.6mm thick aluminum plate shown in Figure 1, the Lamb wave modes of the received signal are identified. As Figure 4 shows, after marginal electromagnetic interference, the first-arrival pulse burst of the sensor signal is identified as the symmetric S_0 mode and the second-arrival pulse burst as the anti-symmetric A_0 mode, followed by the pulse bursts of reflected S_0 mode and A_0 mode from boundary. The S_0 mode signals detected by sensor have much higher peak amplitude than that of A_0 mode signals. However, both the waveforms of S_0 mode and A_0 mode received by the sensor are distorted, due to the dispersive nature of the Lamb wave modes and the interference from the holes in the wave propagation process. In particular, it is found that the S_0 mode is large in amplitude and small in waveform distortion at the actuation central frequency 400 kHz.

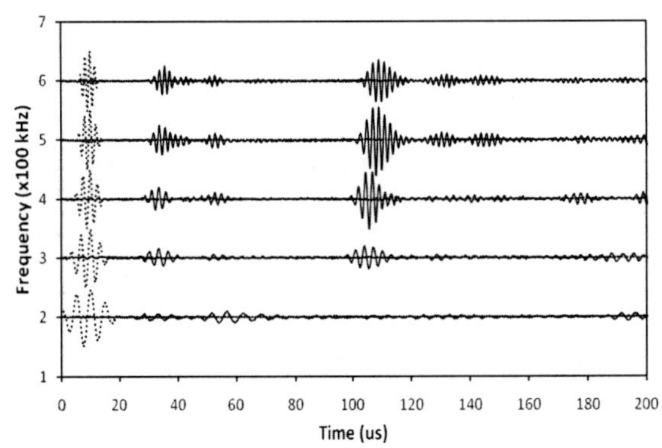

Fig. 4 The Lamb wave signals recorded at different central frequency.

Experiments for detection of cracks in the plates using the PZT transducer network are then carried out. The test specimens with artificial edge cracks at the holes with different crack length and locations are tested. The Lamb wave signals from the selected sensing paths P4-4 and P4-1 for a 10 mm edge crack at the hole 4 along the center line of the plate are displayed in Figure 5. As expected, the fundamental symmetric mode signals S_0 from the path P4-4 shows higher sensitivity to the crack. With the presence of the crack, noticeable increase in the arrival time and decrease amplitude of the transmitted S_0 signals from the path P4-4 were observed. This is attributed to the fact that since the sensing path P4-4 is normal to and cross through the crack, the transmission of S_0 waves along this path is significantly impaired by crack. In contrast, signals from sensing path P4-1 are less sensitive to the presence of the cracks at hole 4.

Since this sensing path is located relatively distant from the cracks, the lamb waves are only slightly impaired by the cracks. Moreover, it is observed that after the transmitted S_0 signals, the signals induced from the holes and cracks are followed, which are combined with the A_0 signals. Therefore, it can be concluded that for crack detection at such a structure with rivet holes, the diagnostic algorithm should be based on the extracted characters of the transmitted S_0 signals.

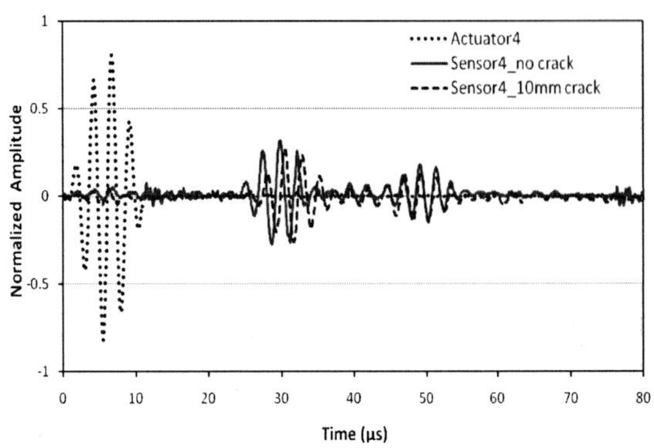

(a) The Lamb wave signals from path 4-4.

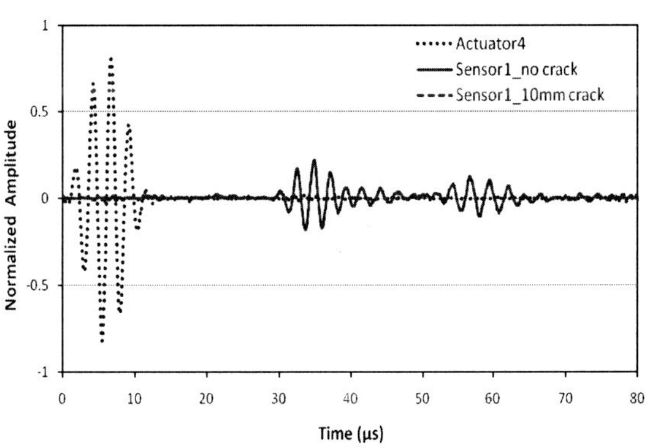

(b) The Lamb wave signals from path 4-1.

Fig. 5 Lamb wave signals from different paths for crack detections.

NUMERICAL MODELING AND SIMULATION

Numerical simulations of the propagation process of Lamb waves were performed using the explicit finite element analysis software, LS-DYNA. The ANSYS preprocessor was used to create the three-dimensional FEM models in which eight-node brick solid elements were employed for the 500 × 500 × 1.6 mm^3 aluminum specimens. The dynamic simulations of the FEM models were accomplished using the explicit LS-DYNA solver, and the results were post processed with the LS-Prepost. The typical element size is less than 1/20 of the smallest wavelength of the Lamb wave modes excited. Since contribution to the mass and stiffness of the specimen structure due to the PZT transducer discs is negligible, the PZT transducer discs are not explicitly modeled but represented with the circular mounting area. The structural defects, such as the through-thickness cracks starting from the holes, are modeled by removing the equivalent elements [8]. To obtain better resolution for simulation of the wave propagation through the regions of holes, cracks and mounting areas of the PZT transducers, the finite element mesh was locally refined. To model the excitation of the PZT actuator, an in-plane uniform radial force proportional to the actuator experimental voltage input signals was applied to all the nodes along the circular boundary of the PZT actuator mounting area. In the sensor model, the output voltage of the PZT sensor was derived from a combination of the strain from the nodes at the circular boundary of the PZT sensor mounting area, and the magnitude was calibrated with the experimental data. The sampling rate to collect the simulated Lamb wave signals was set at 20 MHz, same as that of experiment.

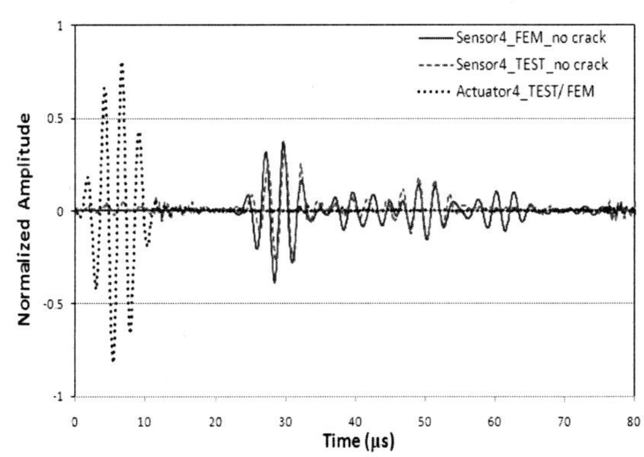

(a) The intact specimen without crack.

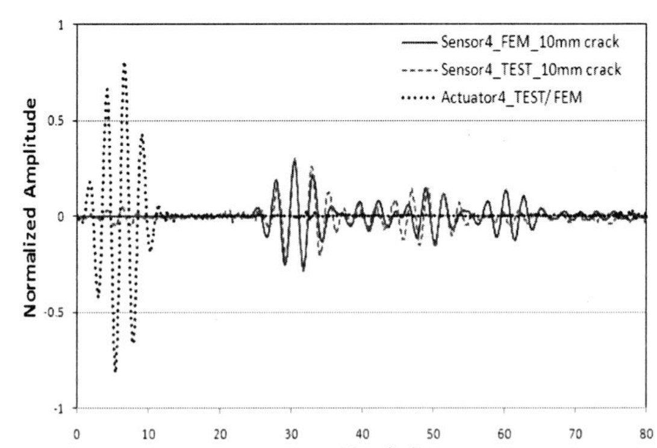

(b) The specimen with 10mm crack at hole 4 along the center line.

Fig. 6 Simulation and experimental wave signals for the path P4-4.

The numerical simulation results have good agreements with the experimental results for the signals of the Lamb waves. The comparison between the simulation results and experimental results for signals from sensing path P4-4 is shown in Figure 6. For the S_0 mode, the numerical simulation results match the experimental results very well in terms of both the arrival time and amplitude of the peaks.

The simulation of the wave propagation through the holes in the intact specimen is shown in Figure 7. Figure 8 and Figure 9 show the simulation of wave propagation in the specimen with a 10 mm crack at the hole 4 and hole 3 along the center line respectively. The propagations of the two fundamental Lamb wave modes display clearly different characteristics. The faster wave is S_0 mode which shows particle displacements that are mostly in the radial direction of the actuator, and symmetric about the mid-plane of the plate. The slower wave is the A_0 mode, which shows particle displacements that are mostly in the out-of plane direction and are anti-symmetric about the mid-plane of the plate. Since the S_0 mode propagates faster than the A_0 mode, it reaches the holes and cracks earlier. It can be seen that after the interaction with the holes and cracks, the S_0 mode is partially transmitted, diffracted, and reflected. The diffracted and reflected S_0 modes are converted into different induced wave modes. The interaction of later A_0 mode with the holes and cracks is more complicated due to its combination with the wave modes induced by the S_0 reflection and diffraction.

With the presence of the crack, decay of the transmitted S_0 mode after the holes and cracks in the relevant paths can be observed. Comparison between the S_0 and A_0 mode waves demonstrates that the S_0 mode propagates through the holes and cracks with higher detectable attenuation than the A_0 mode, which is consistent with the experimental results.

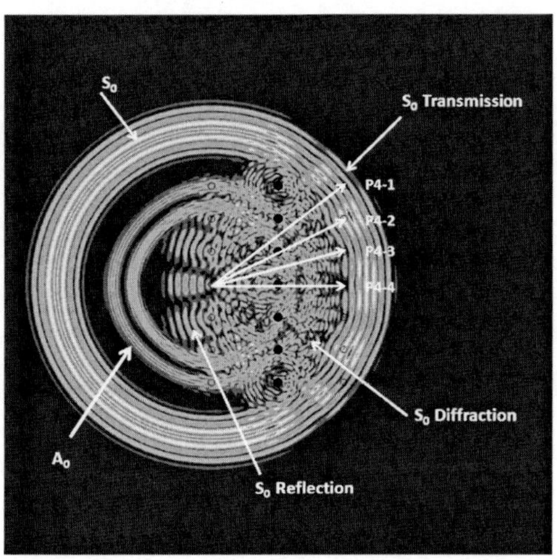

Fig. 7 Wave propagation in the intact specimen without crack.

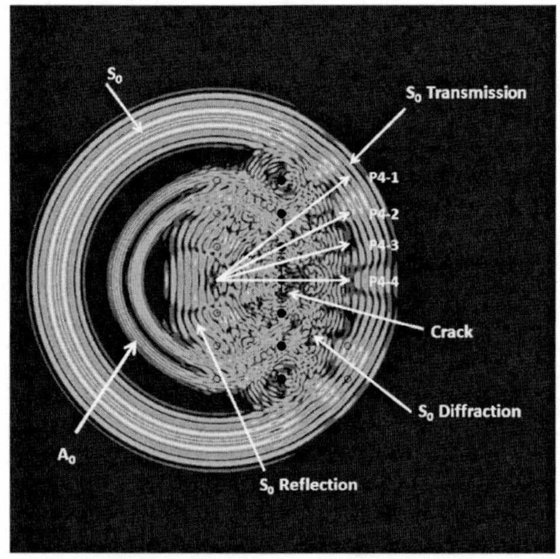

Fig. 8 Wave propagation in the specimen with crack at hole 4 along the center line.

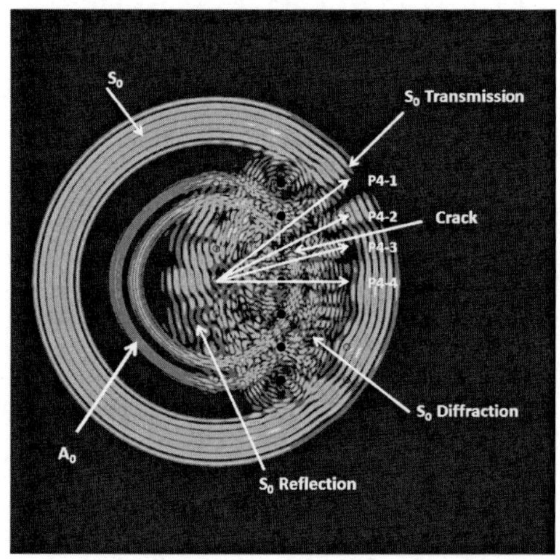

Fig.9 Wave propagation in the specimen with crack at hole 3 along the center line.

SINGNAL PROCESSING AND DAMAGE INDEX

With the LS-DYNA numerical models verified by experiments, the Lamb wave propagation and the wave signals were analyzed for the specimen with different crack dimensions and locations. The time of flight (*TOF*) and the energy transmission ratio of the S_0 wave mode processed from the wave signals in the PZT transducer network are used for diagnosis of cracks in the specimen. The short time Fourier transform (*STFT*) is used in the signal processing to calculate the *TOF* and the transmission energy of S_0 wave mode. The short time Fourier transform of a signal $s(t)$ is defined as [2]:

$$S(\omega,t) = \frac{1}{2\pi}\int_{-\infty}^{\infty} e^{-i\omega\tau} s(\tau)h(\tau - t)d\tau \quad (3)$$

Since the window function $h(t)$ has a short time duration, by moving $h(t)$ with Fourier integrals, the signal's local frequency properties evolving over time can be revealed. The Hanning window is used as a window function. The input signal of an actuator and the corresponding sensor signal, as well as the processed *STFT* spectral amplitude at the central frequency 400 kHz representing the envelope of the time signal are illustrated in the Figure 10.

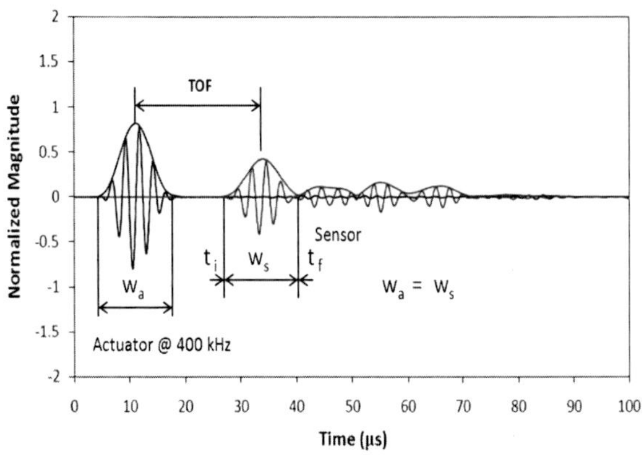

Fig. 10 The Lamb wave signals and the *STFT* envelope

The *TOF* of the transmitted S_0 signal for an actuator sensor path can be obtained by taking the time difference between peak of the first-arrival envelope of the sensor and peak of the first-arrival envelope of the actuator, as shown in Figure 10. Moreover, the energy transmission ratio (*TR*) of the S_0 mode for a specific path from an actuator to a sensor is defined as:

$$TR = \frac{\int_{t_i}^{t_f} |S_s(\omega_0, t)|^2 dt}{\int_{t_i}^{t_f} |S_a(\omega_0, t)|^2 dt} \quad (4)$$

where S_s denotes the time spectral amplitude of the transmitted signals received by the sensor; S_a denotes the time spectral amplitude of input signal of the actuator; ω_0 denotes the selected central frequency for actuator excitation; t_i and t_f denote the upper bound and lower bound of the fundamental symmetric S_0 mode in the time domain respectively.

The normalized variations of the detected *TOF* and *TR* with respect to the TOF_0 and TR_0 of the intact specimen respectively are evaluated as the damage index (DI), formulated as:

$$DI_{TOF} = \left| \frac{TOF - TOF_0}{TOF_0} \right| \quad (5)$$

$$DI_{TR} = \left| \frac{TR - TR_0}{TR_0} \right| \quad (6)$$

The sensitivities of the variations of *TOF* and *TR* to crack length and location are examined with the numerical simulation results for different sensing paths, as demonstrate in Figure 11~12. It is found that variation of the *TOF* will be sensitive to the crack length in the case of the sensing path normal to and cross through the crack. On the other hand, the variation of the *TR* is more robust to indicate the change of crack length in the more general cases of the sensing path close to the crack, and less affected by the crack directions. Future work will focus on the development of the intelligent algorithm such as artificial neural network for precise diagnosis of the crack growth in the structure.

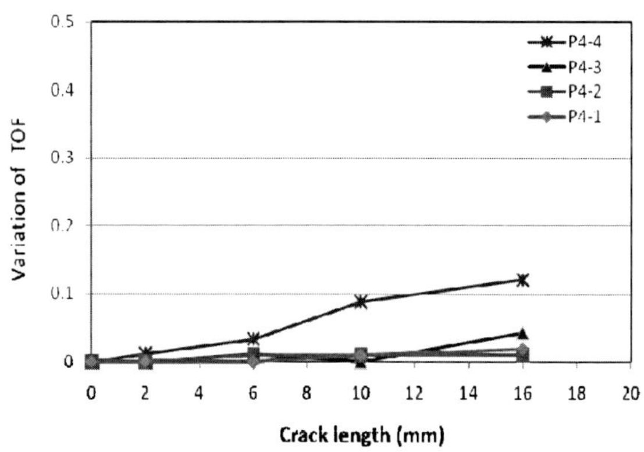

(a) Crack at hole 4 along the center line

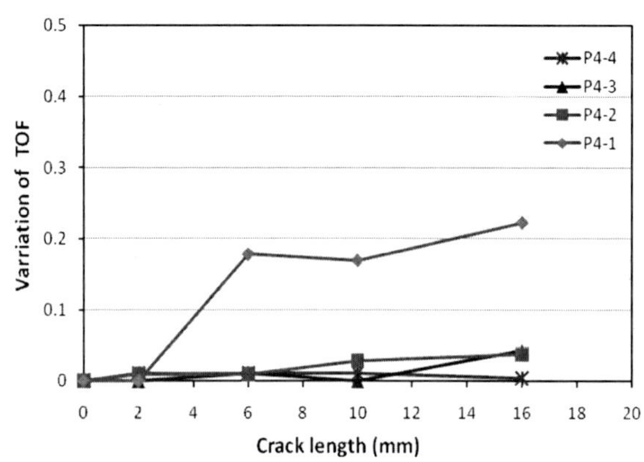

(b) Crack at hole 3 along the center line

Fig. 11 Normalized variation of *TOF* with crack length

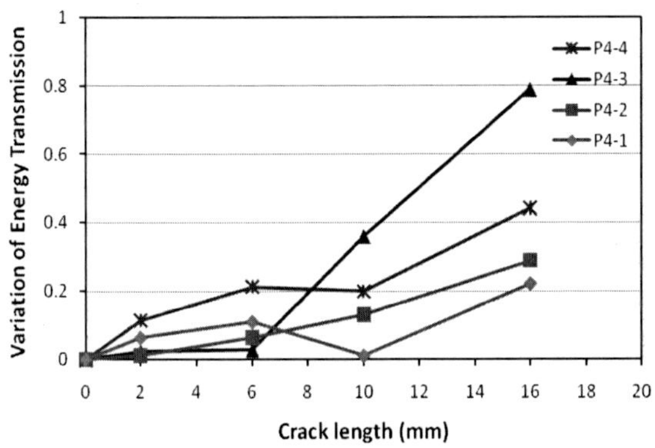

(a) Crack at hole 4 along the center line

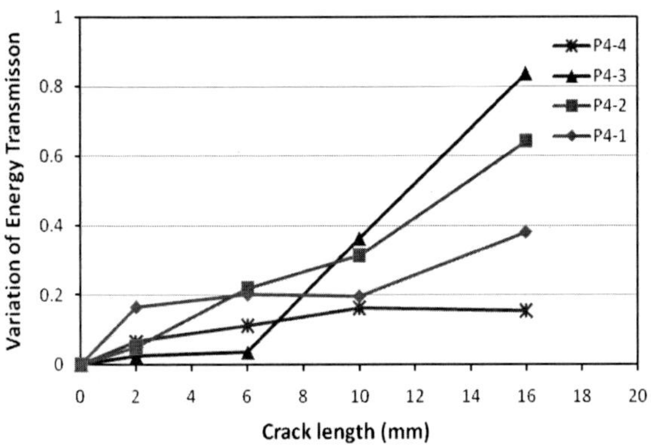

(b) Crack at hole 3 along the center line

Fig. 12 Normalized variation of *TR* with crack length

CONCLUSION

In this paper, the monitoring of crack growth at rivet holes in an aluminum test plate using a PZT transducer network and Lamb waves is investigated. The symmetrical S_0 mode and anti-symmetrical A_0 mode of the Lamb waves in the structures are studied with correlated theoretical, experimental and numerical analysis. The explicit FEM numerical simulations are accomplished with LS-DYNA solver and verified with the experimental results. Analyses of the wave propagation and interaction with the holes and cracks in the structure are carried out. It is found that the faster S_0 mode is partially transmitted, diffracted, and reflected after its interaction with the holes and cracks, and the propagation of the slower A_0 mode through the holes and crack is combined with the waves induced by the S_0 reflection and diffraction. The S_0 mode propagates through the holes and cracks with higher detectable attenuation than the A_0 mode. The short time Fourier transform (*STFT*) is used in processing of the Lamb wave signals in the PZT transducer network. It is demonstrated that the time of flight and the energy transmission ratio of the S_0 waves are sensitive to the cracks in the structure.

REFERENCES

[1] S. Grondel, C. elebarre, J. Assad, J. P. Dupuis, L. Reithler, "Fatiguage crack monitoring of riveted aluminium strap joints by Lamb wave analysis and acoustic emmision measurement techniques", DNT & E International, 35, pp137-146, 2002 .

[2] J. B. Ihn and F. K. Chang, "Detection and monitoring of hidden fatigue crack growth using a built-in piezoelectric sensor/actuator network: I. Diagnostics", Smart Materials and Structures, 13, pp 609-620, 2004.

[3] J. B. Ihn and F. K. Chang, "Detection and monitoring of hidden fatigue crack growth using a built-in piezoelectric sensor/actuator network: II. Validation using riveted joints and repair patches", Smart Materials and Structures, 13, pp 621-630, 2004.

[4] S. Park, C. B. Yun, Y. Roh, "Damage diagnostics on a welded zone of a steel truss member using an active sensing network system", NDT & E International, 40, pp 71-76, 2007.

[5] D. Wang, L Ye, Y. Lu, Z. Su, "Probability of the presence of damage estimated from an active sensor network in a composite pannel of multiple stiffeners", Composites Science and Rechnology, 69, pp 2054-2063, 2009.

[6] J. L. Rose, Ultrasonic Waves in Solid Media, Cambridge University Press, 1999.

[7] K. Diamanti, C. Soutis, J. M. Hodgkinson, "Piezoelectric transducer arrangement for the inspection of large composite structures", Composites: Part A 38, pp 1121–1130, 2007.

[8] Y. Lu, L. Ye, Z. Q. Su, C. H. Yang, "Quantitative assessment of through-thickness crack size based on Lamb wave scattering in aluminum plates" , NDT&E International, 41, pp59–68, 2008.

Proceedings of the ASME 2010 International Mechanical Engineering Congress & Exposition
IMECE2010
November 12-18, 2010, Vancouver, British Columbia, Canada

IMECE2010-38553

ANALYSIS of FREE FLEXURAL VIBRATIONS of COMPOSITE MINDLIN BASE PLATES or PANELS REINFORCED by THREE BONDED PLATE STRIPS

Umur Yuceoglu [#] Jaber Javanshir [+] Aydin Amireghbali [+] Özen Güvendik [+]

Dept. of Aerospace Engineering
Middle East Tech University
Ankara 06531, TURKEY
Phone: (++90)-(312)-2102471
Fax: (++90)-(312)-2101272
e-mail: yuceoglu@ae.metu.edu.tr
e-mail: jaberjavanshir@hotmail.com
e-mail: ogunendik@tai.com.tr

ABSTRACT

The present study is about the theoretical analysis and the numerical solutions of the free vibrations of a certain type of the "Bonded and Stiffened System". In other words, the "Free Bending Vibrations of Composite Mindlin Base Plates or Panels Reinforced by Three Bonded Stiffening Plate Strips" are investigated. The plate elements of the entire "System" are considered as the dissimilar "Orthotropic Mindlin Plates" with the transverse shear deformations and the transverse and the rotatory moments of inertia. The relatively very thin and linearly elastic adhesive layers between the "Stiffening Plate Strips" and the "Base Plate or Panel" are assumed to be subjected to transverse normal and shear stresses. Also, the dissimilar adhesive layers with unequal thicknesses and the different material characteristics are to be taken into account in the theoretical formulation. The dynamic equations, the stress resultant-displacement expressions and the adhesive layer equations of the "System" under study are combined together in each "Part" (or "Region") of the "Bonded and Stiffened System". Following some manipulations and combinations, they are, eventually, reduced to a set of the "Governing System of the First Order Ordinary Differential Equations" in "state vector" forms. The aforementioned system of equations are very suitable for the direct application of the present "Method of Solution" developed for the problem. This semi-analytical and the numerical solution technique is the "Modified Transfer Matrix Method(MTMM) (with Interpolation Polynomials)". The system of equations in the "state vector" forms are numerically integrated and the mode shapes and the natural frequencies are presented for various sets of the "Boundary Conditions". It was observed that, in the "Hard" (or relatively "Stiff") adhesive layers case, the regions with the "Bonded Stiffening Plate Strips" are "almost stationary" in all modes up to the sixth mode, while the rest of the "System" vibrates in a way dependent on the "Support Conditions" of the "Base Plate or Panel". However, in the "Soft" (or relatively "Flexible") adhesive layer cases, there is no "almost stationary" region anywhere in the "System". Additionally, some important parameters such as "Thickness Ratio" on the natural frequencies are investigated and graphically presented for a set of the "Support Conditions".

[#]Professor and PhD and Member -ASME,
[+]Graduate student

INTRODUCTORY REMARKS

In general, the "Bonded and Stiffened Plate or Panel Systems" of various configurations are utilized in flight vehicle or rather air and space vehicle structures and sub-structures [1-3]. They are also employed in light-weight, high-speed, hydrodynamic marine vehicles (such as motor boats, yachts, light and fast coast guard ships for shore protection, etc)

Aforementioned "Systems" offer certain desirable advantages in terms of their light-weight, high-strength, high resistance to dynamic crack propagation and fatigue-fracture. Also, they are free of rivets, welds and other mechanical connectors which may cause high stress concentrations and eventually resulting in cracks of various sizes in the structural system. Another practical advantage they provide is that the "Bonded Stiffening Plate Strips" can stop, or at least, slow down the crack propagation emanating from micro-cracks by reinforcing the main or primary structural element or "Base Plate or Panel" under them [4].

Generally speaking, the advanced "Composite (and/or Metal-Alloy) Stiffened Plate or Panel Systems" may be categorized in terms of the "Four Main Groups" as shown in Figure 1. Furthermore, "Group II Systems" may be categorized in terms of the several "Classes" based on the number of "Bonded Stiffening Plate Strips" as presented in Figure 2.

In the "Group I Systems" that is the "Integrally-Stiffened Plates or Panels" are considered in Yuceoglu et al [5,6]. On the "Bonded and Stiffened Systems", there are relatively few number of research studies available in the world-wide scientific and engineering publications and literature. In this connection, in "Group II Systems" (with "Bonded Stiffening Plate Strips"), one may mention some recent studies by Yuceoglu, Javanshir and Guvendik [7,8] and also Yuceoglu and Ozerciyes [9-16]. In the case of the "Bonded Joints", some important older studies [17-24], and the more recent investigations by Yuceoglu et al [25-35] can be mentioned.

Considering the "Group II and Class 3" cases of Figure 2, there are no published studies the present authors are aware of in the open engineering and scientific literature.

Therefore, the main objectives of this study are (i) theoretically analyze and (ii) numerically solve the "Free Bending Dynamic Response of Composite Mindlin Plates or Panels Reinforced by Three Bonded Stiffening Plate Strips" (which is the "Group II and Class 3" problem as can be seen in Figures 2 and 3). This study is a further extension and generalization of the "Group II and Class 2" case (see also Figure 2.) recently studied by Yuceoglu et al [7,8].

ANALYSIS and DEVELOPMENT of GOVERNING EQUATIONS

As stated in the preceding section, the "Group II and Class 3" case of the "Bonded and Stiffened System" of Figure 3 is the subject matter of the present study. The general configuration, the geometry, the coordinate and material directions of the "System" are presented in Figure 3.a . The longitudinal cross –section is of the "System" is shown in Figure 3.b.

In the theoretical analysis, the "Base Plate or Panel" and the "Three Bonded Stiffening Plate Strips" are assumed to be dissimilar "Orthotropic Mindlin Plates" with unequal thicknesses. They are adhesively bonded together by means of the adhesive layers. Then, in the dynamic equations of the individual plate elements of the "System", the "Mindlin Plate Theory" (which is one of the "First Order Shear Deformation Plate Theories" (FSDPT) [36-38]) are utilized.

The in-between adhesive layers are assumed to be relatively very thin and linearly elastic continua. Also, the "Hard" (or relatively "Stiff") and the "Soft" (or relatively "Flexible") adhesive material characteristics are included in the theoretical analysis and numerical calculations.

It is a well known fact that, in multi-layer plates (here, in the present case, bonded two layer plates), significant effects of transverse shear deformations in the plate elements as well as in the very thin adhesive layers of the entire "System" on the natural frequencies are shown by Pagano [39], Whitney and Pagano [40], Librescu et al [41] and other authors [42-44].

Also, Librescu et al [41] showed that the natural frequencies obtained by (FSDPT's) and by the "Higher Order Shear Deformation Plate Theories (HSDPT's)" [36-38] are not significantly different. Therefore, the (FSDPT) such as the "Mindlin Plate Theory" [36,37] is employed in the present study.

In the development of the theoretical analysis, the preliminary or the first step is to apply the "Domain Decomposition Technique" to the "Bonded and Stiffened System" as shown in Figures 3 a and b. Hence, the base plate and the stiffening plate strips with the in-between adhesive layers are considered to form Part I, Part II, Part III "Regions" and the rest of the one-layer base plate or panel are defined as Part IV,..., Part VII, "Regions" (or "Domains") as indicated in Figure 3.b.

In the second step, in each "Part" (or "Region" or "Domain"), the dynamic equations of the dissimilar "Orthotropic Mindlin Plates", the adhesive layer equations (in Parts I, II, III , only) and the stress resultant-displacement expressions of each plate element from Appendix A are combined and manipulated and reorganized. Thus, the "Governing System of P.D.E's" in a special form as given in Appendix B are developed for each "Part"(or "Region" or "Domain").

In the third step, the "Non-Dimensionalization" is in order. For this purpose, the stiffening plate strip in Part I region is chosen as the "reference plate", consequently, "B_{11}", "ρ_{11}", "h_1" and "**a**" are chosen as the "reference quantities (or parameters)". Additionally, ℓ_I, ℓ_{II}, ℓ_{III},..., ℓ_{VII} in the y-directions are selected as the "length reference (or parameters)"in each corresponding "Part" (or "Region"). Thus, all other quantities are non-dimensionalized with respect to these "parameters". Then, the following expressions are obtained:

The dimensionless coordinates,

$$\begin{aligned}
&\xi_I = y_I/\ell_I, \eta = x/a && \text{(in Part I region)}\\
&\xi_{II} = y_{II}/\ell_{II}, \eta = x/a && \text{(in Part II region)}\\
&\xi_{III} = y_{III}/\ell_{III}, \eta = x/a && \text{(in Part III region)}\\
&\xi_{IV} = y_{IV}/\ell_{IV}, \eta = x/a && \text{(in Part IV region)}\\
&\xi_V = y_V/\ell_V, \eta = x/a && \text{(in Part V region)}\\
&\xi_{VI} = y_{VI}/\ell_{VI}, \eta = x/a && \text{(in Part VI region)}\\
&\xi_{VII} = y_{VII}/\ell_{VII}, \eta = x/a && \text{(in Part VII region)}
\end{aligned} \tag{1}$$

The dimensionless parameters related to the densities and the geometries of the base plate or panel and the stiffening plate strips,

$$\begin{aligned}
&\bar{\rho}_1 = \rho_1/\rho_1 = 1, \bar{\rho}_2 = \rho_2/\rho_1, \bar{\rho}_3 = \rho_3/\rho_1, \bar{\rho}_4 = \rho_4/\rho_1\\
&\bar{h}_1 = h_1/h_1 = 1, \bar{h}_2 = h_2/h_1, \bar{h}_3 = h_3/h_1, \bar{h}_4 = h_4/h_1\\
&\bar{h}_{a1} = h_{a1}/h_1, \bar{h}_{a2} = h_{a2}/h_1, \bar{h}_{a3} = h_{a3}/h_1\\
&\bar{L}_I = \ell_I/a, \bar{L}_{II} = \ell_{II}/a, \bar{L}_{III} = \ell_{III}/a,\\
&\bar{L}_{IV} = \ell_{IV}/a, \bar{L}_V = \ell_V/a, \bar{L}_{VI} = \ell_{VI}/a, \bar{L}_{VII} = \ell_{VII}/a
\end{aligned} \tag{2}$$

where "**a**" is the width of the entire "Bonded and Stiffened System".

The dimensionless material parameters related to the plates and the adhesive layer elastic constants are given as,

$$\begin{aligned}
&\bar{B}_{ik}^{(j)} = B_{ik}^{(j)}/B_{11}^{(1)} \quad (j = 1,2,3,4 \text{ and } i,k = 1,2),\\
&\bar{B}_{\ell\ell}^{(j)} = B_{\ell\ell}^{(j)}/B_{11}^{(1)} \quad (j = 1,2,3,4 \text{ and } \ell = 4,5,6)\\
&\bar{G}_{a1} = G_{a1}/B_{11}^{(1)}, \bar{G}_{a2} = G_{a2}/B_{11}^{(1)}\\
&\bar{E}_{a1} = E_{a1}/B_{11}^{(1)}, \bar{E}_{a2} = E_{a2}/B_{11}^{(1)}
\end{aligned} \tag{3}$$

where the subscript "a" indicates the adhesive layer, while "a" by itself is the width of the entire "System".

The dimensionless natural frequency (or the circular frequency) parameter $\bar{\omega}_{mn}$ of the entire "System",

$$\begin{aligned}
&\bar{\omega}_{mn} = \rho_1 a^4 \omega_{mn}^2 / h_1^2 B_{11}^{(1)} \qquad (m, n=1,2,3...)\\
&\bar{\Omega} = \bar{\omega}_{mn}
\end{aligned} \tag{4}$$

The substitution of (1), (2) and (3) and the expressions in Appendix C into the "Governing PDS's" yield the "Governing System of the First Order Ordinary Differential Equations" of the present problem. Then,

In Part I region (Left Two-Layer Plate and Adhesive Layer),

$$\frac{d}{d\xi_I}\begin{Bmatrix}\bar{Y}_{mn}^{(1)}\\ \bar{Y}_{mn}^{(4)}\end{Bmatrix} = \begin{bmatrix}\bar{\mathcal{C}}_{1,1} & \bar{\mathcal{C}}_{1,2}\\ \bar{\mathcal{C}}_{2,1} & \bar{\mathcal{C}}_{2,2}\end{bmatrix}\begin{Bmatrix}\bar{Y}_{mn}^{(1)}\\ \bar{Y}_{mn}^{(4)}\end{Bmatrix}, \tag{5}$$

$(0 < \xi_I < 1)$ $(j = 1)$ (Left Stiffening Plate Strip)

$(0 \le \xi_I \le 1)$ $(j = 4)$ (Lower Base Plate)

with the "Appropriate Boundary Conditions" at $\xi_I=0$ and $\xi_I=1$ and the "Continuity Conditions" at $\xi_{IV}=1$ and $\xi_{VI}=0$.

In Part II region (Middle Two-Layer Plate and Adhesive Layer),

$$\frac{d}{d\xi_{II}}\begin{Bmatrix}\overline{Y}_{mn}^{(2)}\\ \overline{Y}_{mn}^{(4)}\end{Bmatrix}=\begin{bmatrix}\overline{\mathcal{C}}'_{1,1} & \overline{\mathcal{C}}'_{1,2}\\ \overline{\mathcal{C}}'_{2,1} & \overline{\mathcal{C}}'_{2,2}\end{bmatrix}\begin{Bmatrix}\overline{Y}_{mn}^{(2)}\\ \overline{Y}_{mn}^{(3)}\end{Bmatrix}, \quad (6)$$

$(0 < \xi_{II} < 1)$ $(j = 2)$ (Middle Stiffening Plate Strip)

$(0 \le \xi_{II} \le 1)$ $(j = 3)$ (Lower Base Plate)

with the "Appropriate Boundary Conditions" at

$\xi_{II}=0$ and $\xi_{II}=1$ and the "Continuity Conditions" at $\xi_{VI}=1$ and $\xi_{VII}=0$.

In Part III region (Right Two-Layer Plate and Adhesive Layer),

$$\frac{d}{d\xi_{II}}\begin{Bmatrix}\overline{Y}_{mn}^{(3)}\\ \overline{Y}_{mn}^{(4)}\end{Bmatrix}=\begin{bmatrix}\overline{\mathcal{C}}''_{1,1} & \overline{\mathcal{C}}''_{1,2}\\ \overline{\mathcal{C}}''_{2,1} & \overline{\mathcal{C}}''_{2,2}\end{bmatrix}\begin{Bmatrix}\overline{Y}_{mn}^{(3)}\\ \overline{Y}_{mn}^{(4)}\end{Bmatrix}, \quad (7)$$

$(0 < \xi_{III} < 1)$ $(j = 3)$ (Right Stiffening Plate Strip)

$(0 \le \xi_{III} \le 1)$ $(j = 4)$ (Lower Base Plate)

with the "Appropriate Boundary Conditions" at

$\xi_{III}=0$ and $\xi_{III}=1$ and the "Continuity Conditions" at $\xi_{VII}=1$ and $\xi_{V}=0$.

In Part. IV region (Far Left Base Plate)

$$\frac{d}{d\xi_{IV}}\left\{\overline{Y}_{mn}^{(4)}\right\}=\left[\overline{\mathcal{D}}\right]\left\{\overline{Y}_{mn}^{(4)}\right\}, \qquad (0 < \xi_{IV} \le 1) \quad (8)$$

with the "Far Left End Boundary Conditions" at $\xi_{IV}=0$

and the "Continuity Conditions" $\xi_{IV}=1$.

In Part. V region (Far Right Base Plate),

$$\frac{d}{d\xi_{V}}\left\{\overline{Y}_{mn}^{(4)}\right\}=\left[\overline{\mathcal{E}}\right]\left\{\overline{Y}_{mn}^{(4)}\right\}, \qquad (0 \le \xi_{V} < 1) \quad (9)$$

with the "Far Right End Boundary Conditions" at $\xi_{V}=1$

and the "Continuity Conditions" at $\xi_{V}=0$.

In Part. VI region (Left Middle Base Plate),

$$\frac{d}{d\xi_{VI}}\left\{\overline{Y}_{mn}^{(4)}\right\}=\left[\overline{\mathcal{F}}\right]\left\{\overline{Y}_{mn}^{(4)}\right\} \qquad (0 \le \xi_{VI} \le 1) \quad (10)$$

with the "Continuity Conditions" at $\xi_{VI}=0$ and $\xi_{VI}=1$.

In Part. VII region (Right Middle Base Plate),

$$\frac{d}{d\xi_{VII}}\left\{\overline{Y}_{mn}^{(4)}\right\}=\left[\overline{\mathcal{G}}\right]\left\{\overline{Y}_{mn}^{(4)}\right\}, \qquad (0 \le \xi_{VII} \le 1) \quad (11)$$

with the "Continuity Conditions" at $\xi_{VII}=0$ and $\xi_{VII}=1.0$

where, in the above "System of Equations" of (5-11), the "barred" (or "—") elements are the dimensionless matrices and the coefficient matrices $[\overline{\mathcal{C}}]$, $[\overline{\mathcal{C}}']$, $[\overline{\mathcal{C}}'']$, …, $[\overline{\mathcal{F}}]$, $[\overline{\mathcal{G}}]$. The coefficient matrices include dimensionless geometric and material constants and the dimensionless natural frequency parameter $\overline{\omega}_{mn}$ of the entire "System". The dimensionless column matrices or the "state vectors" associated with the three upper stiffening plate strips (j=1, 2,3) and the lower base plate or panel (j=4), are,

$$\left\{\overline{Y}_{mn}^{(j)}(\xi_k)\right\}=\left\{\overline{\Psi}_{mnx}^{(j)},\overline{\Psi}_{mny}^{(j)},\overline{w}_{mn}^{(j)};\overline{M}_{mnyx}^{(j)},\overline{M}_{mny}^{(j)},\overline{Q}_{mny}^{(j)}\right\}^{T}$$

(j=1,2,3,4) (k=I,II,III,…,VII) (12)

which includes as its elements the dimensionless "fundamental dependent variables" of the problem or the "System" under study.

It is important to note that the "Governing System of Equations" as given in (5-11) establishes a "Two Point or rather Multi-Point Boundary Value Problem" of Mechanics and Physics. In other words, the "Initial and Boundary Value Problem" (which was, at the beginning, the free vibrations problem of the "Bonded and Stiffened System" under consideration here) is now finally reduced to a "Two-Point or rather Multi-Point Boundary Value Problem". This is the crucial step in the present theoretical analysis which facilitates the direct application of the "Solution Procedure" developed by the present authors to be employed in this study. Here, in this case, the "Multi-Point Boundary Value Problem" means that, in addition to the far left end and far right end "Support Conditions" of the base plate or panel, and also the left and the right free "Boundary Conditions" in each of the three bonded stiffening plate strips have to be satisfied.

SOLUTION TECHNIQUE and NUMERICAL PROCEDURE

(9)

The coupled "Governing System of the First Order ODE's" as presented in "state vector" forms in (5-11) can be integrated by the "Modified Transfer Matrix Method (MTMM) (with Interpolation Polynomials)". This solution technique is a combination of the "Classical Levy's Method", the "Transfer Matrix Method" and the "Integrating Matrix Method (with Interpolation Polynomials)" developed by Yuceoglu et al [7, 8, 9-16] to handle some free vibrations problems of composite stiffened plates and shallow shells and also the free vibration problems of various types of "Bonded Joints" in plates and shallow shell configurations [25-35] .

The main steps in the several versions of the present semi-analytical and numerical "Solution Method" can be found in more detail, for instance, in Yuceoglu et al [7, 8 , 9-16]. A more recent version is given in Yuceoglu and Özerciyes [9, 31]. And an earlier version is presented in Yuceoglu et al [25].

In the numerical solution procedure, the first step is the discretization of the "state vectors" and the coefficient matrices in (5-11). The discretization process is accomplished by dividing Part I, Part II, … , Part VII regions along $\ell_I, \ell_{II}, \ell_{III}, \ldots, \ell_{VII}$ into a sufficient number of points os stations along the $\ell_I, \ell_{II}, \ell_{III}, \ldots, \ell_{VII}$ regions in the in the ξ_I , ξ_{II} , $\xi_{III}, \ldots, \xi_{VII}$ directions, respectively. After then, in the second step, the discretized versions of the "state vectors" and the coefficient matrices in (5-11) are premultiplied by the "Global Integrating Matrices" " $[\mathcal{L}_I]$, $[\mathcal{L}_{II}]$, $[\mathcal{L}_{III}]$, …, $[\mathcal{L}_{VII}]$, respectively. Aforementioned matrices act as the "Integrating Operators" within the "Global Integrating Matrices" which include "Integrating Sub-Matrices $[\ell]$". Then,

In Part I region (Two-Layer Mindlin Plate and Adhesive Layer),

$$\begin{Bmatrix} \dot{Y}^{(1)}_{mn} \\ \dot{Y}^{(4)}_{mn} \end{Bmatrix} - \begin{Bmatrix} \dot{Y}^{(1)}_{\xi_I=0} \\ \dot{Y}^{(4)}_{\xi_I=0} \end{Bmatrix} = \left[\mathcal{L}_I\right]\left[\dot{C}\right] \begin{Bmatrix} \overline{Y}^{(1)}_{mn} \\ \overline{Y}^{(4)}_{mn} \end{Bmatrix},$$

$$\left[\dot{\mathcal{U}}\right]_{01} = \left((I) - \left[\mathcal{L}_I\right]\left[\dot{C}\right]\right)^{-1} \tag{13}$$

where, in (11), "•" on the matrices indicates discretization along ξ_I - direction and [I] is the "unit matrix".

In the next step, one can obtain relations between the "state vectors" at the "initial end point" (ξ_I , ξ_{II} ,… , ξ_V)=0 and the "final end points" (ξ_I , ξ_{II} ,… , ξ_V) =1 in each Part I, Part II, … , Part V regions, respectively. Dropping "•" sign and "mn" subscript for convenience, this operation yields the following in Part I region,

In Part I region (Two-Layer Mindlin Plate and Adhesive Layer),

$$\begin{Bmatrix} \overline{Y}^{(1)}_{\xi_I=1} \\ \overline{Y}^{(4)}_{\xi_I=1} \end{Bmatrix} = \begin{bmatrix} \tilde{\mathcal{U}}_{1,1} & \tilde{\mathcal{U}}_{1,2} \\ \tilde{\mathcal{U}}_{2,1} & \tilde{\mathcal{U}}_{2,2} \end{bmatrix}_{01} \begin{Bmatrix} \overline{Y}^{(1)}_{\xi_I=0} \\ \overline{Y}^{(4)}_{\xi_I=0} \end{Bmatrix},$$

$$\left[\tilde{\mathcal{U}}\right]_{01} = \left((I) - \left[\mathcal{L}_I\right]\left[\dot{C}\right]\right)^{-1} \tag{14}$$

$$(0 \leq \xi_I \leq 1)$$

here $\left[\tilde{\mathcal{U}}\right]_{01}$, is the discretized version of the "Modified Transfer Matrix" for Part I region. It transfers the "state variables" from the "Initial end point $\xi_I = 0$ " to the "final end point $\xi_I = 1$ " in Part I region.

A similar set of expressions can be written in Part II and Part III regions. Hence,

In Part II region (Two-Layer Mindlin Plate and Adhesive Layer),

$$\begin{Bmatrix} \overline{Y}^{(2)}_{\xi_{II}=1} \\ \overline{Y}^{(3)}_{\xi_{II}=1} \end{Bmatrix} = \begin{bmatrix} \tilde{\mathcal{U}}'_{1,1} & \tilde{\mathcal{U}}'_{1,2} \\ \tilde{\mathcal{U}}'_{2,1} & \tilde{\mathcal{U}}'_{2,2} \end{bmatrix}_{01} \begin{Bmatrix} \overline{Y}^{(2)}_{\xi_{II}=0} \\ \overline{Y}^{(3)}_{\xi_{II}=0} \end{Bmatrix}$$

$$\left[\tilde{\mathcal{U}}'\right]_{01} = \left((I) - \left[\mathcal{L}_{II}\right]\left[\dot{C}'\right]\right)^{-1} \tag{15}$$

$$(0 \leq \xi_{II} \leq 1)$$

where the subscript "01" and "~" in the above mean that the "final form" of the discretized "Modified Transfer Matrix" are transferring the above quantities from the "initial end point 0" to the "final end point 1" in Part II region.

In Part III region (Two-Layer Mindlin Plate and Adhesive Layer),

$$\begin{Bmatrix} \overline{Y}^{(3)}_{\xi_{II}=1} \\ \overline{Y}^{(4)}_{\xi_{II}=1} \end{Bmatrix} = \begin{bmatrix} \tilde{\mathcal{U}}''_{1,1} & \tilde{\mathcal{U}}''_{1,2} \\ \tilde{\mathcal{U}}''_{2,1} & \tilde{\mathcal{U}}''_{2,2} \end{bmatrix}_{01} \begin{Bmatrix} \overline{Y}^{(3)}_{\xi_{II}=0} \\ \overline{Y}^{(4)}_{\xi_{II}=0} \end{Bmatrix}$$

$$\left[\tilde{\mathcal{U}}''\right]_{01} = \left((I) - \left[\mathcal{L}_{II}\right]\left[\dot{C}''\right]\right)^{-1} \tag{16}$$

$$(0 \leq \xi_{III} \leq 1)$$

In the single layer base plate or panel regions (or rather ℓ_{IV}, ℓ_V ℓ_{VI}, ℓ_{VII} regions) one may write similar expressions with $[\tilde{V}][\tilde{W}][\tilde{S}]$ and, $[\tilde{T}]$, as the "Modified Transfer Matrices" in "Part IV", "Part V", "Part VI", "Part VII" regions, respectively.

In the final step, by making use of the "Continuity Conditions", followed by some matrix manipulations, one can write the following matrix expression:

For the entire "Bonded and Stiffened System"

$$\begin{Bmatrix} \overline{Y}^{(1)}_{\xi_I=1} \\ \overline{Y}^{(2)}_{\xi_{II}=1} \\ \overline{Y}^{(3)}_{\xi_{III}=1} \\ \overline{Y}^{(4)}_{\xi_V=1} \end{Bmatrix} = \begin{bmatrix} \tilde{Q}_{1,1} & \tilde{Q}_{1,2} & \tilde{Q}_{1,3} & \tilde{Q}_{1,4} \\ \tilde{Q}_{2,1} & \tilde{Q}_{2,2} & \tilde{Q}_{2,3} & \tilde{Q}_{2,4} \\ \tilde{Q}_{3,1} & \tilde{Q}_{3,2} & \tilde{Q}_{3,3} & \tilde{Q}_{3,4} \\ \tilde{Q}_{4,1} & \tilde{Q}_{4,2} & \tilde{Q}_{4,3} & \tilde{Q}_{4,4} \end{bmatrix}_{01} \begin{Bmatrix} \overline{Y}^{(1)}_{\xi_I=0} \\ \overline{Y}^{(2)}_{\xi_{II}=0} \\ \overline{Y}^{(2)}_{\xi_{III}=0} \\ \overline{Y}^{(2)}_{\xi_{VI}=0} \end{Bmatrix} \tag{17}$$

where $[\tilde{Q}]_{01}$ in (17) is the "final form" of the "Overall Modified Transfer Matrix" which transfers the above "state variables" from the "initial end point (or far left end support) at $\xi_{IV} = 0$" to the "final end point (or far right end support) at $\xi_V = 1$" of the entire "Bonded and Stiffened System" under consideration here.

The above "Overall Modified Transfer Matrix" $[\tilde{Q}]_{01}$ can be reduced to a smaller size by using or inserting the "Boundary Conditions" of the base plate or panel at the far left end support $\xi_{IV} = 0$ and the far right end support at $\xi_V = 1$, and also the "Free Support Conditions" of the bonded stiffening plate strips at $\xi_I = 1$, $\xi_I = 1$, $\xi_{II} = 0$ and $\xi_{II} = 1$, and $\xi_{III} = 0$ and $\xi_{III} = 1$, respectively.

These operations yield, the following matrix equation in which the "Coefficient Matrix $[C_0]$" implicitly includes the unknown "dimensionless frequency parameter" $\bar{\omega}_{mn}$ of the entire "System". Therefore,

$$[\mathbf{C_0}(\bar{\omega}_{\mathbf{mn}})]\{\mathbf{Y_0}\} = \{\mathbf{0}\} \Rightarrow \tag{18}$$

Determinant of "Coefficient Matrix" $|[\mathbf{C_0}]| = 0$

where $\{\mathbf{Y_0}\}$ is not necessarily zero, then the "Coefficient Matrix" in (18) must be equal to zero. Thus, the roots which are the dimensionless natural frequencies of the "Bonded and Stiffened System" are,

$$\overline{\mathbf{\Omega}} = \overline{\boldsymbol{\omega}}_{\mathbf{mn}} \quad (\mathbf{m}, \mathbf{n} = \mathbf{1,2,3...}) \\ \overline{\mathbf{\Omega}}_1 < \overline{\mathbf{\Omega}}_2 < \overline{\mathbf{\Omega}}_3 < .... \tag{19}$$

where the dimensionless natural frequencies are computed by searching the roots numerically based on the given "m" in "Fourier Series" and, later, "n" values are assigned accordingly. Then, they are reordered or sequenced according to their magnitudes as $\overline{\mathbf{\Omega}}_1 < \overline{\mathbf{\Omega}}_2 < \overline{\mathbf{\Omega}}_3 < ....$ as shown in (19).

SOME NUMERICAL RESULTS and BRIEF CONCLUSIONS

The present theoretical formulation and the present method of solution are applied to the general configuration and the geometry shown in detail in Figure 3. The material directions and the geometric characteristics of the "Composite Mindlin Base Plates or Panels Reinforced by Three Bonded Stiffening Plate Strips" and also the adhesive layer elastic constants (i.e. the "Hard" (or relatively "Stiff") and the "Soft" (or relatively "Flexible") adhesive material properties) are all given in Table 1.

The "Boundary Conditions" given on all Figures are, first, the "Support Conditions" for the three bonded stiffening plate strips followed by the far left end and the far right end "Boundary Conditions" of the base plate or panel. All conditions read from left to right, in the y-directions, for the upper stiffening plate strips and followed the "Support Conditions" for the lower base plate or panel. Hence, they are defined as C=Clamped, S=Simple Support and F=Free "Support Conditions".

In the Figures 4 a, b and up to 6 a, b, the mode shapes with their dimensionless natural frequencies are presented for various "Boundary Conditions" of the base plate or panel. In Figures, first, the "Hard" adhesive case followed by the "Soft" adhesive case for the same or identical "Support Conditions" are shown. Hence, for instance, in Figures 4.a ("Hard" adhesive case) and in Figure 4.b ("Soft" adhesive case) both for the "(FFFFFFCC) Boundary Conditions", the mode shapes up to the sixth mode are presented. In a similar manner, in Figure 5.a ("Hard" adhesive case) and in Figure 5.b ("Soft" adhesive case), the mode shapes up to the sixth mode with their dimensionless natural frequencies are presented both for the same "(FFFFFFCF) Boundary Conditions", etc, etc.

A simple comparison of Figure 4.a ("Hard" adhesive case) and Figure 4.b ("Soft" adhesive case) with the "(FFFFFFCC) Support Conditions" clearly indicates that the mode shapes and their dimensionless natural frequencies are drastically different. One may reach similar conclusions by comparing Figure 5.a and Figure 5.b both for the "(FFFFFFSS) Boundary Condition"s. These conclusions also hold for Figure 6.a and Figure 6.b for the "(FFFFFFCF) Boundary Conditions".

It is also very interesting to observe that, when the "complete symmetry" (i.e. symmetric boundaries, material and support conditions hold), then the "symmetric" and "skew-symmetric" modes follow each other. This is the case in Figures 4.a and 4.b, Figures 5.a and 5.b.

Another important conclusion is that, by the simple comparisons of the dimensionless natural frequencies in Figure 4.a and 4.b, and also in Figures 5.a and 5.b, clearly show that the corresponding natural frequencies in the "Hard" adhesive case, are very much higher than the "Soft" adhesive case with the same "Support Conditions". These results obviously indicate that very significant rather very large differences occur in the natural frequencies as well as in the mode shapes caused by the "Hard" and the "Soft" adhesive material constants.

Again another significant observation regarding the mode shapes is that, in the "Hard" adhesive cases of Figure 4.a and also of Figures 5.a and of Figure 6.a, in all modes up to the sixth mode, one may see that the three bonded stiffening plate strips and the base plate or panel region under them, experience "relatively small motion". In other words, they become relatively "almost stationary" in the modes up to the sixth mode.

In figures 6.a and 6.b, the "complete symmetry" condition does not hold because of the (FFFFFFCF) support conditions. Thus, one can see that in both Figures 6.a ("Hard" adhesive case) and 6.b ("Soft" adhesive case), the mode shapes do exhibit neither "symmetric" nor "skew-symmetric" properties. However, the natural frequencies in the "Hard" adhesive case, are again almost 10 times higher than the "Soft" adhesive case for the same "(FFFFFFCS) Boundary Conditions".

As a further study, the effects of some important "parameters" on the dimensionless natural frequencies of the "System" are also investigated. As an example, one of the parameters such as the "Stiffeners' Thickness Ratio h_4/h_1 $(= h_2 = h_3)$" versus the natural frequencies are presented in Figure 7.a, b both for "(FFFFFFCS) Support Condition"s. In Figure 7.a, the "Hard" adhesive case is shown and in Figure 7.b the "Soft" adhesive case is given. By comparing aforementioned Figures, one may conclude that the "Thickness Ratio" drastically influences the natural frequencies of the "System". Another observation is that, in both the "Hard" and the "Soft" adhesive cases, the natural frequencies change almost linearly with respect to the "Thickness Ratio" of the "Stiffening Plate Strips" as seen in Figures 7.a, b.

SOME REMARKS ABOUT ACCURACY of NUMERICAL RESULTS

At this point, it is worthwhile to say a few words about the "Accuracy" of the "numerical results" presented in this study. One starts with the natural frequencies of the "System", which are numerically searched and located by making use of Eqs.(18) and (19) for a given set (far left and far right end and in-between) "Support Conditions". In Equation (18) and (19), the geometry and material characteristics are also to be taken into account. Once, one gets the natural frequencies accurately (for the given set of the "Boundary Conditions" and the given geometry and the materials), we now conduct a few "numerical experiments".

For this purpose, for each set of the "Support Conditions" (far left and far right end and in-between) "Support Conditions", we consider the number of discretization points along y-direction in each region (see Figures 1 and 2) and the appropriate degree of the "Interpolation Polynomial" in each region. Then, we use Eq.(17) (in the text). One repeates these "numerical experiments" until far right "Support Conditions" are numerically satisfied.

Now, starting with the exact "Support Conditions" at the far left support, one gets from Eq.(17) numerically the far right "Support Conditions", then, one checks if they are reasonably and numerically good. For instance, a "Simple Support Condition" at the far right support means that $M_x = 0,\ w = 0,\ \psi_x = 0$. Then, one numerically calculate these quantities by making use of Eq.(17) and the exact far left end "Support Conditions". Thus, if those quantities are numerically satisfied or rather they are in the order of $O(10^{-15}) \cong 0$ or so, then the natural frequency obtained from Eq.(18,19) are quite accurate. If not, one again considers the "discretization procedure" in each region and the appropriate "Interpolation Polynomial" and make necessary modifications or adjustments to obtain the desired "numerical results". Following a few "numerical experiments" (which can be put into a table) for each set of the "Boundary Conditions", one gets the correct "numerical results".

REFERENCES

[1] Hoskins, B.C. and Baker, A.A., 1986, Composite Materials for Aircraft Structures, AIAA Education Series, Washington D.C.

[2] Niu, M.C.-Y, 1988, "Airframe Structural Design" Comilit Press, Ltd., Hong Kong (see also Niu, M.C., -Y, 2005, "Composite Airframe Structures", Comilit Press, Hong Kong).

[3] Marshall, I.H. and Demuts, H. (Editors), 1988, "Supportability of Composite Airframes and Aerostructures", Elsevier Applied Science Publishers, New York

[4] Baker AA., and Jones R., 1988, "Bonded Repair of Aircraft Structures". Martinus Nijhoff Publishers, New York , (see also, Baker A.A., Rose. L.R.F., and Jones R., 2002, "Advances in Bonded Composite Repair of Metallic Aircraft Structures", Elsevier Science, Oxford, England).

[5] Yuceoglu U., Javanshir J., Güvendik Ö., 2008, "Free Vibrations of Integrally Stiffened and/or Stepped-thickness Plates or Panels with Three Integral Plate Stiffeners", 2008, ASME Intr. Mech. Engineering Congress and Exposition (IMECE-2008), Oct. 24-04 Nov., 2008, Boston, MA, "Symposium on Advanced Struct. and Materials for Aerospace and light Weight Design".(ASME Paper No. IMECE-2008-68372)

[6] Yuceoglu U., Gemalmayan N., Sunar O., 2007, "Free Bending Vibrations of Integrally-Stiffened and/or Stepped Thickness Rectangular Plates or Panels with Noncentral Plate Stiffeners", 2007 ASME Inter. Mech. Eng. Congress and Exposition, Nov.11-15, 2007, Seattle, Washington, (ASME Paper No. IMECE-2007-41066)

[7] Yuceoglu, U., Javanshir, J., and Guvendik, 2009, "Free Bending Vibrations Analysis of Composite Mindlin Base Plates or Panels with Two Bonded Stiffening Plate Strips", 50th AIAA/ASME/ASCE/AHS/ASC , Structures, Struc. Dynamics and Materials (SDM) Conference and Exhibit", May 4-7, 2009, Palm Springs, California (AIAA Paper No. AIAA-2009-2375)

[8] Yuceoglu U., Javanshir J., Güvendik Ö., 2008, "On a General Approach to Free Vibrations Response of Integrally-Stiffened and/or Stepped Thickness Rectangular Plates or Panels", 2008, ASME Intr. Mech. Engineering Congress and Exposition (IMECE-2008), Oct. 24- Nov. 04, 2008, Boston, MA, "Symposium in Honor and Memory of Prof. L. Librescu" (ASME Paper No. IMECE-2008-66980)

[9] Yuceoglu, U. and Özerciyes, V., 2003, "Free Flexural Vibrations of Orthotropic Composite Base Plates or Panels with a Bonded Non-central (or Eccentric) Stiffening Plate Strip," ASME Jour. of Vibration and Acoustics, Vol. 125, pp. 1-16.

[10] Yuceoglu, U. and Özerciyes, V., 1996, "Free Bending Vibrations of Partially- Stiffened, Stepped-Thickness Composite Plates," ASME-Noise Control and Acoustics Div. Proceed. "Advanced Materials for Vibro-Acoustic Applications NCAD- Vol. 23, pp. 191-202, (an ASME Publication).

[11] Yuceoglu, U., and Özerciyes, V., 1997, "Natural Frequencies and Mode Shapes of Composite Plates or Panels with a Bonded Central Stiffening Plate Strip," ASME-Noise Control and Acoustics Div. Symposium on Vibroacoustic Methods in Processing and Characterization of Advanced Materials and Structures, NCAD-Vol. 24, pp.185-196, (an ASME Publication)

[12] Yuceoglu, U., and Özerciyes, V., 1998, "Free Bending Vibrations of Composite Base Plates or Panels Reinforced with a Bonded Noncentral Stiffening Plate Strip," ASME-Noise Control and Acoustics Div. Vibroacoustic Characterization of Advanced Materials and Structures NCAD-Vol.25, pp 233-243, (an ASME Publication).

[13] Yuceoglu, U. and Özerciyes, V., 1999, "Sudden Drop Phenomena in Natural Frequencies of Partially Stiffened, Stepped-Thickness, Composite Plates or Panels," Proceed. of the "40th AIAA/ASME/ASCE/AHS/ASC Structures and Struct. Dyn. and Materials (SDM) Conference and Exhibit", (AIAA Paper No. AIAA-1999-1483), pp. 2336-2347.

[14] Yuceoglu, U., and Özerciyes, V., 2000, "Natural Frequencies and Modes in Free Transverse Vibrations of Stepped-Thickness and/or Stiffened Plates and Panels," Proceed. of the "41st AIAA/ASME/ASCE/AHS/ASC Structures and Struct. Dyn. and Materials (SDM) Conference and Exhibit", (AIAA Paper No. AIAA-2000-1348).

[15] Yuceoglu, U. and Özerciyes, V., 2000, "Sudden Drop Phenomena in Natural Frequencies of Composite Plates or Panels with a Central Stiffening Plate Strip," Inter. Jour. of Computers and Structures, 76(1-3) (Special Issue), pp. 247-262.

[16] Yuceoglu et al, 1990, "Vibrations of Two-Layer, Orthotropic Beam-Like Strips Connected by Mechanical Springs," ASME Aerospace Division Proceedings AD-Vol.19 (also AMD-Vol.113), pp. 67-72, (an ASME Publication).

[17] Saito, H., and Tani, H., 1984, "Vibrations of Bonded Beams with a Single Lap Adhesive Joint," Journal of Sound and Vibration, 92(2), pp. 299-309.

[18] Rao, M. D., Croker, M. J., 1990, "Analytical and Experimental Study of the Vibration of Bonded Beams with a Lap Joint," ASME Jour. of Applied Mechanics, 112(4), pp.444-451.

[19] He, S., and Rao, M. D., 1992, "Vibration Analysis of Adhesively Bonded Beam Lap Joint, Part I: Theory," Journal of Sound and Vibration, 152(3), pp. 405-416.

[20] Rao, M. D., He, S., 1992, "Vibration Analysis of Adhesively Bonded Beam Lap Joint, Part II: Numerical Solution," Journal of Sound and Vibration, 152(3), pp. 417-425.

[21] He, S., and Rao, M.D., 1992, "Longitudinal Vibration and Damping Analysis of Adhesively Bonded Double-Strap Joints," ASME Jour. of Vibration and Acoustics, 114(3), pp. 330-337.

[22] Rao, M.D., and He, S., 1992, "Analysis of Natural Frequencies and Modal Loss Factors of Simply Supported Beams with Adhesively Bonded Double-Strap Joints," Jour. of Acoustical Society of America, 92(1), pp. 268-276.

[23] Ko, T.-C., Lin, C.-C. and Chu, R.-C., 1995, "Vibrations of Bonded Laminated Lap Joint (Beamlike) Plates using Adhesive Interface Elements," Journal of Sound and Vibration, 184(4), pp. 567-583.

[24] Gorrepati, K.M. and Rao, M.D., 1996, "Analysis of Modal Parameters of Adhesively Bonded Double-Strap Joints by the Modal Strain Energy Method," ASME Jour. of Vibration and Acoustics, 118(1), pp. 28-35.

[25] Yuceoglu, U., Toghi, F., and Tekinalp, O., 1996, "Free Bending Vibrations of Adhesively Bonded, Orthotropic Plates with a Single Lap Joint," ASME Jour. of Vibration and Acoustics, 118, pp 122-134.

[26] Yuceoglu, U., Guvendik Ö., Özerciyes V., 2003, "Free Flexural (or Bending) Vibrations of Orthotropic Composite Mindlin Plates with a Bonded Non-Central (or Eccentric) Lap Joint", 2003 ASME International Mechanical Engineering Congress & Exposition, Nov.15-21, 2003, Washington", D.C, (ASME-Paper No. IMECE2003-43651)

[27] Yuceoglu, U., Güvendik, Ö. and Özerciyes, V., 2005, "Free Flexural (or Bending) Vibrations of Orthotropic Composite Plates or Panels with a Centrally Bonded Symmetric Lap Joint (or Symmetric Doubler Joint)", "46th AIAA/ASME/ASCE/AHS/ASC Structures, Structural Dynamics and Materials (SDM) Conference and Exhibit", April 18-21 2005 / Austin, TX, (AIAA-Paper No. AIAA-2005-26174).

[28] Yuceoglu, U., Guvendik, O., and Ozerciyes V., 2007, "Free Bending Vibrations of Centrally Bonded Symmetric Double Lap Joint (or Symmetric Double Doubler Joint) with a gap in Mindlin Plates or Panels, "Proceedings of the 2007 ASME Inter. Mech. Engineering Congress and Exposition, "IMECE-2007, Vol.1 Advances in Aerospace Technology", Nov.11-15, 2007, Seattle, Washington, pp.151-167 (ASME Paper No. IMECE-2007-41152).

[29] Yuceoglu U, Özerciyes V., 1997, "Natural Frequencies and Mode Shapes of Composite Plates or Panels with a Bonded Central Stiffening Plate Strips", 1997 ASME Noise Control and Acoustics Div. Symposium on Vibroacoustics characteristics of Advanced Materials and Structures, NCAD. Vol.24, pp. 185-196 (an ASME publication).

[30] Yuceoglu U. , Özerciyes V., 2000, "Sudden Drop Phenomena in Natural Frequencies of Composite Plates or Panels with a Central Stiffening Plate Strip", Intr. Jour. of Computers and StructuresVol.76 (1-3) (Special Issue) pp.247-262

[31] Yuceoglu U. , Özerciyes V., 2005, "Free Vibrations of Adhesively Bonded Single Lap Joint in Shallow Cylindrical Shell Panels" , 2005. AIAA Journal Vol.43, pp.2537-2548.

[32] Yuceoglu U., Guvendik Ö., Özerciyes V., 2005, "Free Flexural (or Bending) Vibrations of Orthotropic Composite Plates and Panels with a Centrally Bonded Symmetric Lap Joint (or Symmetric Doubler Joint)" , "46th AIAA/ASME/ASCE/AHS/ASC Struct. Struct. Dynamics and Materials (SDM) Conference and Exhibit", (AIAA Paper No. AIAA-2005-26174).

[33] Yuceoglu U., Guvendik Ö., 2006, "Free Vibrations of Centrally Bonded Symmetric Lap Joint (or Symmetric Doubler Joint) with a Gap in Mindlin Plates or Panels", 47th AIAA/ASME/ASCE/AHS/ASC Struct. Struct. Dynamics and Materials (SDM) Conference and Exhibit, May 1-4, 2006, New Port , R.I. (AIAA Paper No. AIAA-2006-2109)

[34] Yuceoglu U., Guvendik, Ö, Özerciyes V., 2006, "Free Transverse Vibrations of Orthotropic Composite Mindlin Plates with a Non-Centrally Bonded Symmetric Lap Joint (or Symmetric Doubler Joint)" , 2006, ASME Intr. Mech. Engineering Congress and Exposition (IMECE, Nov. 5-10, 2006, Chicago, Illinois, "Joining Technologies for Advanced Materials and Structure Symposium" (ASME paper No. IMECE-2006-14013).

[35] Yuceoglu U., Guvendik, Ö, Özerciyes V., 2007, "Free Bending Vibrations of Centrally Bonded Symmetric Double Lap Joint (or Symmetric Double Doubler Joint) with a gap in Mindlin Plates or Panels", 2007 ASME Intr. Mech. Engineering Congress and Exposition, Nov.11-15, 2007, Seattle, Washington (ASME Paper No. IMECE-2007-41073)

[36] Mindlin, R. D., 1945, "Effects of Transverse Shear Deformations on the Bending of Elastic Plates", ASME, Jour. of Applied Mechanics, Vol.12, No.2, pp-A.69-A.77.

[37] Mindlin, R. D., 1951, "Influence of Rotatory Inertia and Transverse Shear on Flexural Motions of Isotropic, Elastic Plates," ASME Jour. of Applied Mechanics, 18, pp.31-38.

[38] Reissner, E., 1945, "The Effect of Transverse Shear Deformation on the Bending of Elastic Plates," ASME Jour. of Applied Mechanics, 12(2), pp A.69-A.77.

[39] Whitney, J.M., 1969, "The Effect of Transverse Shear Deformation on the Bending of Laminated Plates," Jour. of Composite Materials, 3, pp. 534-546.

[40] Whitney C. M., Pagano N. J., 1970, "Shear Deformation in Heterogeneous Anisotropic Plates," ASME Journal of Applied Mechanics, pp. 1031-1036.

[41] Khdeir, A. A. and Librescu, L., 1988, “Analysis of Symmetric Cross-ply Laminated Elastic Plates using Higher Order Theory - Part II,” Jour. of Composite Structures, 9, pp. 259-277.

[42] Kapania, R. K. and Raciti, S., 1989, “Recent Advances in Analysis of Laminated Beams and Plates, Part II,” AIAA Journal, 27, pp. 935-946.

[43] Reddy, J. N., 1990, “A Review of Refined Theories of Laminated Composite Plates," Shock and Vibration Digest, 22, pp. 3-17.

[44] Reddy, J. N. and Robbins, D. H., 1994, “Theories and Computational Models for Composite Laminates," ASME Applied Mechanics Reviews, 47(6), pp. 147-169.

TABLE 1. MATERIAL PROPERTIES AND DIMENSIONS of "STIFFENED PLATE or PANEL SYSTEM with TWO BONDED STIFFENING PLATE STRIPS"

"Isotropic" "Al-Alloy" System		Adhesive Layer		"Orthotropic" "Composite" System	
Al-Alloy Plate 1= Plate 2=Plate3 (j=1,2,3)	Al-Alloy Plate 4 (j=4)	"Hard"	"Soft"	Graphite-Epoxy Plate 1= Plate 2=Plate 3 (j=1,2,2)	Kevlar-Epoxy Plate 4 (j=4)
$E_x^{(j)}$=72.69 GPa	$E_x^{(j)}$=72.69 GPa	$E_a^{(j)}$=4.0 GPa	$E_a^{(j)}$=1.771MPa	$E_x^{(j)}$=11.71 GPa	$E_x^{(j)}$=5.5 GPa
$E_y^{(j)}$=72.69 GPa	$E_y^{(j)}$=72.69 GPa	$G_a^{(j)}$=1.4 GPa	$G_a^{(j)}$=0.453MPa	$E_y^{(j)}$=137.8 GPa	$E_y^{(j)}$=76.0 GPa
$G_{xy}^{(j)}$=25.78 GPa	$G_{xy}^{(j)}$=25.78 GPa	h_a=0.15x10^{-3} m	h_a=0.15x10^{-3} m	$G_{xy}^{(j)}$=5.51 GPa	$G_{xy}^{(j)}$=2.10 GPa
$G_{xz}^{(j)}$=25.78 GPa	$G_{xz}^{(j)}$=25.78 GPa	ρ_a=neglected	ρ_a=neglected	$G_{xz}^{(j)}$=2.5 GPa	$G_{xz}^{(j)}$=1.5 GPa
$G_{yz}^{(j)}$=25.78 GPa	$G_{yz}^{(j)}$=25.78 GPa			$G_{yz}^{(j)}$=3.0 GPa	$G_{yz}^{(j)}$=2.0 GPa
$\upsilon_{xy}^{(j)}$=0.313	$\upsilon_{xy}^{(j)}$=0.313			$\upsilon_{xy}^{(j)}$=0.0213	$\upsilon_{xy}^{(j)}$=0.024
$\upsilon_{yx}^{(j)}$=0.313	$\upsilon_{yx}^{(j)}$=0.313			$\upsilon_{yx}^{(j)}$=0.25	$\upsilon_{yx}^{(j)}$=0.34
$\rho^{(j)}$=2.796gr/cm^3	$\rho^{(j)}$=2.796gr/cm^3			$\rho^{(j)}$=1.6gr/cm^3	$\rho^{(j)}$=1.3gr/cm^3
$h_1 = h_2$=0,005m	h_3=0,005m			$h_1 = h_2$=0,005m	h_3=0,005 m
a=0.50m	**a**=0.50 m L=1.00 m			**a**=0.50m	**a**=0.50 m L=1.00 m

ℓ_I=Width of Left "Stiffening Plate Strip"
h_I=Thickness ("Stiffening Plate Strip")
h_{a_1} = Left Adhesive Thickness
ℓ_{III}=Far Left "Base Plate"
$\tilde{b}_I$=Mid-Point (Left "Stiffening Plate Strip")
h_4=Thickness ("Base Plate")
L=Total Length (Entire System),
a=Width (Entire System),

ℓ_{II}=Width of Middle "Stiffening Plate Strip"
h_2= Thickness ("Stiffening Plate Strip")
h_{a_2} = Middle Adhesive Thickness
ℓ_{IV}=Far Right "Base Plate"
$\tilde{b}_{II}$=Mid-Point (Middle "Stiffening Plate Strip")
h_4=Thickness ("Base Plate")
ℓ_V=Middle "Base Plate"

ℓ_{III}=Width of Right "Stiffening Plate Strip"
h_3=Thickness ("Stiffening Plate Strip")
h_{a_3} = Right Adhesive Thickness
ℓ_{III}=Far Right "Base Plate"
$\tilde{b}_{III}$=Mid-Point (Right "Stiffening Plate Strip")
h_4=Thickness ("Base Plate")
L=Total Length (Entire System),

APPENDIX A
"Mindlin Plate Theory" as Applied to Two-Layer (with an Adhesive Layer) and One-Layer Orthotropic Plates

- **Stress Resultant-Displacement Relations**

$$M_x = \frac{h^3}{12}\left(B_{11}\frac{\partial \psi_x}{\partial x} + B_{12}\frac{\partial \psi_y}{\partial y}\right)$$

$$Q_x = \kappa_x^2 hB_{55}\left(\psi_x + \frac{\partial w}{\partial x}\right)$$

$$M_y = \frac{h^3}{12}\left(B_{21}\frac{\partial \psi_x}{\partial x} + B_{22}\frac{\partial \psi_y}{\partial y}\right),$$

$$Q_y = \kappa_y^2 hB_{44}\left(\psi_y + \frac{\partial w}{\partial y}\right)$$

$$M_{yx} = \frac{h^3}{12}B_{66}\left(\frac{\partial \psi_x}{\partial y} + \frac{\partial \psi_y}{\partial x}\right)$$

where κ^2 terms are the "Shear Correction Factors" of the "Mindlin Plate Theory", and B's are material coefficients in the orthotropic stress-strain relations (Hooke's Law) such that,

$$B_{11} = \frac{E_x}{1-\nu_{xy}\nu_{yx}},\ B_{44} = G_{yz},\ B_{22} = \frac{E_y}{1-\nu_{yx}\nu_{xy}},$$

$$B_{55} = G_{xz}$$

$$B_{12} = B_{21} = \nu_{yx}B_{11} = \nu_{xy}B_{22},\ B_{66} = G_{xy}$$

$$B_{55} = G_{xz}$$

- **Equations Of Motion Of Mindlin Plates**

$$\frac{\partial M_x}{\partial x} + \frac{\partial M_{yx}}{\partial y} - Q_x + \frac{h}{2}\left(q_{zx}^{(+)} + q_{zx}^{(-)}\right) = \frac{\rho h^3}{12}\frac{\partial^2 \psi_x}{\partial t^2}$$

$$\frac{\partial M_{yx}}{\partial x} + \frac{\partial M_y}{\partial y} - Q_y + \frac{h}{2}\left(q_{zy}^{(+)} + q_{zy}^{(-)}\right) = \frac{\rho h^3}{12}\frac{\partial^2 \psi_y}{\partial t^2}$$

$$\frac{\partial Q_x}{\partial x} + \frac{\partial Q_y}{\partial y} + \left(q_z^{(+)} - q_z^{(-)}\right) = \rho h\frac{\partial^2 w}{\partial t^2}$$

- **Adhesive Layer Equations**

$$\gamma_{xz}^{(j)} = \underline{\frac{1}{2h_{ak}}\left(h_m\psi_x^{(m)} + h_n\psi_x^{(n)}\right)} - \underline{\underline{\frac{\partial w^{(j)}}{\partial x}}}$$

$$\gamma_{yz}^{(j)} = \underline{\frac{1}{2h_{ak}}\left(h_m\psi_y^{(m)} + h_n\psi_y^{(n)}\right)} - \underline{\underline{\frac{\partial w^{(j)}}{\partial y}}}$$

$$\varepsilon_z^{(j)} = \underline{\frac{w^{(n)} - w^{(m)}}{h_{ak}}}$$

for Part I
(m=1, n=4, k=1, for Adhesive Layer between "Left Plate Stiffener" and "Base Plate"

for Part II
(m=2, n=4, k=2, for Adhesive Layer between "Middle Plate Stiffener" and "Base Plate"

for Part III
(m=3, n=4, k=3, for Adhesive Layer between "Right Plate Stiffener" and "Base Plate"

- **Equations of Motion for "Plate Stiffeners" and "Base Palte" ("Bonded Joint" Region)**

$$\frac{\partial M_x^{(j)}}{\partial x} + \frac{\partial M_{yx}^{(j)}}{\partial y} - Q_x^{(j)} - \underline{\frac{h_k G_{ak}}{4h_{ak}}\left(h_m\psi_x^{(m)} + h_n\psi_x^{(n)}\right)} - \underline{\underline{\frac{h_1 G_{a1}}{2}\frac{\partial w^{(1)}}{\partial x}}} = \frac{\rho_j h_j^3}{12}\frac{\partial^2 \psi_x^{(j)}}{\partial t^2}$$

$$\frac{\partial M_{yx}^{(j)}}{\partial x} + \frac{\partial M_y^{(j)}}{\partial y} - Q_y^{(j)} - \underline{\frac{h_k G_{ak}}{4h_{ak}}\left(h_m\psi_y^{(m)} + h_n\psi_y^{(n)}\right)} - \underline{\underline{\frac{h_1 G_{a1}}{2}\frac{\partial w^{(1)}}{\partial y}}} = \frac{\rho_j h_j^3}{12}\frac{\partial^2 \psi_y^{(j)}}{\partial t^2}$$

$$\frac{\partial Q_x^{(j)}}{\partial x} + \frac{\partial Q_y^{(j)}}{\partial y} + \underline{\frac{E_{ak}}{h_{ak}}\left(w^{(n)} - w^{(m)}\right)} = \rho_j h_j\frac{\partial^2 w^{(j)}}{\partial t^2}$$

For Part I,
(m=1, n=4, k=1,and j=1 for "Left Plate Stiffener")
(m=4, n=1, k=1 and j=4 for "Base Plate")

For Part II,
(m=2, n=4, k=2,and j=2 for "Middle Plate Stiffener")
(m=4, n=2, k=2 and j=4 for "Base Plate")

For Part III,
(m=3, n=4, k=3,and j=3 for "Right Plate Stiffener")
(m=4, n=3, k=3 and j=4 for "Base Plate")

- **Equations of Motion for "Base Plate" (outside of "Bonded Joint" Region)**

$$\frac{\partial M_x^{(j)}}{\partial x} + \frac{\partial M_{yx}^{(j)}}{\partial y} - Q_x^{(j)} = \frac{\rho_j h_j^3}{12}\frac{\partial^2 \psi_x^{(j)}}{\partial t^2}$$

$$\frac{\partial M_{yx}^{(j)}}{\partial x} + \frac{\partial M_y^{(j)}}{\partial y} - Q_y^{(j)} = \frac{\rho_j h_j^3}{12}\frac{\partial^2 \psi_y^{(j)}}{\partial t^2}$$

$$\frac{\partial Q_x^{(j)}}{\partial x} + \frac{\partial Q_y^{(j)}}{\partial y} = \rho_j h_j\frac{\partial^2 w^{(j)}}{\partial t^2}$$

For Parts IV, V, VI, VII
(j=4 for "Base Plate")

- **Mindlin Boundary Conditions**

(F) (free) $M_{nt}=M_n=Q_n=0$
(S) (simply supported) $w=\psi_t=M_n=0$
(C) (clamped) $w=\psi_n=\psi_t=0$
where n and t are normal and tangential directions

APPENDIX B

- **Special Form of "Governing PDE's" in Part I and Part III,...,Part VII Regions:**

In Part I, Part II and Part III Regions (or Left, Middle, Right "Bonded Joint Regions")

$$\frac{1}{\ell_k}\frac{\partial \psi_y^{(j)}}{\partial \xi_k}=\frac{1}{B_{22}^{(j)}}\left(\frac{12}{h_j^3}M_y^{(j)}-B_{12}^{(j)}\frac{1}{a}\frac{\partial \psi_x^{(j)}}{\partial \eta}\right)$$

$$\frac{1}{\ell_k}\frac{\partial \psi_x^{(j)}}{\partial \xi_k}=\frac{12}{h_j^{(3)}B_{66}^{(j)}}M_{yx}^{(j)}-\frac{1}{a}\frac{\partial \psi_y^{(j)}}{\partial \eta}$$

$$\frac{1}{\ell_k}\frac{\partial w^{(j)}}{\partial \xi_k}=\frac{1}{\kappa_y^2 h_j B_{44}^{(j)}}Q_y^{(j)}-\psi_y^{(j)}$$

$$\frac{1}{\ell_k}\frac{\partial M_{yx}^{(j)}}{\partial \xi_k}=\frac{\rho_j h_j^3}{12}\frac{\partial^2 \psi_x^{(j)}}{\partial t^2}-\frac{1}{a}\frac{\partial M_x^{(j)}}{\partial \eta}+Q_x^{(j)}-\frac{h_j}{2}\left(q_{zx}^{(+)}+q_{zx}^{(-)}\right)$$

$$\frac{1}{\ell_k}\frac{\partial M_y^{(j)}}{\partial \xi_k}=\frac{\rho_j h_j^3}{12}\frac{\partial^2 \psi_y^{(j)}}{\partial t^2}-\frac{1}{a}\frac{\partial M_{yx}^{(j)}}{\partial \eta}+Q_y^{(j)}-\frac{h_j}{2}\left(q_{zy}^{(+)}+q_{zy}^{(-)}\right)$$

$$\frac{1}{\ell_k}\frac{\partial Q_y^{(j)}}{\partial \xi_k}=\rho_j h_j\frac{\partial^2 w^{(j)}}{\partial t^2}-\frac{1}{a}\frac{\partial Q_x^{(j)}}{\partial \eta}-\left(q_z^{(+)}-q_z^{(-)}\right)$$

$$\left|\begin{array}{l}(j=1,4)\ ,(k=\mathrm{I})\\(j=2,4)\ ,(k=\mathrm{II})\\(j=3,4),(k=\mathrm{III})\end{array}\right. \qquad \text{(B.1)}$$

In Part IV, Part V, Part VI, Part VII Regions (or "One-Layer Base Plate Regions")

$$\frac{1}{\ell_k}\frac{\partial \psi_y^{(j)}}{\partial \xi_k}=\frac{1}{B_{22}^{(j)}}\left(\frac{12}{h_j^3}M_y^{(j)}-B_{12}^{(j)}\frac{1}{a}\frac{\partial \psi_x^{(j)}}{\partial \eta}\right)$$

$$\frac{1}{\ell_k}\frac{\partial \psi_x^{(j)}}{\partial \xi_k}=\frac{12}{h_j^{(3)}B_{66}^{(j)}}M_{yx}^{(j)}-\frac{1}{a}\frac{\partial \psi_y^{(j)}}{\partial \eta}$$

$$\frac{1}{\ell_k}\frac{\partial w^{(j)}}{\partial \xi_k}=\frac{1}{\kappa_y^2 h_j B_{44}^{(j)}}Q_y^{(j)}-\psi_y^{(j)}$$

$$\frac{1}{\ell_k}\frac{\partial M_{yx}^{(j)}}{\partial \xi_k}=\frac{\rho_j h_j^3}{12}\frac{\partial^2 \psi_x^{(j)}}{\partial t^2}-\frac{1}{a}\frac{\partial M_x^{(j)}}{\partial \eta}+Q_x^{(j)}-\frac{h_j}{2}\left(q_{zx}^{(+)}+q_{zx}^{(-)}\right)$$

$$\frac{1}{\ell_k}\frac{\partial M_y^{(j)}}{\partial \xi_k}=\frac{\rho_j h_j^3}{12}\frac{\partial^2 \psi_y^{(j)}}{\partial t^2}-\frac{1}{a}\frac{\partial M_{yx}^{(j)}}{\partial \eta}+Q_y^{(j)}-\frac{h_j}{2}\left(q_{zy}^{(+)}+q_{zy}^{(-)}\right)$$

$$\frac{1}{\ell_k}\frac{\partial Q_y^{(j)}}{\partial \xi_k}=\rho_j h_j\frac{\partial^2 w^{(j)}}{\partial t^2}-\frac{1}{a}\frac{\partial Q_x^{(j)}}{\partial \eta}-\left(q_z^{(+)}-q_z^{(-)}\right)$$

$$\left|\begin{array}{l}(j=4)\ ,(k=\mathrm{IV})\\(j=4)\ ,(k=\mathrm{V})\\(j=4),(k=\mathrm{VI})\\(j=4),(k=\mathrm{VII})\end{array}\right. \qquad \text{(B.2)}$$

where q's are surface loads or the stresses on the upper (-) and lower (+) surfaces of the plates. They correspond, where appropriate, to the adhesive layer stresses as surface loads in bonded regions. Also, note that the superscript and the subscript "j" denotes the "Base Plate" for (j=4) and the "Stiffening Plate Strips" for (j=1, 2,3) (k=I,II,III) respectively.

APPENDIX C

- **"Classical Levy's Solutions" in Part I, Part II and Part III Regions:**

Displacements In Part I, Part II and Part III regions,

$$w^{(j)}(\eta,\xi_k,t)=h_1\sum_{m=1}^{\infty}\sum_{n=1}^{\infty}\overline{W}_{mn}^{(j)}(\xi_k)\sin(m\pi\eta)e^{i\omega_{mn}t}$$

$$\psi_x^{(j)}(\eta,\xi_k,t)=\sum_{m=1}^{\infty}\sum_{n=1}^{\infty}\overline{\Psi}_{mnx}^{(j)}(\xi_k)\cos(m\pi\eta)e^{i\omega_{mn}t}$$

$$\psi_y^{(j)}(\eta,\xi_k,t)=\sum_{m=1}^{\infty}\sum_{n=1}^{\infty}\overline{\Psi}_{mny}^{(j)}(\xi_k)\sin(m\pi\eta)e^{i\omega_{mn}t}$$

$$\begin{matrix}(0<\xi_I<1)\\(0<\xi_{II}<1)\\(0<\xi_{III}<1)\end{matrix}\quad\left|\begin{matrix}j=1,4 \text{ for Part I}\\ j=2,4 \text{ for Part II}\\ j=3,4 \text{ for Part III}\\ k=I \text{ for Part I}\\ k=II \text{ for Part II}\\ k=III \text{ for Part III}\end{matrix}\right.\qquad (C.1)$$

Stress Resultants In Part I, Part II and Part III regions,

$$M_x^{(j)}(\eta,\xi_k,t)=\frac{h_1^5B_{11}^{(1)}}{a^3}\sum_{m=1}^{\infty}\sum_{n=1}^{\infty}\overline{M}_{mnx}^{(j)}(\xi_k)\sin(m\pi\eta)e^{i\omega_{mn}t}$$

$$M_{yx}^{(j)}(\eta,\xi_k,t)=\frac{h_1^5B_{11}^{(1)}}{a^3}\sum_{m=1}^{\infty}\sum_{n=1}^{\infty}\overline{M}_{mnyx}^{(j)}(\xi_k)\cos(m\pi\eta)e^{i\omega_{mn}t}$$

$$M_y^{(j)}(\eta,\xi_k,t)=\frac{h_1^5B_{11}^{(1)}}{a^3}\sum_{m=1}^{\infty}\sum_{n=1}^{\infty}\overline{M}_{mny}^{(j)}(\xi_k)\sin(m\pi\eta)e^{i\omega_{mn}t}$$

$$Q_y^{(j)}(\eta,\xi_k,t)=\frac{h_1^4B_{11}^{(1)}}{a^3}\sum_{m=1}^{\infty}\sum_{n=1}^{\infty}\overline{Q}_{mny}^{(j)}(\xi_k)\sin(m\pi\eta)e^{i\omega_{mn}t}$$

$$Q_x^{(j)}(\eta,\xi_k,t)=\frac{h_1^4B_{11}^{(1)}}{a^3}\sum_{m=1}^{\infty}\sum_{n=1}^{\infty}\overline{Q}_{mnx}^{(j)}(\xi_k)\cos(m\pi\eta)e^{i\omega_{mn}t}$$

$$\begin{matrix}(0<\xi_I<1)\\(0<\xi_{II}<1)\\(0<\xi_{III}<1)\end{matrix}\quad\left|\begin{matrix}j=1,4 \text{ for Part I}\\ j=2,4 \text{ for Part II}\\ j=3,4 \text{ for Part III}\\ k=I \text{ for Part I}\\ k=II \text{ for Part II}\\ k=III \text{ for Part III}\end{matrix}\right.\qquad (C.2)$$

where "barred" or $(-)$ quantities are the nondimensional displacements, angles of rotations and stress resultants, and "m" is the number of half-waves in Fourier Series in the x-direction, and ξ_I is defined as y_I/ℓ_I and ξ_{II} is defined as y_{II}/ℓ_{II} and y_{III}/ℓ_{III} (see also Fig.3). whereas "*i*"$=\sqrt{-1}$.

Similar expressions can be written in Part IV, Part V, Part VI and Part VII (all of them one-layer, "Base Plate or Panel "), with (j=4) and (k=IV,V,VI,VII), respectively.

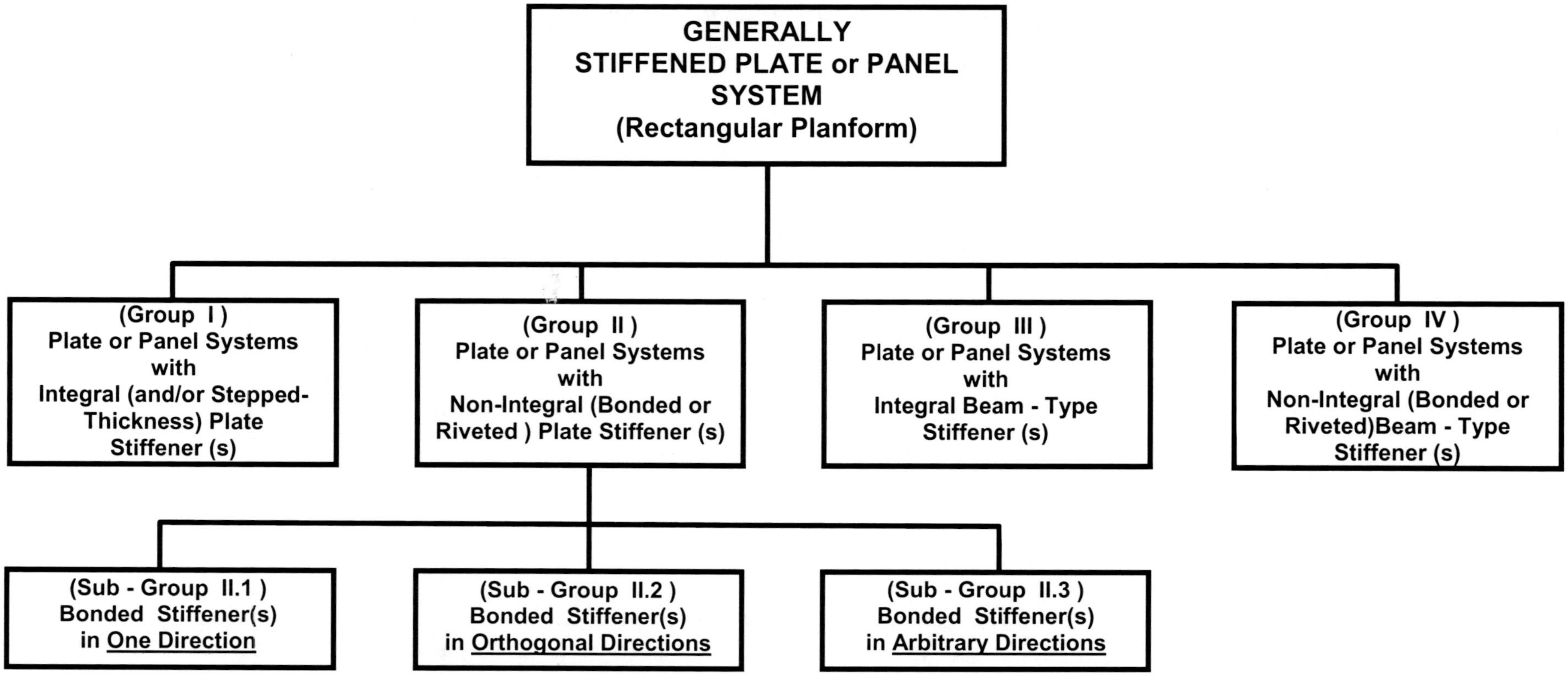

Figure 1. “Groups” of “Stiffened Plate or Panel Systems” (Ref. Yuceoglu et al [5,6])

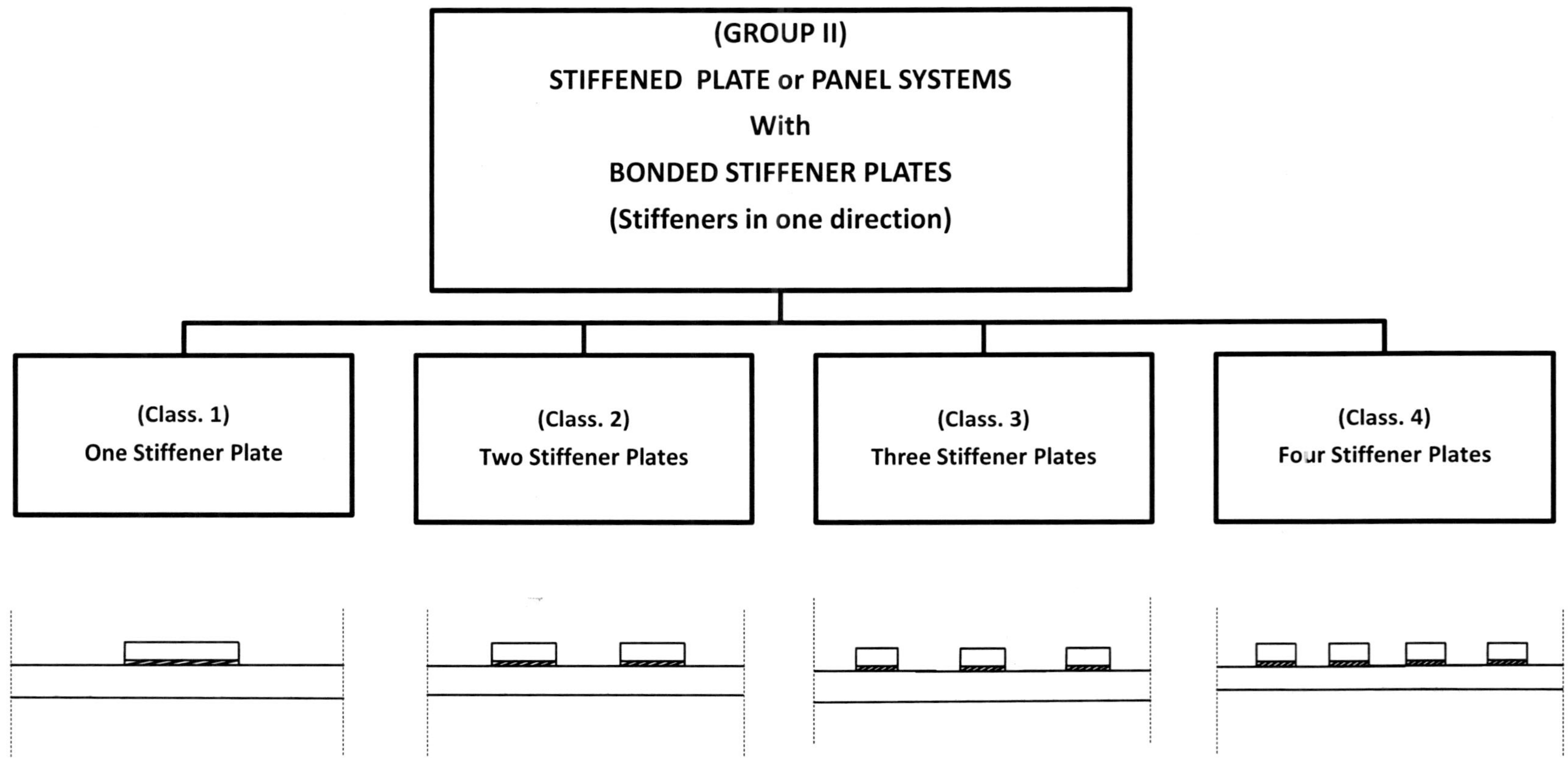

Figure 2. "Classes" of "Stiffened Plate or Panel Systems with Bonded Stiffener Plates" (Ref. Yuceoglu et al [5,6])

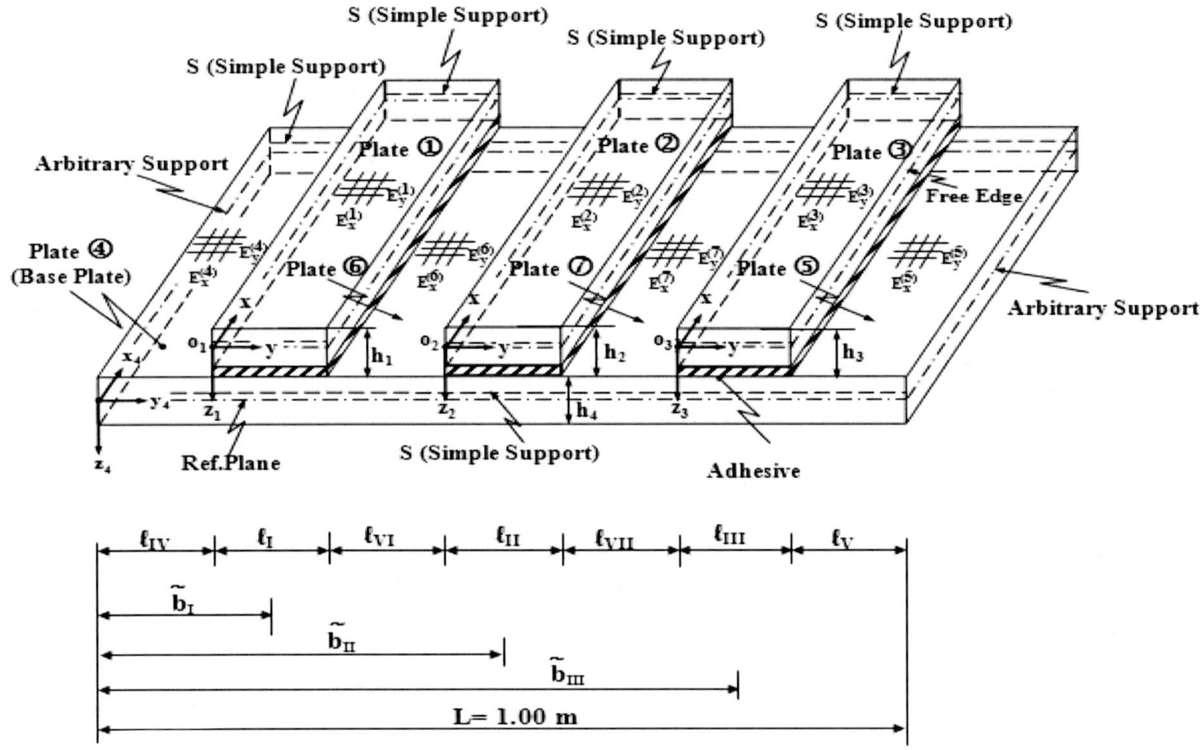

(a) General Configuration, Geometry, Coordinate Systems and Material Directions

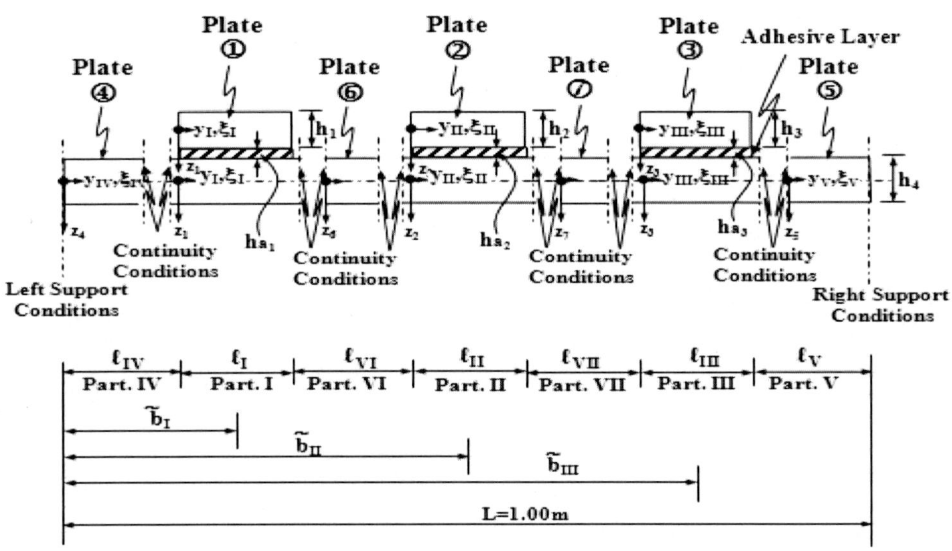

(a) Longitudinal Cross-Section with Parts I,II, III, IV, V, VI, VII and Coordinate Systems

Figure 3. "Stiffened Composite Base Plate or Panel System with Three Bonded Stiffening Plate Strips"

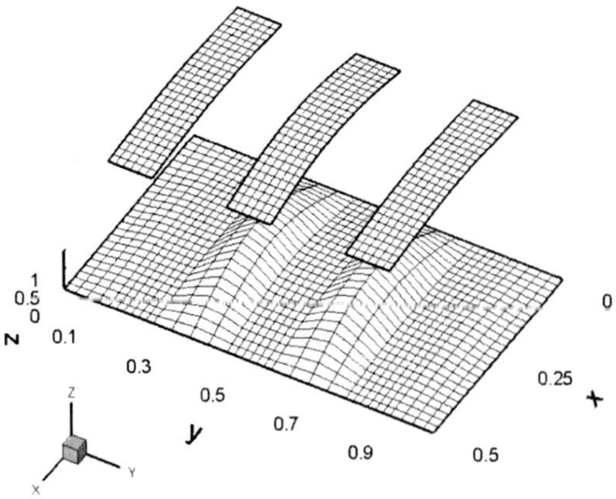

First Mode With $\bar{\Omega}_1=\bar{\omega}_{11}$ =5738.572

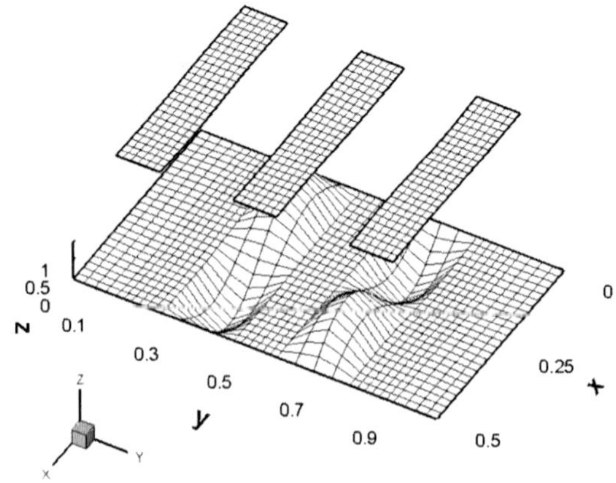

Fourth Mode With $\bar{\Omega}_4=\bar{\omega}_{22}$ =8946.250

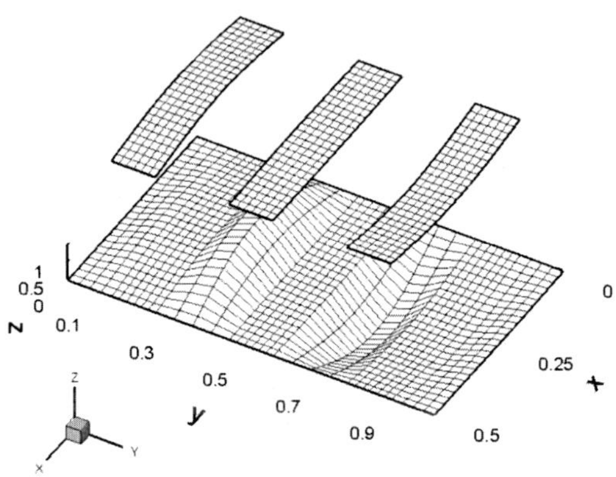

Second Mode With $\bar{\Omega}_2=\bar{\omega}_{12}$ = 7210.463

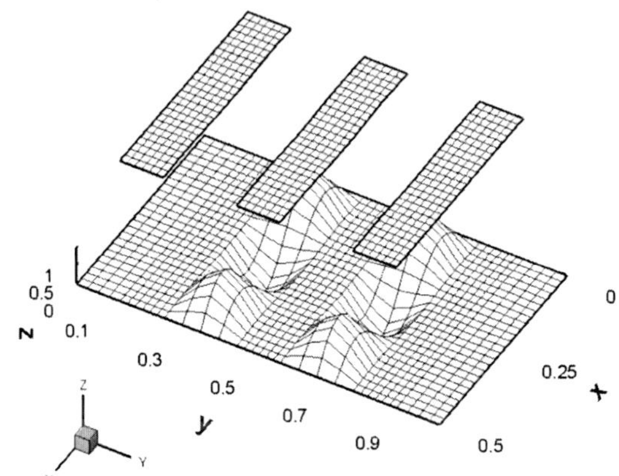

Fifth Mode With $\bar{\Omega}_5=\bar{\omega}_{31}$ =9415.105

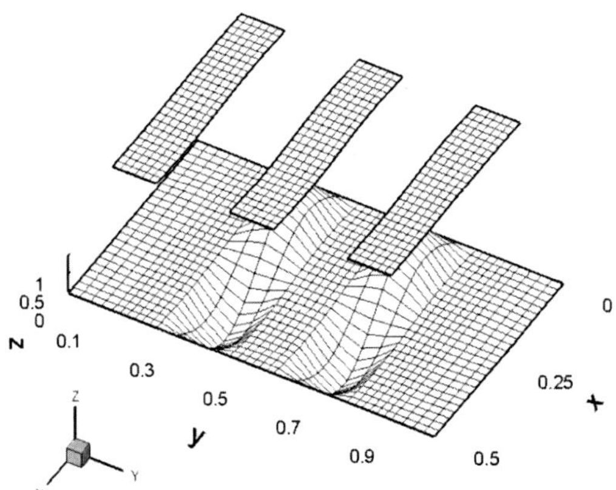

Third Mode With $\bar{\Omega}_3=\bar{\omega}_{21}$ =8399.761

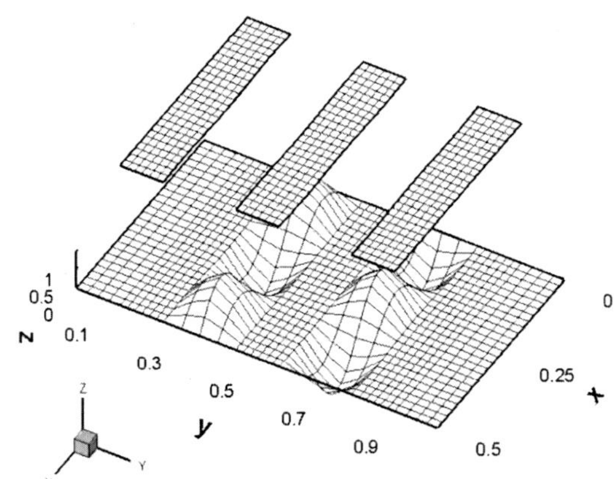

Sixth Mode With $\bar{\Omega}_6=\bar{\omega}_{32}$ =9738.124

Figure 4 .a Mode Shapes and Dimensionless Nat. Freqs of "Composite Base Plate or Panel with Three Bonded Stiffening Plate Strips"
("Orthotropic" case with "Hard" adhesive)
(Plate Strip ①=Plate Strip ②= Plate Strip ③=Graphite-Epoxy, Base Plate=Plate ④= Kevlar-Epoxy)
($\ell_I=\ell_{II}=\ell_{III}=\ell_{IV}=\ell_V$=0.12m, $\ell_{VI}=\ell_{VII}$=0.2m)
(a=0.5m, $h_1=h_2=h_3$=0.005m, h_4=0.005m, $h_{a1}=h_{a2}=h_{a3}=0.15\times10^{-3}$ m , L=1.00m)
(Boundary Conditions in y-direction FFFFFFCC)

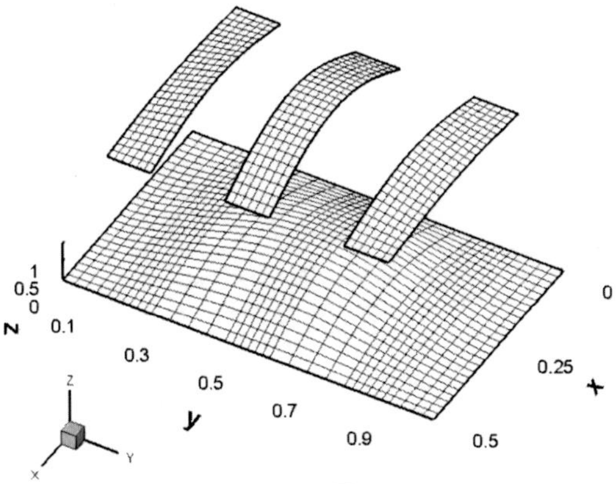

First Mode With $\bar{\Omega}_1=\bar{\omega}_{11}=76.985$

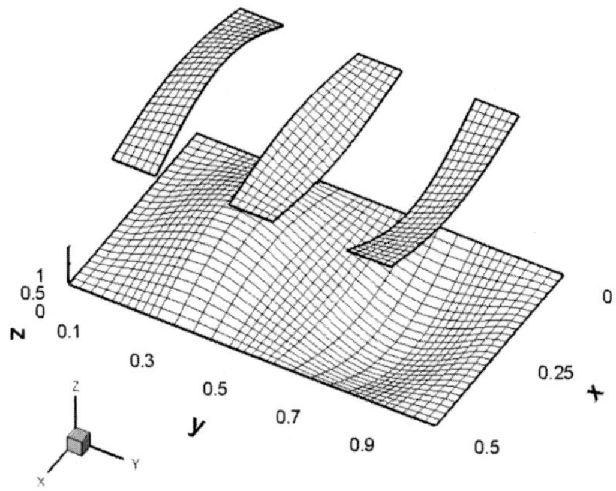

Second Mode With $\bar{\Omega}_2=\bar{\omega}_{12}=240.439$

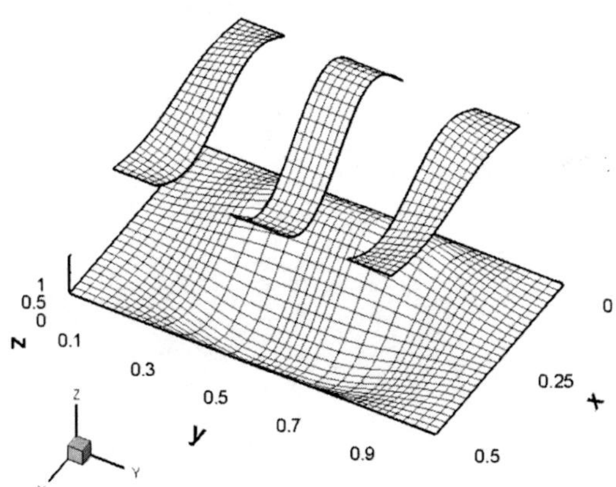

Third Mode With $\bar{\Omega}_3=\bar{\omega}_{21}=288.474$

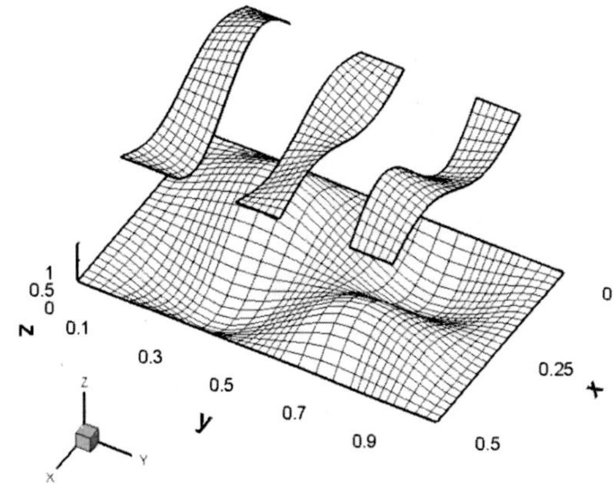

Fourth Mode With $\bar{\Omega}_4=\bar{\omega}_{22}=475.055$

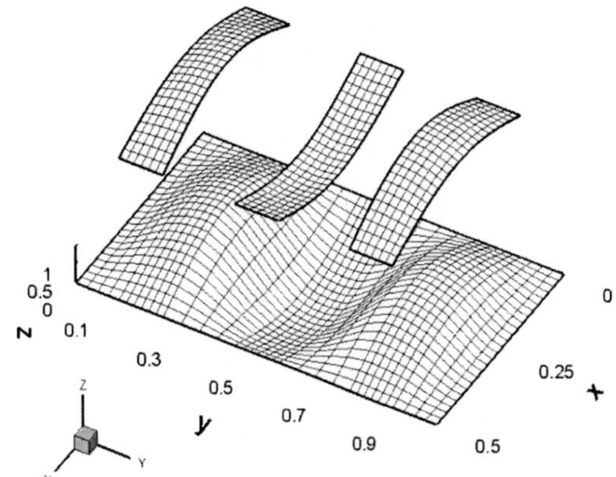

Fifth Mode With $\bar{\Omega}_5=\bar{\omega}_{13}=527.957$

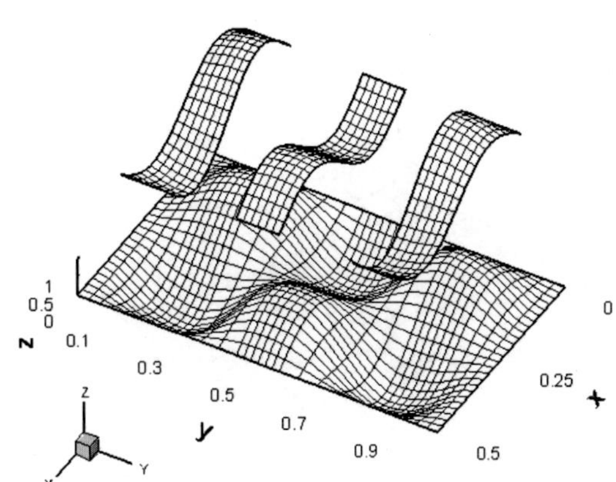

Sixth Mode With $\bar{\Omega}_6=\bar{\omega}_{23}=833.047$

Figure 4.b Mode Shapes and Dimensionless Nat. Freqs of "Composite Base Plate or Panel with Three Bonded Stiffening Plate Strips"
("Orthotropic" case with "Soft" adhesive)
(Plate Strip ①=Plate Strip ②= Plate Strip ③=Graphite-Epoxy, Base Plate=Plate ④= Kevlar-Epoxy)
($\ell_I=\ell_{II}=\ell_{III}=\ell_{IV}=\ell_V=0.12$m, $\ell_{VI}=\ell_{VII}=0.2$m)
(a=0.5m, $h_1=h_2=h_3=0.005$m, $h_4=0.005$m, $h_{a1}=h_{a2}=h_{a3}=0.15\times10^{-3}$ m , L=1.00m)
(Boundary Conditions in y-direction FFFFFFCC)

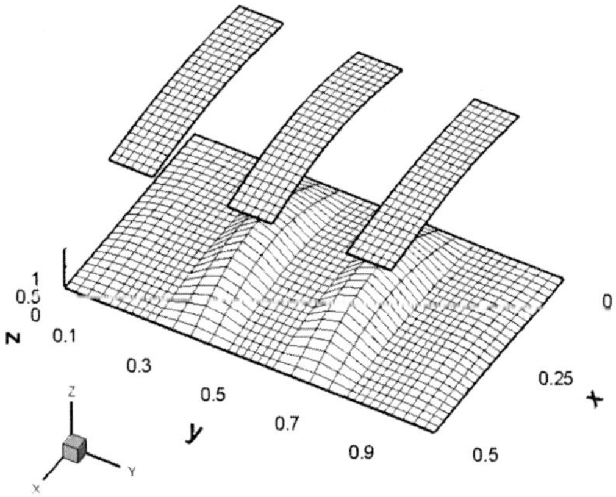

First Mode With $\bar{\Omega}_1=\bar{\omega}_{11}=5659.121$

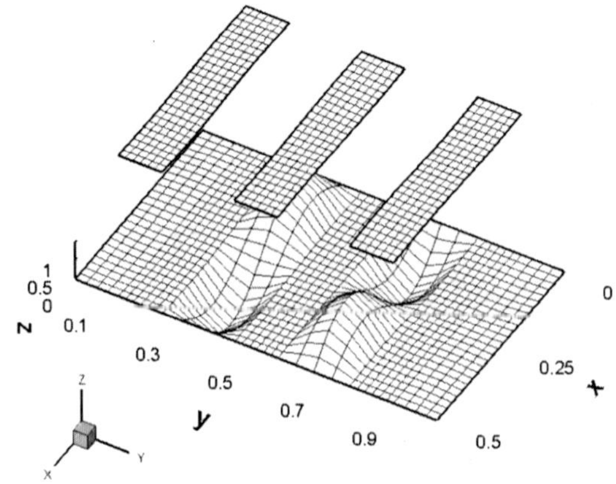

Fourth Mode With $\bar{\Omega}_4=\bar{\omega}_{22}=8935.610$

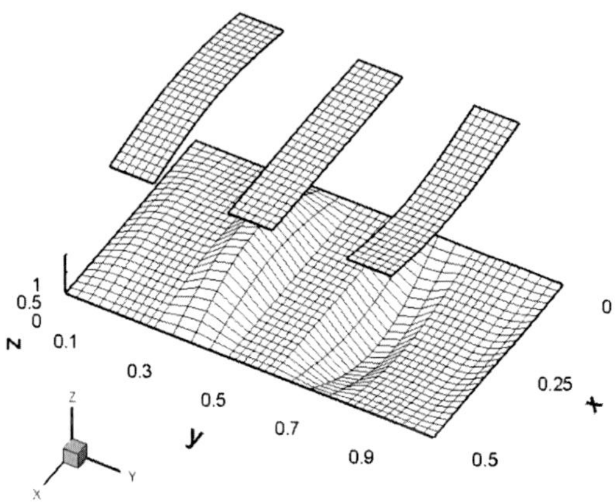

Second Mode With $\bar{\Omega}_2=\bar{\omega}_{12}=6952.001$

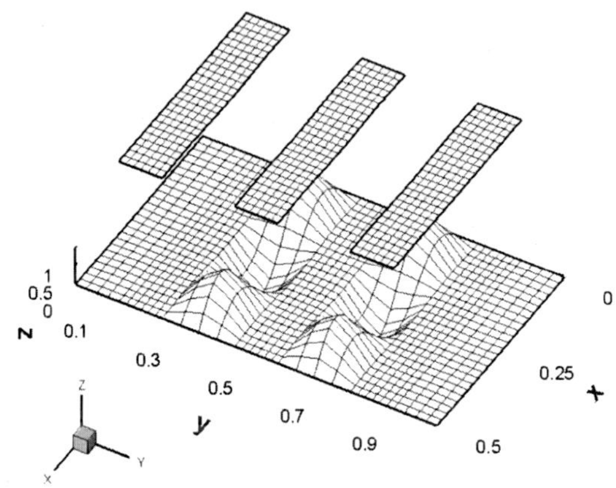

Fifth Mode With $\bar{\Omega}_5=\bar{\omega}_{31}=9412.487$

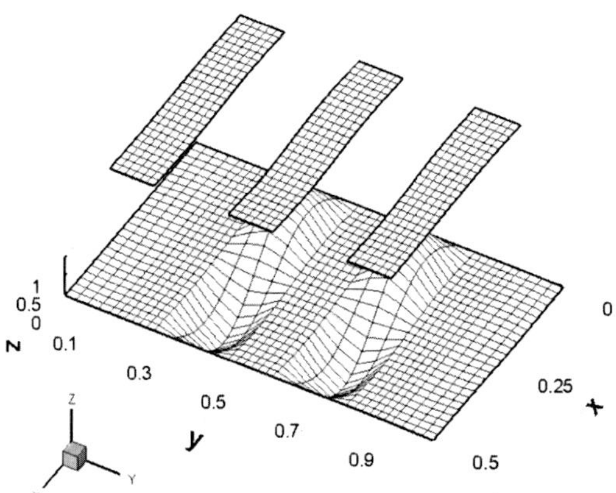

Third Mode With $\bar{\Omega}_3=\bar{\omega}_{21}=8391.412$

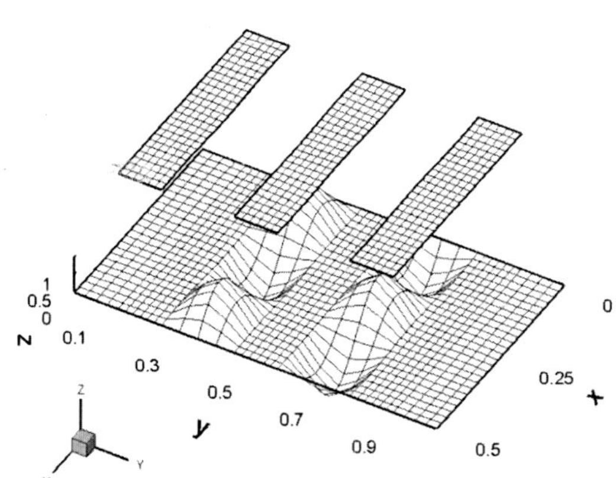

Sixth Mode With $\bar{\Omega}_6=\bar{\omega}_{32}=9735.242$

Figure 5 .a Mode Shapes and Dimensionless Nat. Freqs of "Composite Base Plate or Panel with Three Bonded Stiffening Plate Strips"
("Orthotropic" case with "Hard" adhesive)
(Plate Strip ①=Plate Strip ②= Plate Strip ③=Graphite-Epoxy, Base Plate=Plate ④= Kevlar-Epoxy)
($\ell_I=\ell_{II}=\ell_{III}=\ell_{IV}=\ell_V=0.12m$, $\ell_{VI}=\ell_{VII}=0.2m$)
($a=0.5m$, $h_1=h_2=h_3=0.005m$, $h_4=0.005m$, $h_{a1}=h_{a2}=h_{a3}=0.15x10^{-3}$ m , L=1.00m)
(Boundary Conditions in y-direction FFFFFFSS)

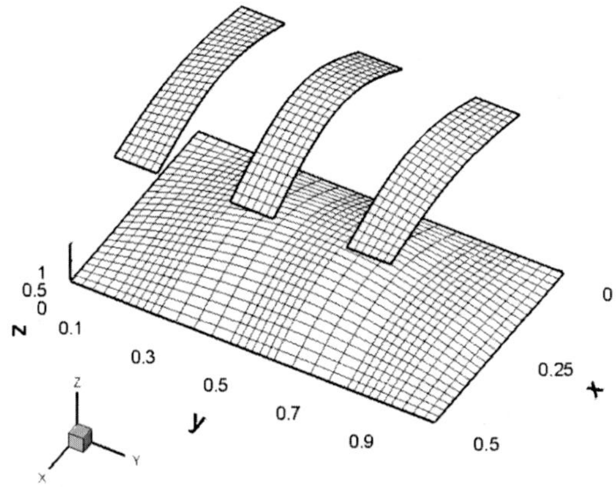

First Mode With $\bar{\Omega}_1=\bar{\omega}_{11}$ =59.540

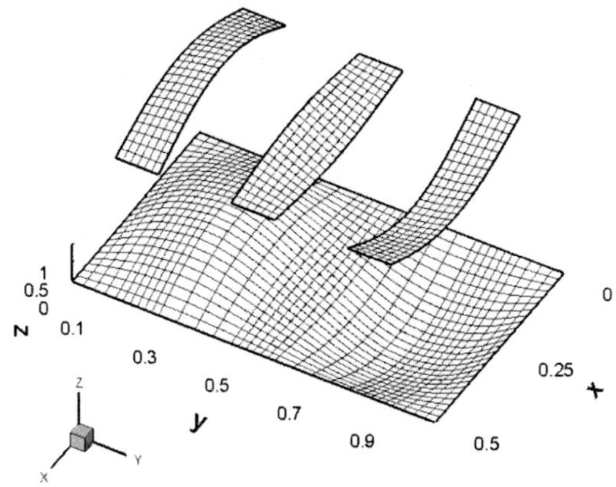

Second Mode With $\bar{\Omega}_2=\bar{\omega}_{12}$ = 139.426

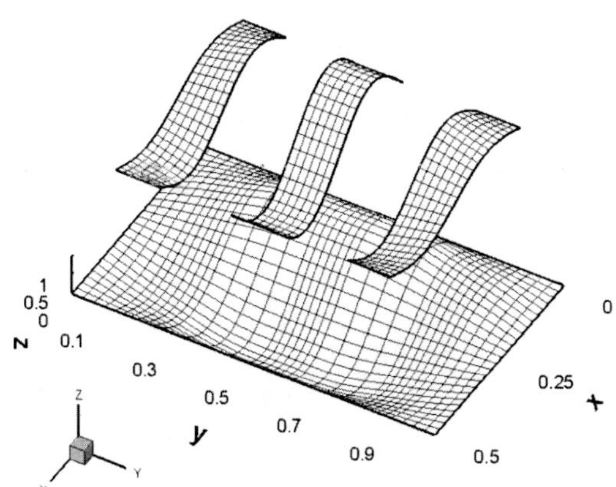

Third Mode With $\bar{\Omega}_3=\bar{\omega}_{21}$ =270.6121

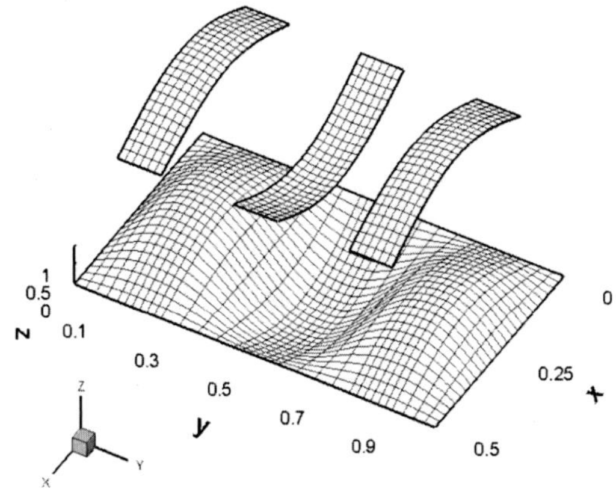

Fourth Mode With $\bar{\Omega}_4=\bar{\omega}_{13}$ =335.259

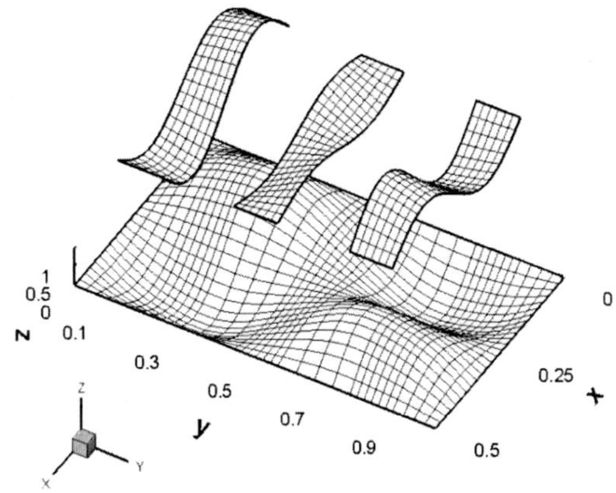

Fifth Mode With $\bar{\Omega}_5=\bar{\omega}_{22}$ =378.223

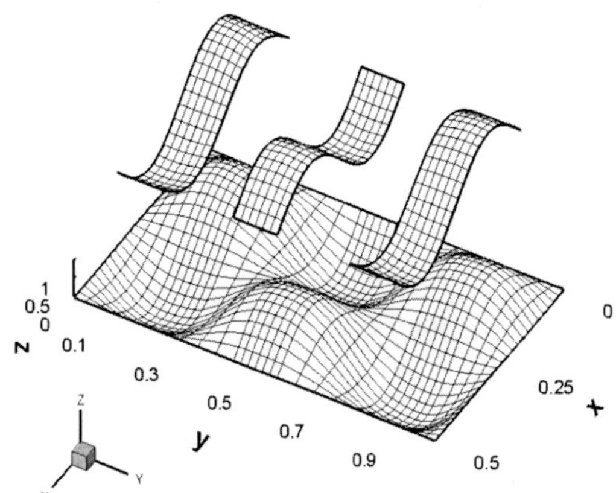

Sixth Mode With $\bar{\Omega}_6=\bar{\omega}_{23}$ =633.163

Figure 5 .b Mode Shapes and Dimensionless Nat. Freqs of "Composite Base Plate or Panel with Three Bonded Stiffening Plate Strips"
("Orthotropic" case with "Soft" adhesive)
(Plate Strip ①=Plate Strip ②= Plate Strip ③=Graphite-Epoxy, Base Plate=Plate ④= Kevlar-Epoxy)
($\ell_I = \ell_{II} = \ell_{III} = \ell_{IV} = \ell_V$=0.12m, $\ell_{VI} = \ell_{VII}$=0.2m)
(a=0.5m, $h_1=h_2=h_3$=0.005m, h_4=0.005m, $h_{a1}=h_{a2}=h_{a3}=0.15\times10^{-3}$ m , L=1.00m)
(Boundary Conditions in y-direction FFFFFFSS)

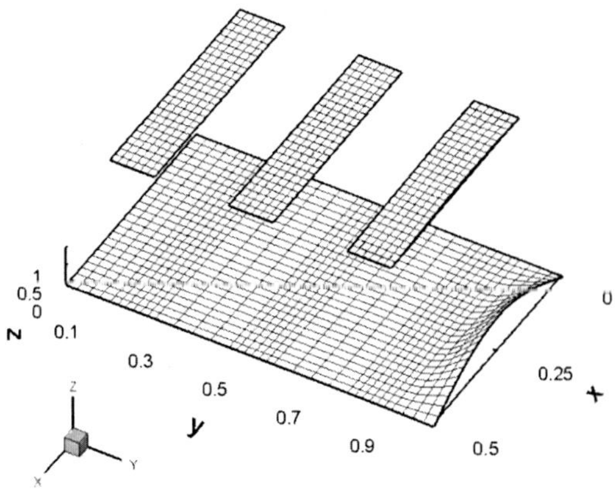

First Mode With $\bar{\Omega}_1=\bar{\omega}_{11}$ =1885.347

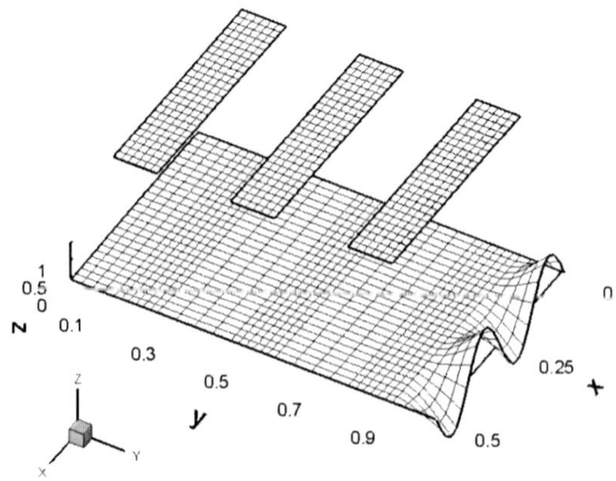

Fourth Mode With $\bar{\Omega}_4=\bar{\omega}_{41}$ =3817.117

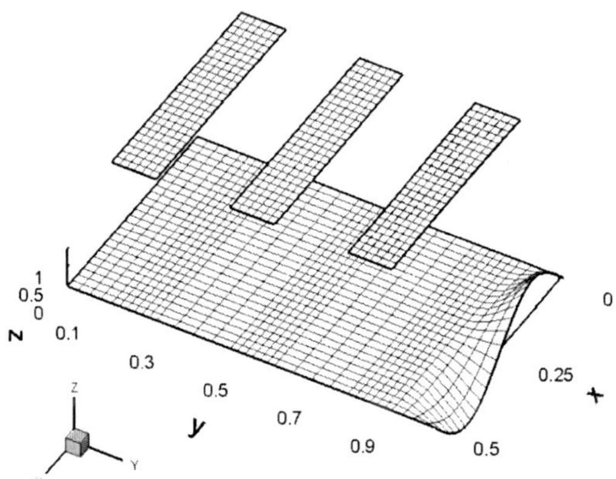

Second Mode With $\bar{\Omega}_2=\bar{\omega}_{21}$ = 2163.944

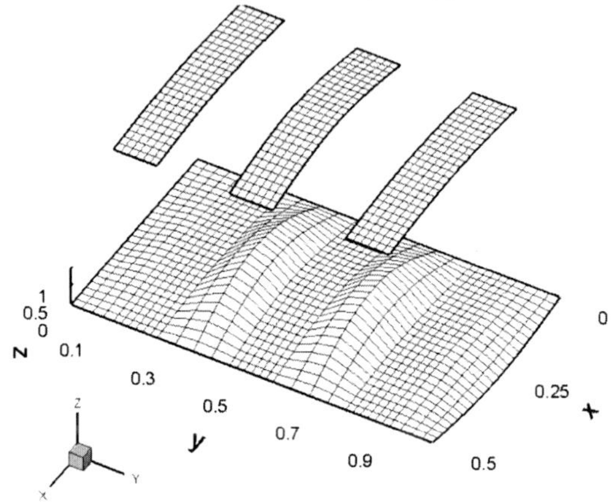

Fifth Mode With $\bar{\Omega}_5=\bar{\omega}_{12}$ =5737.679

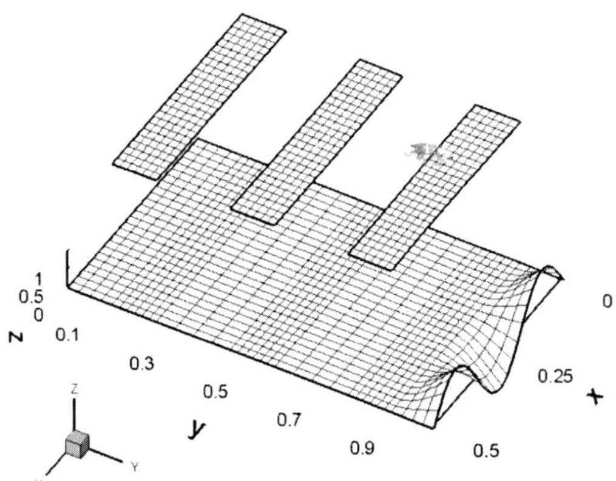

Third Mode With $\bar{\Omega}_3=\bar{\omega}_{31}$ =2698.198

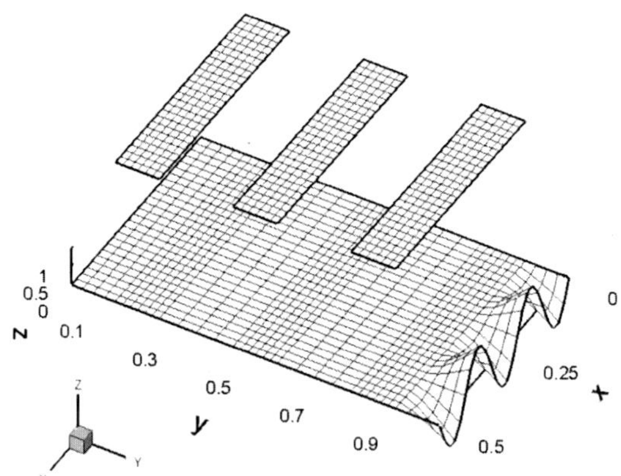

Sixth Mode With $\bar{\Omega}_6=\bar{\omega}_{51}$ =5903.169

Figure 6.a Mode Shapes and Dimensionless Nat. Freqs of "Composite Base Plate or Panel with Three Bonded Stiffening Plate Strips"
("Orthotropic" case with "Hard" adhesive)
(Plate Strip ①=Plate Strip ②= Plate Strip ③=Graphite-Epoxy, Base Plate=Plate ④= Kevlar-Epoxy)
($\ell_I=\ell_{II}=\ell_{III}=\ell_{IV}=\ell_V$=0.12m, $\ell_{VI}=\ell_{VII}$=0.2m)
(a=0.5m, $h_1=h_2=h_3$=0.005m, h_4=0.005m, $h_{a1}=h_{a2}=h_{a3}=0.15\times10^{-3}$ m , L=1.00m)
(Boundary Conditions in y-direction FFFFFFCF)

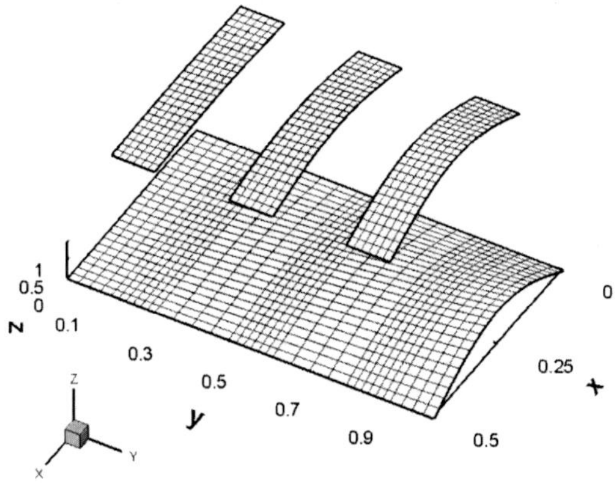

First Mode With $\bar{\Omega}_1=\bar{\omega}_{11}$ =47.275

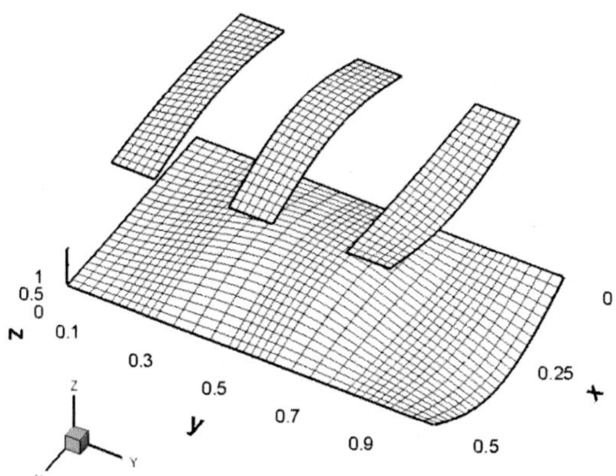

Second Mode With $\bar{\Omega}_2=\bar{\omega}_{12}$ = 93.767

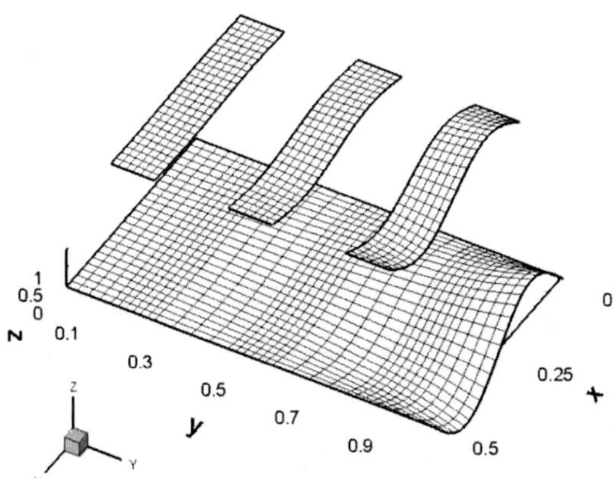

Third Mode With $\bar{\Omega}_3=\bar{\omega}_{21}$ =235.616

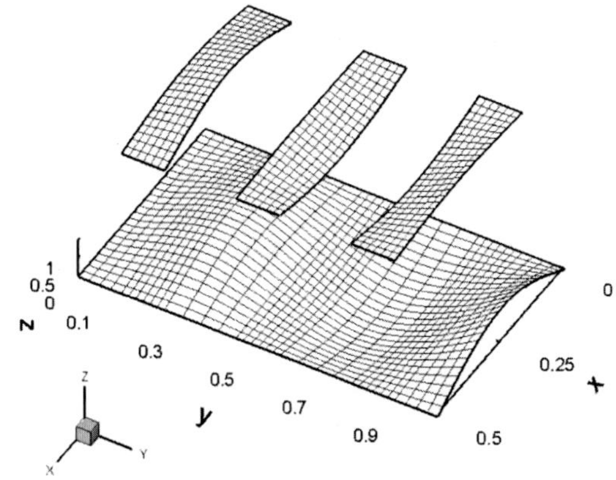

Fourth Mode With $\bar{\Omega}_4=\bar{\omega}_{13}$ =288.941

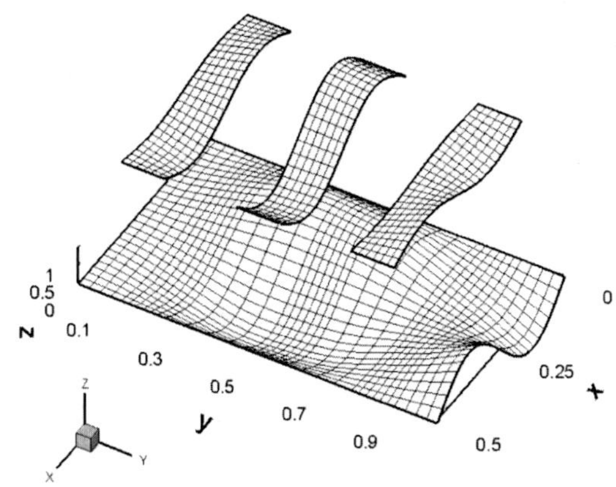

Fifth Mode With $\bar{\Omega}_5=\bar{\omega}_{22}$ =302.163

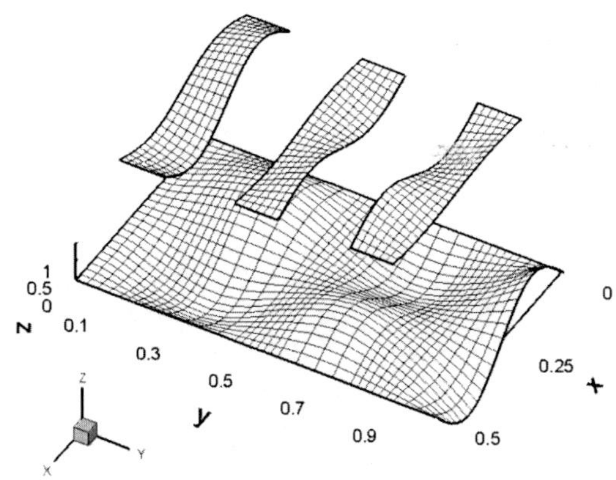

Sixth Mode With $\bar{\Omega}_6=\bar{\omega}_{23}$ =520.225

Figure 6 .b Mode Shapes and Dimensionless Nat. Freqs of "Composite Base Plate or Panel with Three Bonded Stiffening Plate Strips"
("orthotropic" case with "Soft" adhesive)
(Plate Strip ①=Plate Strip ②= Plate Strip ③=Graphite-Epoxy, Base Plate=Plate ④= Kevlar-Epoxy)
($\ell_I= \ell_{II} = \ell_{III} = \ell_{IV}= \ell_V$=0.12m, $\ell_{VI}= \ell_{VII}$=0.2m)
(a=0.5m, $h_1=h_2= h_3$ =0.005m, h_4=0.005m, $h_{a1}=h_{a2}= h_{a3}$ =0.15x10^{-3} m , L=1.00m)
(Boundary Conditions in y-direction FFFFFFCF)

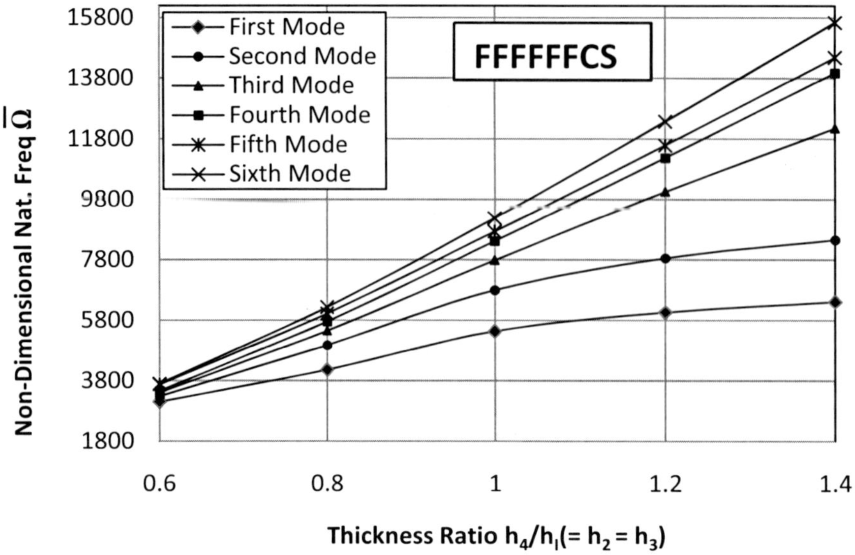

a) "Hard" Adhesive Case

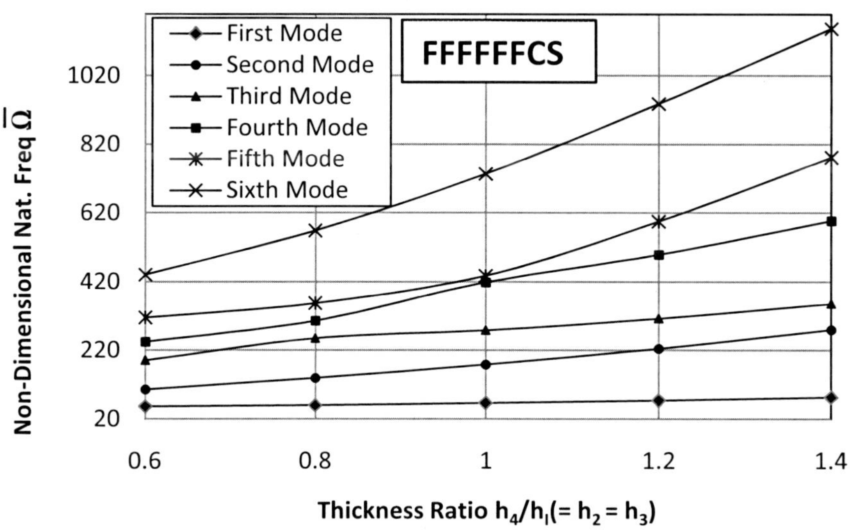

b) "Soft" Adhesive Case

Figure 7a,b Dimensionless Nat. Freqs versus "Thickness Ratio $h_4/h_i(=h_2=h_3)$" in "Orthotropic Base Plate with Three Bonded Stiffening Plate Strips".
("Orthotropic" Case)
(Plate Strip ①=Plate Strip ②= Plate Strip ③=Graphite-Epoxy, Base Plate=Plate ④= Kevlar-Epoxy)
($\ell_I=\ell_{II}=\ell_{III}=\ell_{IV}=\ell_V=0.12$m, $\ell_{VI}=\ell_{VII}=0.2$m)
(a=0.5m, $h_1=h_2=h_3=0.005$m, h_4=Variess, $h_{a1}=h_{a2}=h_{a3}=0.15$x10-3 m , L=1.00m)
(Boundary Conditions in y-direction FFFFFFCS)

Proceedings of the ASME 2010 International Mechanical Engineering Congress & Exposition
IMECE2010
November 12-18, 2010, Vancouver, British Columbia, Canada

IMECE2010-38851

NOVEL MORPHING WING DESIGN USING ANTAGONISTIC SHAPE MEMORY ALLOY ACTUATION

AYN Sofla
University of Toronto
Toronto, Ontario, Canada

SA Meguid
University of Toronto
Toronto, Ontario, Canada

KT Tan
Defence Science Organization
Singapore, Singapore

ABSTRACT

A wing shear concept is adopted here to design and fabricate morphing wing for an unmanned aerial vehicle. The concept uses a parallelogram wing-box that consists of several composite rib shells that are hinged to two active spars. Antagonistic shape memory actuation is used to flex the spars and consequently shape morph the wing between straight and curved shapes.

INTRODUCTION

The wing of conventional plane is equipped with high lift devices such as flaps or slats that are deployed when higher lift is required or when the speed is too low to generate sufficient lift. Examples for such flight conditions are take-off or landing. These devices are mainly used in larger planes where the weight of accessory components such as hydraulic/electric actuators or power transmission mechanisms or overlapping aerodynamic surfaces are negligible compared to the gross weight of the plane. For smaller aircraft especially unmanned aerial vehicles (UAVs) such luxury is not affordable. As a result many UAVs require a form of launching system to provide the required lift for the take-off by the initial speeding of the plane. Such small planes, however, would be more robust by adapting to different flight conditions.

Smart materials such as shape memory alloy (SMA) or piezoelectric materials have unveiled exciting opportunities in the field of adaptive structures in the past decade. In the area of aeronautics, many researchers have been exploring smart materials enabled *shape morphing* vehicles [1]. The main reason is that the multifunctional smart material generates actuation and yet carries aerodynamic loads. This is superior to the conventional actuating devices such as electromagnetic or hydraulic motors that the bulk of their body does not carry aerodynamic load. On the other hand, the smart materials can be distributed within the shape morphing device and therefore eliminate the need for power transmission systems such as linkages or lead/screw systems [2]. The use of smart materials, therefore, overcomes the undesirable extra weight associated with the conventional actuation systems.

The wing morphing designs can be classified into three major types: plan-form alternation, out-of-plane transformation, and airfoil adjustment [1]. Here we report on our recent developments in the design and fabrication of a shape morphing wing for UAV. We previously developed a new concept for shape morphing of the wing of small airplanes [3]. The objective of this wing morphing design is to reduce the wing drag by reducing the wing area (plan-form) during flight. The spars and ribs in this design form a parallelogram wing-box that is fixed to the fuselage. Equal and in-phase flexing of the active spars causes the ribs to slide in the wing-plane. As a result, the area of the resulted curved wing is less than its original straight shape. Each adaptive spar in this wing morphing design is actuated by means of a pair of one one-way SMA actuator in a setup called Antagonistic Flexural Cell (AFC) [4].

SHAPE MORPHING WING CONCEPT

The novel morphing wing design in this study uses a shearing approach to reduce the wing drag by reducing its area during flight as shown in the bottom picture of Fig. 1. The wing shear concepts [5-8] use several ribs that are hinged to few (in this research two) parallel spars to form parallelogram wing-boxes (one wing-box at either side of the plane). A major problem with the shear wing concept, shown in the top of Fig. 1, is that at least one of the spars needs to be shifted laterally (and consequently longitudinally) to reduce the area of the wing. The linkage that carries out the above task at the wing root is under extreme aerodynamic load and therefore can be complex to build. Additionally, large stroke linear displacement actuators are needed to provide such large movements. As a result, shear concepts have not yet been seriously pursued in the past for the shape morphing of aircraft wing.

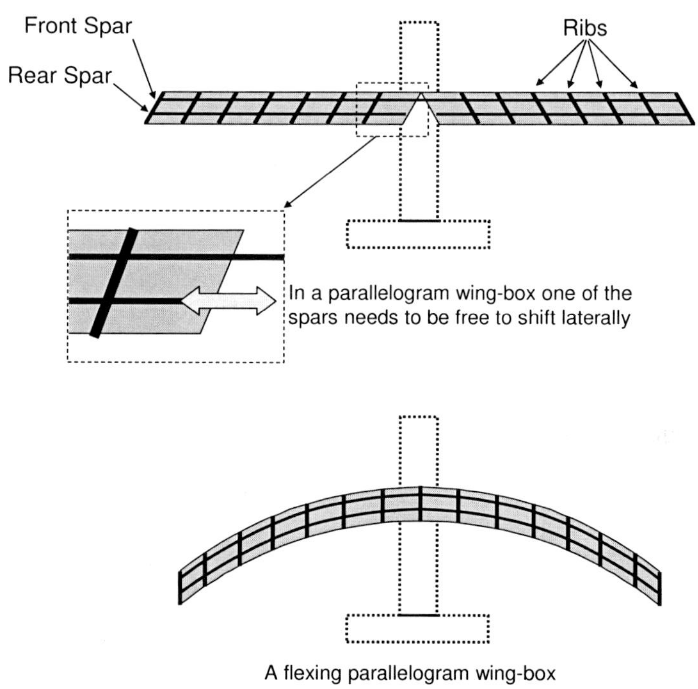

Figure 1: Parallelogram concept to change the wing area. Top) regular parallelogram, bottom) flexing parallelogram.

In this research we have developed a new wing shear concept called flexing parallelogram that uses smart materials and also addresses the spar-to-fuselage connection problem. The spars and ribs in this design form a parallelogram wing-box that is fixed to the fuselage to eliminate the need for pivots or sliding links at the joints (Fig. 1). Equal and in-phase flexing of the spars in the plane of the wing (upon actuation) causes the ribs to slide in the wing-plane (Fig. 1). As a result, the area of the resulted curved wing is less than its original straight shape. Proportionately, the drag and lift forces are also reduced. The concept is shown in more detail in Fig.2.

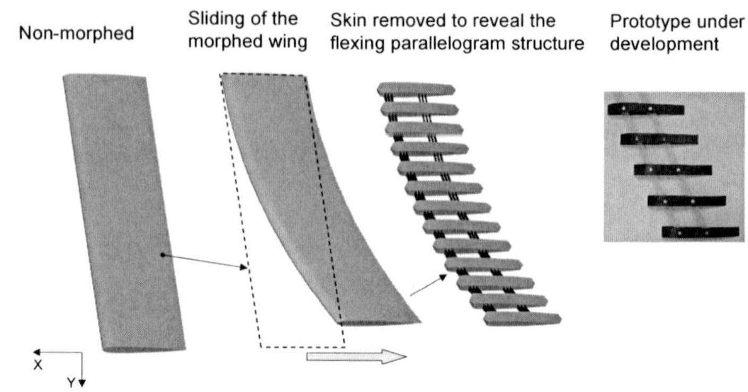

Figure 2: The flexing parallelogram wing box: identical bending of the wing box in the plane of the wing box slides the ribs to form a curved wing.

For small aircraft and most UAVs, application of shape memory alloy (SMA) actuated beams is a viable approach to carry out the flexing of the spars. For instance, Sofla et al. studied antagonistic shape memory alloy (SMA) actuated flexural structure suitable for the shape morphing of small aircraft [9, 10], Fig. 3.

Figure 3: Shape memory alloy (SMA) actuated beam can be used as the spar of flexing parallelogram morphing wing box [9].

The shape morphing of an originally swept wing via the flexing parallelogram concept is illustrated in Fig. 4. The sweep angle of the wing before morphing is represented by Λ. The final curved shape of the morphed wing is defined by the angle $(\Lambda + \beta)$, in which the shape morphing extent of the beam (sweep change) is quantified by the angle β in this paper. This sweep change (β) is defined here as the angle of rotation of a straight line that connects the wing-tip to the root upon morphing; see Fig. 4 for details.

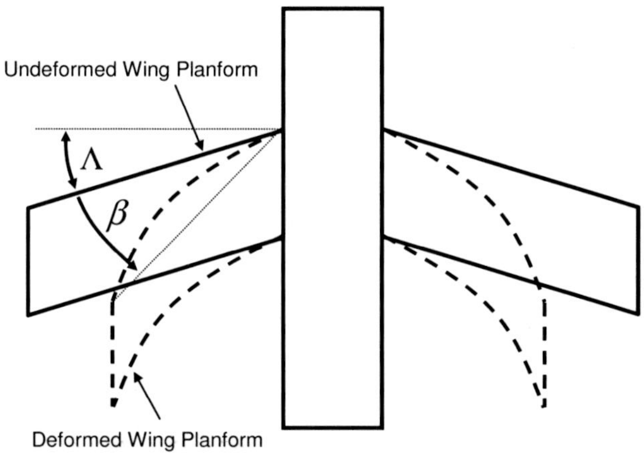

Figure 4: The wing with sweep angle (Λ) is transformed to a new shape that is identified by the angle ($\Lambda + \beta$). The area of the curved wing in the picture is 75% of the area of the wing prior to the morphing.

The relative reduction in the wing area for the configuration shown in Fig. 4 is related to the sweep angle Λ and the sweep change β by the following geometric relationship:

$$\frac{\Delta A}{A_{unmorphed}} = 1 - \frac{COS(\Lambda + \beta)}{COS(\Lambda)} \quad (1)$$

Where $A_{unmorphed}$ denotes to the wing area before morphing. The sweep change is usually limited by the actuation authority and the stiffness of the wing skin. The sweep angle Λ is a design parameter and can be arbitrarily selected from a range of straight wing (e.g. $\Lambda = 0°$) to highly swept values (e.g. $\Lambda = 20°$).

The wing skin in the flexing parallelogram takes part in supporting the aerodynamic loads on the wing. As demonstrated in Fig. 4, the change in the area is achieved by large in-plane shear deformations. Both top and bottom surface of the wing undergo simultaneous shearing. Therefore, the chord-length and the length of curves parallel to the spars, remain unchanged during the shape morphing. As a consequence, skin systems made of conventional elastomers, which are stiffened with commercial nets, would make ideal skin for the flexing parallelogram shear wing designs. The fibres of the net are aligned parallel with the rib and spars in order to enhance the bending stiffness of the wing in the span and chord directions. Because the net-stiffened elastomer does not restrict the in-plane shear deformation of the skin, it allows high authority for the shape morphing of the wing. Such skin material resembles the textiles that covered the early aircraft wing. The elastomer can be stiffened with rods instead of fabric net to increase the flexural stiffness of the skin material in order to carry higher wing loading.

We performed aerodynamic and performance analysis [3] and found that the morphing wing can reduce the fuel consumption of a typical UAV up to 14%.

PROOF OF CONCEPT

A prototype wing-box is designed and fabricated to replace the wing of an existing UAV (Fig. 5). The morphing wing was designed to change between a straight wing and a 30 degree swept wing. The right half-wing in the figure shows a straight shape after the heating (and subsequent cooling) of the front side SMA ribbons of both spars. The left half-wing shows the curvature of the wing after the heating (and then cooling) of the rear side SMA ribbons. The shapes in Fig. 5 are extended to a full length wing to illustrate that if the prototype was made to the full length, the final design would have a 30 degrees curved shape. The excellent matching of the shape morphing of the wing-box with the target design supports the success of the methods and analysis that was adopted in this research in the design and fabrication of a shape morphing wing [11].

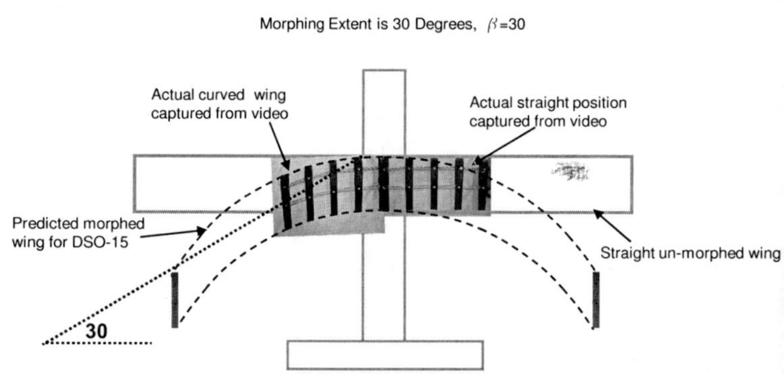

Figure 5: The prototype meets its objective wing shapes of straight shape and a 30 degrees swept shape.

SUMMARY

The shape morphing wing for small aircraft that is presented in this paper uses shape memory alloy (SMA) for actuation. The SMA ribbons are distributed in the spars; therefore no bulky or heavy power transmission means are needed for the shape morphing of the wing. On the other hand, the SMA actuators take part in the supporting of the aerodynamic load, unlike the conventional actuating devices such as electromagnetic motors that their body is fixed somewhere in the aircraft without carrying load. An antagonistic approach was used in this work to create two-way response of the spars by means of the one-way shape memory effect of the SMA ribbon. Although heating is needed to transform the wing from one shape to another, the main advantage of the antagonistic method is that these new shapes for the wing are retained without additional external power (i.e. heat). Therefore, the approach is suitable for the shape morphing of the wing to adapt to changes in condition of flight.

Prototype of a flexing parallelogram wing-box was made in this work. The prototype is fully capable of meeting the design objectives for the morphing wing.

REFERENCES

[1] AYN Sofla, SA Meguid, KT Tan, and WK Yeo, Shape morphing wing design: current status and challenges, Materials and Design, 2009 Volume 31, Issue 3, 1284-1292.

[2] G Reich, and B Sanders, Introduction to morphing aircraft research. Journal of Aircraft. 2007; 44: 1059-1059.

[3] AYN Sofla, VP Galantai and SA Meguid, Flexing parallelogram concept for the shape morphing of aircraft wing, submitted elsewhere, 2010.

[4] AYN Sofla, DM Elzey and HNG Wadley, Two-way antagonistic shape actuation based on the one-way shape memory effect, Journal of Intelligent Material Systems and Structures, 2008, Vol. 19, 1017-1027.

[5] JJ Joo, B Sanders, T Johnson, MI Frecker, Optimal actuator location within a morphing wing scissor mechanism configuration, In proceeding of SPIE: Smart structures and materials: modeling, signal processing, and control. 2006.Vol. 6166 616603-1.

[6] S Bharti, MI Frecker, G Lesieutre, J Browne, Tendon actuated cellular mechanisms for morphing aircraft wing. In Proceeding of of SPIE: Modeling, signal processing, and control for smart structures. 2007. Vol. 6523 652307-1.

[7] JS Flanagan, RC Strutzenberg, RB Myers, and JE Rodrian, Development and flight testing of a morphing aircraft, the NextGen MFX-1. In proceedings of 48th AIAA/ASME/ASCE/AHS/ASC structures, structural dynamics and materials conference, Honolulu, Hawaii, 2007.

[8] Y Yu, X Li, W Zhang, J Leng, Investigation on adaptive wing structure based on shape memory polymer composite hinge, In proceeding of SPIE: International conference on smart materials and nanotechnology in engineering. China; 2007. Vol. 6423 64231D-5.

[9] DM Elzey, AYN Sofla, and HNG Wadley. A bio-inspired, high-authority actuator for shape morphing structures. In proceedings of SPIE: Smart Structures and Materials 2003: Active Materials: Behavior and Mechanics, Vol. 5053.

[10] DM Elzey, AYN Sofla, and HNG Wadley, A shape memory-based multifunctional structural actuator panel, International J. of Solids and Structures, 42 (2005) 1943-1955.

[11] AYN Sofla, Development of a novel morphing wing for unmanned aerial vehicle, submitted elsewhere, 2010.

Proceedings of the ASME 2010 International Mechanical Engineering Congress & Exposition
IMECE2010
November 12-18, 2010, Vancouver, British Columbia, Canada

IMECE2010-39011

DESIGN OF MICROSTRUCTURE RESPONSE USING A COMPLEX STEP PLASTICITY APPROACH

Aswath Rangarajan
PhD student, Department of Aerospace Engineering
University of Michigan
Ann Arbor, MI 48109, USA

Veera Sundararaghavan
Assistant Professor, Department of Aerospace Engineering
University of Michigan
Ann Arbor, MI 48109, USA
veeras@umich.edu

ABSTRACT

Microstructure-sensitive design is performed using a newly developed Multiscale Complex Step Method. Algorithmic implementation of the complex-step method is based on conventional multiscale Taylor or FE2 direct model, with the difference being that the microstructural deformation gradient is modeled as a complex number with a small imaginary component. We introduce methods to calculate derivatives of microstructural fields with respect to loading parameters. The sensitivities are used to identify optimal microstructures with desired elastic and plastic properties.

1 INTRODUCTION

Design of materials with optimal property distributions is important to address the critical performance needs of hardware components in aerospace, naval and automotive applications. In metallic materials, properties can be tailored by controlling the evolution of features of underlying polycrystalline microstructures through controlled plastic deformation. Simulation of microstructure evolution in polycrystals has been well studied in the past. Many of the related works apply the Taylor-type (e.g. for a review see Kocks et al. [1]) or finite element homogenization [2-5] methods that link the kinematics of the macro- and microscopic scales. The success of such approaches has allowed us to efficiently compute the effect of macroscopic parameters (such as strain rates) on the microstructural response. Microstructure-sensitive design methods employ these techniques to address inverse/optimization problems such as: computation of optimal crystal orientation distributions that lead to desired elasto-plastic properties (Adams et al. [6], Kalidindi et al. [7]), identification of processing paths that lead to optimal microstructures using spectral (Li et al. [8]) or finite element representations (Sundararaghavan and Zabaras [9-10]. Many of these methods involve gradient optimization, where the derivative of the objective function with respect to macro-scale design variables is needed to iterate for the best solution. This involves a sensitivity analysis procedure to compute changes in microstructure sensitive properties (such as elastic modulus and yield strength) due to infinitesimal changes in design variables (such as macroscopic strain rates and deformation modes).

In choosing a computational approach for finding sensitivities, one is mainly concerned with its accuracy, computational expense and ease of implementation. Computational expense is all the more important in multiscale design problems where a large number of degrees of freedom (at the macro- and microscales) are handled. One method that is very commonly used is finite differencing. In this approach, the two different problems with small changes in a process variable are executed and the difference in microscale property is used as a measure of sensitivity. The approach is is not known for being accurate (due to cancellation errors) or computationally efficient (due to the need to solve two non-linear problems for each sensitivity). Direct differentiation methods, such as the continuum sensitivity analysis, developed in Badrinarayanan and Zabaras [11], overcome the issues associated with finite differencing. In the continuum sensitivity approach, the sensitivities were obtained by differentiating the governing equations and then by discretizing and solving the resulting equations. The approach is several times faster than finite differencing due to the use of linearized sensitivity equations and is thus ideal for large multiscale problems (in Ref. [12]). However, the ease of implementation is low and the method needs significant additions to the source code.

In this paper, a new computational materials design technique based on the complex-step method is presented. The approach involves perturbation of macroscopic design variables with a small imaginary component. In the computational implementation, the original multiscale program is mostly unchanged while all integer and floating point declarations are converted to complex number declarations. The most significant advantage is that cancellation errors that occur in finite differencing schemes do not occur in the complex step method. This enables the use of the complex step approach in problems involving remeshing and discontinuities (eg. contact/impact problems) with minimal changes to the source code. One important assumption is that the function whose derivative is calculated must be a real valued function, i.e. the domain and codomain of the function must be real values. This is true for most computational material science problems (eg. thermal, mechanical, diffusion related properties) but for certain others (eg. electromagnetic impedance design) the approach may not apply.

Complex step method originated with the work of Lyness [13,14] and has been more recently used for sensitivity analysis in computational fluid dynamics problems by Martins [15]. In the present paper, we propose the use of the approach to computational materials design problems. We address the problem of controlling elasto-plastic properties in polycrystalline microstructures using an already available polycrystal plasticity code [12]. Here, the evolution of the crystal orientations during deformation is modelled using continuum representation of texture [16,17] and incorporates crystal elasto-plasticity through the constitutive equations of Anand and Kothari [18]. For the purpose of demonstration, the numerical examples are performed at a material point with few hundred degrees of freedom. The derivatives are directly compared to the finite difference method. The efficiency of the approach would enable future applications in solving large multiscale design problems.

2 Derivation of the Complex step method for obtaining the derivatives

The complex step method is based on the perturbation of a function using a small arbitrary complex step, thereby obtaining the sensitivity as the imaginary part of the function value obtained. If f is an analytic function of z , then the Taylor series of f in the complex domain can be written as

$$f(z+ih) = f(z) + \frac{ihf'(z)}{1!} + \frac{(ih)^2 f''(z)}{2!} + \frac{(ih)^3 f'''(z)}{3!} + \dots \quad (1)$$

Now, considering that $z = x + 0i$, i.e. considering that the domain is real, we can write,

$$f(x+ih) - f(x) + \frac{ihf'(x)}{1!} + \frac{(ih)^2 f''(x)}{2!} + \frac{(ih)^3 f'''(x)}{3!} + \dots \quad (2)$$

Equating the real and imaginary parts of the Eq. 2,

$$f(x) = Re\{f(x+ih)\} + \frac{(h)^2 f''(x)}{2!} + HOT\dots \quad (3)$$

$$f'(x) = \frac{Imf(x+ih)}{h} + \frac{(h)^2 f'''(x)}{3!} + HOT\dots \quad (4)$$

where HOT stands for the higher order terms in the expansion.

However, when the value of h is chosen to be sufficiently low , the HOT can be neglected. Hence we obtain the following relations for arbitrarily small h.

$$f(x) = Re\{f(x+ih)\} \quad (5)$$

$$f'(x) = \frac{Im\{f(x+ih)\}}{h} \quad (6)$$

One of the main advantages of this method is that one single evaluation of the function is required in order to find the function value as well as the derivative of the function with respect to a variable. The real part of the function obtained with a small complex perturbation does not vary much when compared to the value of the function without such a perturbation. Hence, for an arbitrarily small complex perturbation of a variable, the real part can be assumed to represent the function value evaluated at the real value of the variable comsidered. Hence, the amount of computation is reduced by significant amount when comparing the derivative obtained from a finite differencing scheme (which needs two function evaluations in order to obtain the derivative). Care has to be taken during the implementation of the complex step method while dealing with functions which have discontinuities. In particular, two functions were modified in the implementation of the complex step method for the polycrystal optimization problem.

abs - The absolute function is modified as follows.

$$abs(x+iy) = \begin{cases} -x-iy & \text{if } x<0 \\ x+iy & \text{if } x>0 \\ 0 & \text{if } x=0 \end{cases}$$

signum - The signum function is modified as follows.

$$sign(x+iy) = \begin{cases} -1 & \text{if } x<0 \\ 1 & \text{if } x>0 \\ 0 & \text{if } x=0 \end{cases}$$

3 COMPLEX STEP METHOD FOR MICROSTRUCTURE-SENSITIVE DESIGN

A problem of interest to manufacturing engineers is to identify improved processing parameters that would closely achieve desired properties in materials. We define the design problem of interest as identification of the right combination of process modes such as tension (drawing), plane strain tension/compression (rolling), shear and rotation, and the corresponding strain rates α that would lead to a desired property that is a function of the given microstructure. The macro-velocity gradient ($L = \dot{F}F^{-1}$) is decomposed uniquely for 3D microstructure analysis as follows

$$\mathbf{L} = \alpha_1 \begin{bmatrix} 1 & 0 & 0 \\ 0 & -0.5 & 0 \\ 0 & 0 & -0.5 \end{bmatrix} + \alpha_2 \begin{bmatrix} 0 & 0 & 0 \\ 0 & 1 & 0 \\ 0 & 0 & -1 \end{bmatrix} + \alpha_3 \begin{bmatrix} 0 & 1 & 0 \\ 1 & 0 & 0 \\ 0 & 0 & 0 \end{bmatrix} + \alpha_4 \begin{bmatrix} 0 & 0 & 1 \\ 0 & 0 & 0 \\ 1 & 0 & 0 \end{bmatrix} + \alpha_5 \begin{bmatrix} 0 & 0 & 0 \\ 0 & 0 & 1 \\ 0 & 1 & 0 \end{bmatrix} + \alpha_6 \begin{bmatrix} 0 & -1 & 0 \\ 1 & 0 & 0 \\ 0 & 0 & 0 \end{bmatrix} + \alpha_7 \begin{bmatrix} 0 & 0 & -1 \\ 0 & 0 & 0 \\ 1 & 0 & 0 \end{bmatrix} + \alpha_8 \begin{bmatrix} 0 & 0 & 0 \\ 0 & 0 & -1 \\ 0 & 1 & 0 \end{bmatrix} \quad (7)$$

Each of the matrices in Eq.(7) specify a particular type of loading/deformation process. For example, the first matrix represents uniaxial tension/compression (depending on the sign of α_1), the second matrix represents plane strain tension/compression. The 3rd,4th and 5th matrices represent shear modes, while the last three matrices represent rotation modes.

The macroscopic deformation gradient at time step $t+1$ (F) can be computed using an exponential map ($F = exp(Lt)$). Alternatively, for small strain increments used in this work, an Euler approximation is employed as,

$$F = F_t(I + L\Delta t) \quad (8)$$

Computational scheme for computing microstructural response using this applied deformation gradient is described next.

3.1 Microstructure Evolution Direct Problem

A rate-independent Taylor model of the single-crystal plasticity model developed in Kothari and Anand [is used to compute the effect of macroscopic strain on the polycrystal. In the Taylor model it is assumed that all crystals are subjected to the same macroscopic deformation gradient. To compute the stress in each crystal due to this deformation gradient, we shall employ the (now) classical single-crystal plasticity theory (e.g., [19-27]) based on the notion that plastic flow takes place through slip on prescribed slip systems.

For a material with $\alpha = 1,\ldots,N$ slip systems defined by ortho-normal vector pairs (m^α,n^α) denoting the slip direction and slip plane normal respectively, the constitutive equations relate the following basic fields: the deformation gradient F which can be decomposed into elastic and plastic parts as $F = F^eF^p$, the Cauchy stress T ($= \frac{1}{\det F_r} P_r F_r^T$) and the slip resistances $s^\alpha > 0$. In the constitutive equations (intended to characterize small elastic strains) to be defined below, the Green elastic strain measure $\tilde{E}^e = \frac{1}{2}\left(F^{eT}F^e - I\right)$ defined on the relaxed configuration (plastically deformed, unstressed configuration) is utilized. The conjugate stress measure is then defined as $\tilde{T} = \det F^e (F^e)^{-1} T (F^e)^{-T}$ where T is the Cauchy stress for the crystal in the sample reference frame.

The constitutive relation, for stress, is given by $\tilde{T} = \mathcal{L}^e\left[\tilde{E}^e\right]$ where $\mathcal{L}^e$ is the fourth-order anisotropic elasticity tensor. It is assumed that deformation takes place through dislocation glide and the evolution of the plastic flow is given by

$$L^p = \dot{F}^p(F^p)^{-1} = \sum_\alpha \dot{\gamma}^\alpha S_0^\alpha sign(\tau^\alpha) \quad (9)$$

where $S_0^\alpha = m^\alpha \otimes n^\alpha$ is the Schmid tensor and $\dot{\gamma}^\alpha$ is the plastic shearing rate on the αth slip system. The resolved stress on the αth slip system is given by $\tau^\alpha = \tilde{T}\cdot S_0^\alpha$. The resolved shear stress τ^α attains a critical value s^α on the systems where slip occurs ($\dot{\gamma}^\alpha > 0$). Further, the resolved shear stress does not exceed s^α on the inactive systems with $\dot{\gamma}^\alpha = 0$. The hardening law for the slip resistance s^α is taken as,

$$\dot{s}^\alpha(t) = \sum_\beta h^{\alpha\beta}\dot{\gamma}^\beta,\ s^\alpha(0) = s_0^\alpha \quad (10)$$

3.2 Direct and Sensitivity analysis of texture evolution

During deformation, the crystals in the microstructure reorient and this leads to changes in material properties. Orientation distribution function (ODF) [1], the probability density function for orientations, is employed for the quantification of such changes in crystallographic texture. We employ the axis-angle parametrization of the orientation space proposed by Rodrigues [16]. The Rodrigues' parametrization is created by scaling the axis of rotation $\mathbf{n}$ as $\mathbf{r} = \mathbf{n}tan(\frac{\theta}{2})$, where θ is the rotation angle. A proper rotation matrix $\mathbf{R}$ relates the lattice orientation to a reference orientation. Given the Rodrigues parametrization $\mathbf{r}$, the rotation $\mathbf{R}$ can be obtained as,

$$\mathbf{R} = \frac{1}{1+\mathbf{r}.\mathbf{r}}(\mathbf{I}(1-\mathbf{r}.\mathbf{r}) + 2(\mathbf{r}\otimes\mathbf{r} + \mathbf{I}\times\mathbf{r})) \quad (11)$$

The fundamental region represents a region of the orientation space such that each crystal orientation is represented uniquely within the space. Fundamental region for the cubic symmetry group used in this paper is a truncated cube (as shown later in the numerical examples to quantify texturing). If the orientation-dependent property for a single crystal $\chi(r,t)$ is known, any polycrystal property can be expressed as an expectation value or average given by:

$$<\chi> = \int_{\mathscr{R}} \chi(r,t)\mathscr{A}(r,t)dv \quad (12)$$

Texture Evolution Direct Analysis:
For computing texture evolution during deformation, the reorientation velocity is found as follows (with $\mathbf{r}$ being the Rodrigues parametrization of crystal rotation - Refer [16]):

$$v = \frac{\partial r}{\partial t} = \frac{1}{2}(\omega + (\omega\cdot r)r + \omega\times r) \quad (13)$$

where r is the orientation (Rodrigues' parametrization) and ω represents the spin vector defined as

$$\omega = \mathrm{vect}\left(\dot{R}^e R^{eT}\right) = \mathrm{vect}(\Omega) \quad (14)$$

where R^e is evaluated through the polar decomposition of the elastic deformation gradient F^e as

$$F^e = R^e U^e \quad (15)$$

Considering the Euler approximation of $\dot{R}^e R^{eT} = \Omega$, where Ω is the spin tensor, leads to the following (alternatively matrix exponential $R^e = exp(\Omega t)$ could be used):

$$\Omega = \left\{\frac{R^e_{n+1} - R^e_n}{\Delta t}\right\} R^{eT}_n \quad (16)$$

From the elastic deformation gradients, R^e_{n+1} and R^e_n are evaluated and one can evaluate the spin tensor Ω using Eq. (16) and then the re-orientation velocity from Eq. (13).

Texture Evolution: Sensitivity analysis:
Once $\overset{\star}{F}{}^e_{n+1}$ has been evaluated from the previous sub-sections, $\overset{\star}{R}{}^e_{n+1}$ is obtained from complex step analysis of the polar decomposition function. Sensitivities of spin vector and the spin tensor are obtained as:

$$\overset{\star}{\omega} = \mathrm{vect}\left(\overset{\star}{\Omega}\right) \quad (17)$$

$$\overset{\star}{\Omega} = \overset{\star}{\dot{R}^e} R^{eT} + \dot{R}^e \overset{\star}{\overline{R^{eT}}} \quad (18)$$

Consider the re-orientation velocity defined in Eq. (13). The following equation results in the sensitivity of the re-orientation velocity:

$$\overset{\star}{r}_{n+1} = \overset{\star}{v}\Delta t + \overset{\star}{r}_n \quad (19)$$

3.3 DEFINITION OF THE GRADIENT OPTIMIZATION PROBLEM

Using the computed sensitivity fields, one can evaluate the gradients of any microstructure-dependent property $\langle\chi\rangle$ with respect to the components of a design vector α of size m . The gradient is represented as $\nabla\langle\chi\rangle = \left(\frac{\partial\langle\chi\rangle}{\partial\alpha_1}, \frac{\partial\langle\chi\rangle}{\partial\alpha_2}, \ldots, \frac{\partial\langle\chi\rangle}{\partial\alpha_m}\right)$ where

$$\frac{\partial\langle\chi\rangle}{\partial\alpha_i} = \frac{\overset{\star}{\langle\chi\rangle}(r,t;\alpha_1,\ldots,\alpha_m,0,\ldots,\overset{\star}{\alpha}_i,\ldots,0)}{\overset{\star}{\alpha}_i} \quad (20)$$

The ith sensitivity problem is driven by $\overset{\star}{\alpha}_i$ with $\overset{\star}{\alpha}_j = 0$ for $j \neq i$.

The objective function for the gradient optimization problem where it is desirable to obtain a set of properties $\langle\chi\rangle^{\mathrm{desired}}$ can be defined as follows:

$$\min_{\alpha} F(\alpha) = \frac{1}{2N}\sum_{i=1}^{N}\left(\langle\chi\rangle_i(\alpha) - \langle\chi\rangle_i^{\mathrm{desired}}\right)^2 \quad (21)$$

where N is the total number of sampling points (for e.g. Young's moduli along different loading directions) and $\langle\chi\rangle_i$ defines the property computed at the ith sampling point. Alternately, the objective function can be expressed in vector notation as

$$\min_{\alpha} F(\alpha) = \frac{1}{2N}(\Delta\langle\chi\rangle)^T(\Delta\langle\chi\rangle) \tag{22}$$

where $\Delta\langle\chi\rangle = \langle\chi\rangle(\alpha) - \langle\chi\rangle^{\text{desired}}$. An iterative approach is followed to solve for α which would give the desired property. Let α^r be the solution at the rth optimization iteration step. The descent direction d^r of $F(\alpha)$ can be computed from Eq. (22) as

$$d^r = -\nabla_{\alpha} F(\alpha) = -\frac{1}{N} S^T_{\alpha^r}(\Delta\langle\chi\rangle) \tag{23}$$

where the sensitivity matrix S_{α} of $\langle\chi\rangle$ is defined as $S_{\alpha} \equiv \partial\langle\chi\rangle/\partial\alpha$. S_{α} is of the order $N \times m$, where N is the dimension of $\langle\chi\rangle$ (number of sampling points) and m is the dimension of the design vector α. After each iteration, the design parameters are updated as follows:

$$\beta^{r+1} = \alpha^r + \delta_r d^r, \quad r = 0, 1, \ldots \tag{24}$$

where the optimal scalar δ_r is computed from a line search procedure. The iterative process is repeated until the prescribed convergence criterion is satisfied.

4 NUMERICAL EXAMPLES

The slip system hardening model used in the examples is given as:

$$h^{\alpha\beta} = [q + (1-q)\delta^{\alpha\beta}]h^{\beta} \text{ (no sum on } \beta) \tag{25}$$

where h^{β} is a single slip hardening rate, q is the latent-hardening ratio and $\delta^{\alpha\beta}$ is the Kronecker delta function. The parameter q is taken to be 1.0 for coplanar slip systems and 1.4 for non-coplanar slip systems. For the single-slip hardening rate, the following specific form is adopted:

$$h^{\beta} = h_o(1 - \frac{s^{\beta}}{s_s})^a \tag{26}$$

where h_o, a, and s_s are slip hardening parameters taken to be identical for all slip systems, with values $h_o = 180$ MPa, $s_s = 148$ MPa and $a = 2.25$ for FCC copper single crystals. The initial value of slip system resistance is taken as $s_o = 16MPa$ (Anand and Kothari [18]). Values of elastic parameters for copper crystal are taken as $C_{11} = 170$ GPa, $C_{12} = 124$ GPa and $C_{44} = 75$ GPa. Slip is assumed to occur in the 12 $\{111\} < 110 >$ slip systems in the FCC crystal. As a validation of the micro-scale texture evolution model, results are compared with the numerical example of Anand and Kothari [18]. The experiment corresponds to an x-axis compression with a strain rate of 0.001 s^{-1} of FCC copper polycrystal. The initial texturing of the material is assumed to be random, and this corresponds to a constant Lagrangian ODF of 2.435. The reference fundamental region is discretized into 448 tetrahedral elements with cubic symmetry enforced in the solution procedure. Comparison of results with Ref. [18] with the present ODF-Taylor model is shown in Fig. 1.

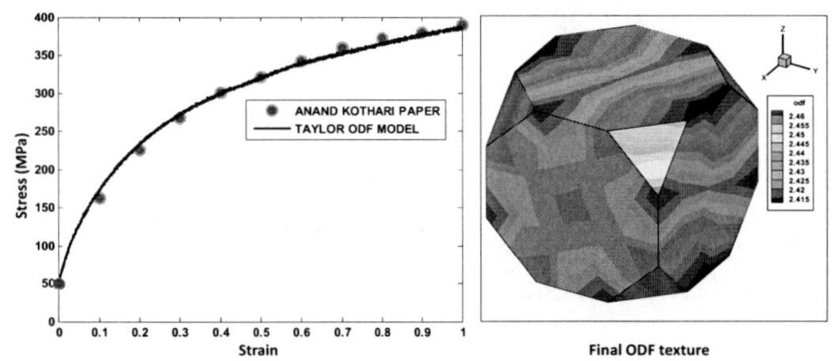

FIGURE 1. Comparison of results of current model with those in Anand and Kothari [18]

The parameter sensitivities in the design problem, as discussed later, are computed with respect to a perturbation of 10^{-7} units to the design parameters.

4.1 Example 1:Comparison of complex step method Vs finite difference method with respect to discretization size.

The first example shows the advantage of the complex step method as to obtain the proper derivative value for any arbitrarily small discretization. In this example, the deformation gradient is given small perturbations in decreasing orders of magnitude (i.e. from 10^{-4} to 10^{-10}) and the convergence of the derivative obtained is plotted in Fig. 2 (difference between the value of derivative for discretization using two consecutive powers of 10). A similar procedure is followed for obtaining the derivative using the complex step method, and the results are compared. The function whose derivative is evaluated is the stress following 2.5 seconds of unidirectional compression at a strain rate of 0.001 s^{-1}. The result shows the considerable advantage of complex step method over a finite differencing approach. In finite differencing, the derivative computed never converges with respect to the perturbation magnitude due to cancellation errors. On the other hand, the result shows that the derivative computed by the complex step method converges and is relatively independent of the magnitude of perturbation provided the perturbation is small.

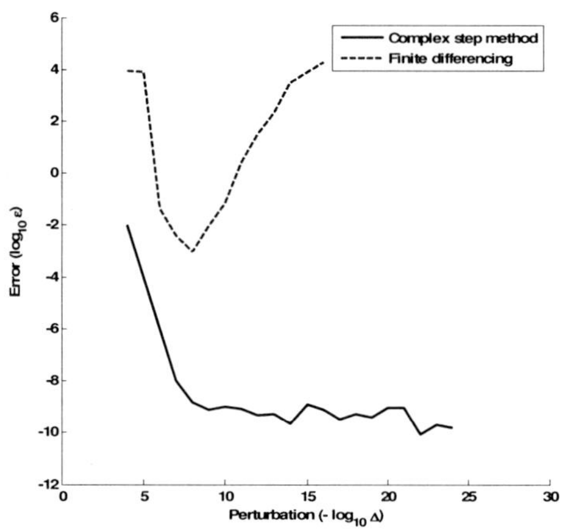

FIGURE 2. Comparison of finite differencing and complex step method where ε is the error and Δ is the applied perturbation

4.2 Example 2:Comparison of ODF sensitivities obtained using complex step method with those obtained using finite difference method

This example shows the comparison of the sensitivities obtained using the complex step method versus the sensitivities obtained using the finite difference method. The data is taken at a perturbation of 10^{-7} where finite differencing errors are minimal. The plots are obtained for a plane strain compression process at a strain rate of 0.001 for 12.5 seconds. The comparison of the sensitivities of the ODFs are shown in Fig. 3. It can be seen that the sensitivities of the ODF obtained using the complex step method compares quite well with those obtained using the finite difference method.

For computing sensitivities using the finite differencing approach, the required simulation time (wall clock time) was twice ($2x$) that of a single direct problem. Total time taken to compute sensitivities using the complex step method was 1.26 times the time needed by the direct problem that did not have complex operations. This additional time is the performance penalty associated with conversion of all integer and floating point operations to complex arithmetic.

In rest of the design examples, sensitivities are computed with a perturbation of $\overset{\star}{\alpha}_i = 10^{-7}$ for each process mode i. The optimization problem was executed until the normalized objective function showed less than 10^{-4} improvement between iterations.

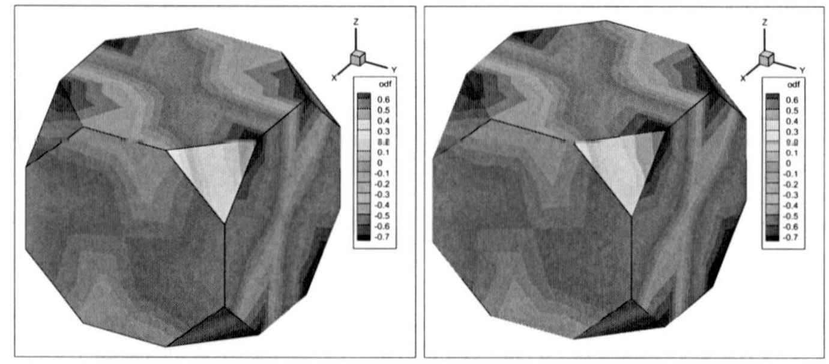

FIGURE 3. Comparison of sensitivities of the ODF obtained using finite difference and complex step method

4.3 Example 3: Optimization of strain rates for required stress

In this example, the variation of homogenized stresses with respect to time is optimized in a process comprising of unidirectional compression along X axis and rotation in YZ plane. The equivalent stress for the microstructure is represented using the von-Mises norm as,

$$\bar{\sigma}_{eff} = \sqrt{\frac{3}{2}\bar{T}'.\bar{T}'} \tag{27}$$

The response was computed for a total time of 11 seconds with a varying time step which is scaled as the value of α changes to maintain computational accuracy. The desired response is shown in Fig. 4. The initial strain rates were taken to be -0.001 sec^{-1} for both processes. The required stresses at times of 0.3, 3, 7 and 11 seconds were taken as 54.8, 84.8, 126 and 156 MPa. The final strain rates for achieving the stresses were identified as $\alpha_1 = -0.00217sec^{-1}$ and $\alpha_8 = -0.00446sec^{-1}$. The plots of convergence of the objective function are shown in Fig. 4. Convergence was achieved with 8 iterations and the objective function shows smooth decrease with respect to design iterations showcasing the efficiency of the complex step method in computing accurate sensitivities.

4.4 Example 4: Optimization of Young's Modulus

In this example, the optimization of the Young's Modulus is performed so as to obtain a desired variation with respect to the rolling direction as shown in Fig. 5. To compute the Young's moduli, the polycrystal stiffness, $\langle C \rangle$, is first computed through a weighted average (over $\mathscr{A}$) of the stiffness of individual crystals expressed in the sample reference frame. The elastic modulus is then computed through this polycrystal stiffness as

$$E = \frac{1.0}{(\langle C \rangle)^{-1}_{(11)}} \tag{28}$$

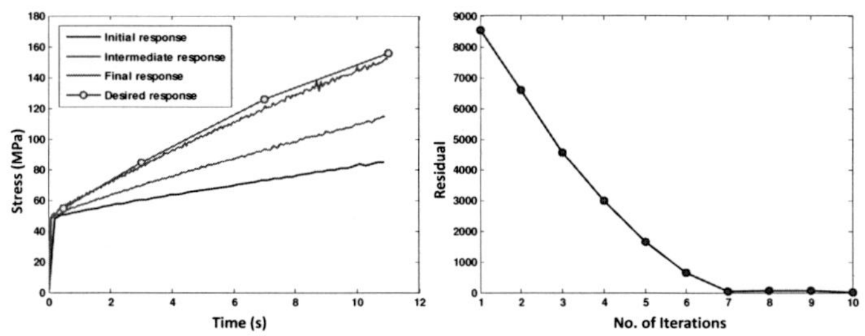

FIGURE 4. Design for desired material response (a) Change in the microstructure response over various iterations of the optimization problem. Desired response of the material given by the 4 desired coordinates indicated. (b) Change in objective function over various design iterations of the gradient optimization algorithm,

Furthermore, the elastic modulus about an angle with the rolling direction (RD) can be evaluated using the above equation, but after a coordinate transformation of $\langle C \rangle$.

The deformation is effected up to a strain of 0.2 starting from a random texture. The deformation modes considered in the problem are unidirectional compression/tension along X axis and shear in XY plane. The initial strain rates are $\alpha_1 = \alpha_3 = -0.001$. The final strain rates obtained after optimization were $\alpha_1 = -0.00142$ and $\alpha_3 = -0.00075$. The initial and final distributions of the Young's moduli with respect to the rolling direction are given in Fig. 5. Convergence was obtained to within 1% of the desired distribution in three iterations showcasing the efficiency of the complex step method.

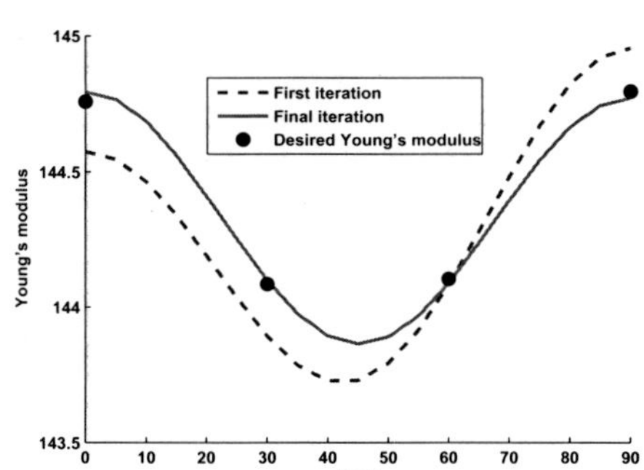

FIGURE 5. Optimization of Young's modulus distribution (with respect to direction of loading): Comparison of the desired distribution, initial and final (optimized) distribution.

5 CONCLUSIONS

Complex step method is a new technique for computational materials design. The approach is motivated by its surprising simplicity and ease of implementation. The approach involves perturbation of macroscopic design variables with a small imaginary component. In the computational implementation, the original multiscale program is mostly unchanged while all integer and floating point declarations are converted to complex number declarations. The sensitivities computed by the complex step method do not depend on the amount of perturbation applied given that the perturbation is sufficiently small. This is a significant advantage over finite differencing methods where such convergence is not seen. The technique is applied to identify optimal strain rates in single and two-stage processes (with intermediate unloading stage) that would lead to a desired microstructure response. The algorithm is computationally efficient and is found to converge to the desired response within a few iterations. Work presented here was focused on the material point problem of controlling microstructures to obtain desired response. In the future, the present analysis will be linked to a continuum model and complex step-based techniques will be extended in a multiscale framework to control macro-parameters to achieve desired property *distribution* in the final product. In this paper, certain special algorithmic treatments to model discontinuous functions that arise in polycrystal plasticity were indicated. The approach needs to be further studied for problems involving remeshing and discontinuities (eg. contact/impact problems, phase change, chemical reactions) where further changes to the algorithm may be warranted.

References

[1] Kocks, UF, Tome, CN, Wenk, HR,, Texture and Anisotropy: Preferred Orientations in Polycrystals and their Effect on Materials Properties, Cambridge University Press, 1998.

[2] S. Harren, R. Asaro, Nonuniform deformations in polycrystals and aspects of the validity of the taylor model, Journal of the Mechanics and Physics of Solids 37 (1980) 191 232.

[3] C. A. Bronkhorst, S. R. Kalidindi, L. Anand, Polycrystalline plasticity and the evolution of crystallographic texture in fcc metals, Philosophical Transactions: Physical Sciences and Engineering 341 (1662) (1992) 443 477.

[4] A. Beaudoin, H. Mecking, U. Kocks, Development of localized orientation gradients in fcc polycrystals, Philosophical Magazine A 73 (6) (1996) 1503 1518.

[5] S. J. Miehe, C, J. Schotte, Computational homogenization analysis in finite plasticity simulation of texture development in polycrystalline materials, Computer Methods in Applied Mechanics and Engineering 171 (3) (1999)

[6] B. L. Adams, A. Henrie, B. Henrie, M. Lyon, S. R. Kalidindi, H. Garmestani, Microstructure-sensitive design of a compliant beam, Journal of the Mechanics and Physics of Solids 49 (8) (2001) 1639 1663.

[7] S. R. Kalidindi, J. R. Houskamp, M. Lyons, B. L. Adams,

Microstructure sensitive design of an orthotropic plate subjected to tensile load, International Journal of Plasticity 20 (8-9) (2004) 1561 1575.

[8] D. Li, II. Garmcstani, B. Adams, A texture evolution model in cubicorthotropic polycrystalline system, International Journal of Plasticity 21 (8) (2005) 1591 1617.

[9] V. Sundararaghavan, N. Zabaras, Linear analysis of texture–property relationships using process-based representations of rodrigues space, Acta Materialia 55 (5) (2007) 1573 1587.

[10] V. Sundararaghavan, N. Zabaras, Design of microstructure-sensitive properties in elasto-viscoplastic polycrystals using multi-scale homogenization, International Journal of Plasticity 22 (10) (2006) 1799 1824.

[11] S. Badrinarayanan, N. Zabaras, A sensitivity analysis for the optimal design of metal-forming processes, Computer Methods in Applied Mechanics and Engineering 129 (4) (1996) 319 348.

[12] V. Sundararaghavan, N. Zabaras, A multi-length scale sensitivity analysis for the control of texture-dependent properties in deformation processing, International Journal of Plasticity 24 (9) (2008) 1581 1605.

[13] J. N. Lyness, C. B. Moler, Numerical differentiation of analytic functions, SIAM Journal on Numerical Analysis 4 (2) (1967) 202210.

[14] J. N. Lyness, Numerical algorithms based on the theory of complex variable, in: ACM 67: Proceedings of the 1967 22nd national conference, ACM, New York, NY, USA, 1967, pp. 125133.

[15] J. R. R. A. Martins, P. Sturdza, J. J. Alonso, The complex-step derivative approximation, ACM Trans. Math. Softw. 29 (3) (2003) 245262.

[16] Dawson, P, Kumar A, Computational modeling of f.c.c. deformation textures over rodrigues space, Acta Materialia 48 (10) (2000) 2719 2736.

[17] S. Ganapathysubramanian, N. Zabaras, Modeling the thermoelasticviscoplastic response of polycrystals using a continuum representation over the orientation space, International Journal of Plasticity 21 (1) (2005) 119

[18] L. Anand, M. Kothari, A computational procedure for rate-independent crystal plasticity, Journal of the Mechanics and Physics of Solids 44 (4) (1996) 525 558.

[19] G. Taylor, Plastic strain in metals, Journal of the Institute for Metals.

[20] J. Mandel, Generalisation de la theorie de plasticite de w. t. koiter, International Journal of Solids and Structures 1 (3) (1965) 273 295.

[21] J. Rice, Inelastic constitutive relations for solids: An internalvariable theory and its application to metal plasticity, Journal of the Mechanics and Physics of Solids 19 (6) (1971) 433 455.

[22] R. Hill, Continuum micro-mechanics of elastoplastic polycrystals, Journal of the Mechanics and Physics of Solids 13 (2) (1965) 89 101.

[23] C. Teodosiu, F. Sidoroff, A theory of finite elastoviscoplasticity of single crystals, International Journal of Engineering Science 14 (2) (1976) 165 176.

[24] R. J. Asaro, Crystal plasticity, Journal of Applied Mechanics 50 (4b) (1983) 921934.

[25] R. J. Asaro, A. Needleman, Texture development and strain hardening in rate dependent polycrystals, Acta Metall 33 (6) (1985) 923953.

[26] M. Rashid, S. Nemat-Nasser, Modeling very large plastic flows at very large strain rates for large-scale computation, Computers and Structures 37 (2) (1990) 119 132, special Issue: Computational Technology for Flight Vehicles.

[27] M. Ortiz, A Cuitino, A material-independent method for extending stress update algorithms from small-strain plasticity to finite plasticity with multiplicative kinematics, Engineering Computationsl 9 (4) (1992) 437 451.

Proceedings of the ASME 2010 International Mechanical Engineering Congress & Exposition
IMECE2010
November 12-18, 2010, Vancouver, British Columbia, Canada

IMECE2010-37014

WIND TUNNEL INVESTIGATION OF THE FLOW FIELD ON SLENDER SHARP-EDGED HALF DELTA WINGS

Sepideh Alibeigi
Research and Science Branch, I.A.U.
Tehran, Iran
alibeigis@yahoo.com

ABSTRACT

An experimental study was conducted in order to investigate the flow field over half delta wings. These wind tunnel tests was carried out over the 70 degree swept half delta wing at angles of attack of 10 to 35 degrees between 1.5×10^5 and 2.5×10^5 Reynolds Numbers. Therefore, the effect of angle of attack's variation and Reynolds Number's variation was perused. The pressure contour at different cross section of the delta wing was obtained by using pitot tubes. Furthermore, by using a hot wire anemometer the turbulence intensity, average velocity, and mean kinetic energy was measured. The results show that as the angle of attack increases, the vortex core become bigger and stronger. Not only does the vortex core become bigger, but also it moves toward inside of the delta wing. The magnitude of pressure coefficient at the center of vortex is lower than the other regions. By increasing of the angle of attack the pressure coefficient at the vortex core declines more. Moreover, the results indicate that the vortex location and its breakdown are not changing with variation of Reynolds number between 1.5 and 2.5×10^5. In addition, the turbulence intensity, average velocity, and mean kinetic energy, obtained by the hot wire, at the vortex core are higher than the other regions.

INTRODUCTION

Many modern fighter aircraft have highly swept, low aspect ratio wings to improve their supersonic cruise performance. Delta wings and their combinations are the most common platforms for fighters in order to be effective in combat arena. The high sweep angle of delta wings provides favorable drag characteristics at high speed. In addition, their sharp leading edges are conducive to flow separation at moderate to high angle of attack, producing substantial non linear lift known as vortex lift required during take off, landing, and maneuvers. The vast majority of delta-winged aircraft spends a great deal of their flight time at subsonic speed, so the low-speed flow field of delta wings has been a subject of much study. This consideration has spawned the current delta wing research.

The flow structure over a delta wing at moderate to high angle of attack is different than that of conventional and low sweep wings. For wings of less sweep, $\Lambda < 40°$ as the angle of attack increases beyond zero degrees, the upper surface flow close to the leading edge separates resulting in an unsteady separation bubble. Beyond a 20 degree angle of attack, the lifting performance of these wings deteriorates rapidly[1] . This behavior of low sweep wings is undesirable in close combat arena.

The delta wing at low angle of attack acts like a flat plate. At zero angle of attack the flow remains entirely attached over both surfaces and no lift is generated. As the angle of attack increases further the pressure difference between the upper and the lower surface causes a flow rounds the leading edge. The flow detached along the leading edge line into a shear layer that curls up into two primary vortices (fig. 1). The dominant feature of this flow is the effect of these two primary vortices. The different regions of vortex are shown in figure 2. The leading-edge vortices are strong and stable. Not only does these vortices' size increase with increasing of the angle of attack, but their strength also enhances. Depending on the delta wing sweep angle, these vortices remain stable through a wide range of angle of attack. Aft of the trailing edge the vortices continue downstream in the direction of the flow[4] .

Delta wing vortices are being a source of high energy increasing with angle of attack. This energy induces additional velocity on the upper surface of the wing, reducing pressure considerably. Consequently, an additional lift which is known as vortex lift increasing nonlinearly with increasing angle of attack. This lift can be as high as 50% of the total lift generated [2]. Depending on the wing sweep angle, these vortices remain stable through a wide rang of angle of attack, up to an angle of 40 degrees for 85 degrees sweep angle [2], and the flow is characteristically steady. However, there is a limitation for all positive effects of delta wing's vortices known as vortex breakdown or vortex bursting. Vortex bursting is associated with an abrupt axial flow deceleration, expansion of the vortex about a stagnant core, and turbulent flow thereafter. At moderate to high angle of attack, primary vortices burst first far downstream of the trailing edge. The burst point moves upstream with increasing the angle of attack. The forward progression of the core breakdown is more rapid when the burst point is in the region of the trailing edge and less rapid at the bursting approaches the apex of the wing [3].

The present study investigates experimentally the flow field of half delta wings by variation of angle of attack and Reynolds number. Furthermore, the turbulence intensity and mean kinetic energy inside and outside of vortex core were obtained; however, it is beyond the scope of this paper to represent all of the wind tunnel test results.

NOMENCLATURE

Λ: Sweep back angle
C_p: Pressure coefficient
K.E.: Kinetic Energy
S: Spanwise distance
A.O.A.: Angle of attack
Re: Reynolds number

EXPERIMENTAL APPARATUS

The experiments were carried out in a subsonic 45 centimeter by 45 centimeter wind tunnel having the maximum velocity of 45 m/sec. Pressure coefficient was measured by a probe with 0.9 millimeter diameter. The pressure coefficient of each position is measured with transducer and the position is measured with potentiometer. By drawing the displacement variation verses voltage and determining its slop, the out put voltage of potentiometer can be stated in displacement. The same work can be carried out for the transducer. The turbulence intensity has measured with hot wire. The model of hot wire is CTA (Constant Temperature Anemometry) which is based on cooling percentage of wire as a result of flow passing. As cooling of wire, the resistance in electronic circuit will change, so output voltage will change. The hot wire probe is joined to the potentiometer, and the output voltage is analyzed by a computer. Therefore, hot wire probe location and output voltage is specified simultaneously each second. Studies show that for the CTA model apparatus, velocity versus voltage will be stated as a forth order polynomial function. The forth order polynomial function for this experiment is:

$$V = -15.66 + 72.9v - 112.77v^2 + 84.25v^3 - 18.71v^4 \tag{1}$$

where V is the wind tunnel velocity, and v is the output voltage.

The delta wing model has a 70 degree sweep back and sharp edge leading edge. The schematic of the model is shown in figure 3. The half delta wing is made of aluminum, and it is attached to a splitter plate.

RESULTS

In figure 4, the pressure contour at different cross sections of delta wing at $\alpha=20°$ is shown. It can be seen as moving toward the leading edge, the vortex becomes bigger, the size of low pressure region increases, and moves upward from the surface. In figure 4 the variation of vortex diameter at different cross section at constant angle of attack is presented. This graph, also, approves that vortex diameter by moving toward the downstream becomes bigger. Furthermore, it illustrates that as the angle of attack increases, the vortex becomes bigger.

By increasing the angle of attack, the size of vortex at each section of delta wing increases, and it moves toward inboard of the surface. It can be observed in figure 5. In addition, it can be observed that the diameter of the vortex increases as moved along the leading edge toward the downstream at constant angle of attack. Moreover, as the angle of attack increases the vortex core moves upward from the surface, and the low pressure region over delta wing increases.

In figure 6, the shape of vortex at different cross section at angle of attack of 30 degrees is presented. The conclusions are the same as conclusion at $\alpha=20°$.

As it can be observed in figure 7, vortex breakdown has been occurred at the angle of attack of 35 degrees. Vortex core becomes big and covers most of the spanwise of the half delta wing.

Figure 8 compares the pressure distribution at the vortex core at different angles of attack. The pressure distribution and pressure coefficient of vortex core decline as angle of attack increases. It can, also, be seen that the pressure coefficient at the vortex core reduce abruptly. As soon as the vortex bursts, occurring at $\alpha=35°$, the situation over the delta wing will be changed.

Figure 9 shows variation of pressure contour at constant angle of attack, cross section, and different Reynolds number. It shows that at the subsonic speed, pressure contour is not very sensitive with Reynolds number variation, and it does not change considerably. The vortex core has not changed its position by variation of Reynolds number.

In figure 10, velocity profile at constant Reynolds number and angle of attack has been shown. It demonstrates that velocity inside of the vortex increases and at the vortex core reaches the maximum value. Y=3 shows the velocity inside the vortex core, and y=4 shows the velocity profile outside the vortex core. It can be seen far from the vortex, the velocity is equal to the free stream velocity. At y=4, the velocity profile is to some extent affected by the vortex; the bulge region shows the affected region. Kinetic Energy profile is alike the velocity profile (fig. 11). Each dot in the figure indicates the kinetic energy at specific height from the surface of the wing.

Figure 12 shows the turbulence intensity inside and outside of the vortex. As it seen, the turbulence at vortex region is higher than the other region of the flow field, and it has its maximum value at the core vortex.

CONCLUSION

To understand angle of attack and Reynolds number effect on flow field over a half delta wing, a series of wind tunnel testing has been conducted over a 70° half delta wing. The present results indicate that as the angle of attack increases, the vortex becomes bigger and stronger. Furthermore, it moves upward from the surface and toward inside of the delta wing, and the pressure coefficient of vortex core declines more. According to the probe results, the pressure coefficient at the vortex core decreases abruptly and it is less than the other regions; moreover, according to the hot wire results, the mean velocity and mean kinetic energy increase near the vortex and at the vortex, and they reach to their maximum value at the core vortex. Furthermore, the turbulence intensity at vortex region increases and at the vortex core is the highest. Probe results, also, indicate that flow field at the subsonic region is insensitive to the Reynolds number variation.

REFERENCES

[1]. Poll, D. I.A., "On the generation and subsequent development of spiral vortex flow over a swept back wing," AGARD-CP-342, April 1983, PP. 6-1 to 6-14.

[2]. Wentz, W. H., and Kohlman, D. L., "Vortex breakdown on slender sharp-edged wings," AIAA paper NO. 69-778.

[3]. Skow, A. M., and Erikson, G. E., "Modern fighter aircraft design for high angle of attack maneuvering," AGARD-LS-121, March 1982, PP. 4-1 to 4-59.

[4]. Soltani, M. R., "An Experimental Study of the Relation between Forces and Moments and Vortex Breakdown on a Pitching Delta Wing," *PhD Thesis*, Univ. of Illinois Urbana-Champaign, 1992.

[5]. Luckring J. M., "Reynolds Number, Compressibility, and Leading-Edge Bluntness Effects on Delta Wing aerodynamics," proceedings of the 24th International Congress of the Aeronautical sciences, 2004.

[6]. Kim H. D., Lee Y. K., Kim T. H., John M. H., "A Study of the Vortical flow over a Delta Wing with a Leading Edge Extension," Journal of Thermal Science, Vol. 12, No. 4. 2003.

[7]. Heron I., "Vortex Burst Behavior of a Dynamically Pitched Delta Wing under the Influence of a Von Karman Vortex Street and Unsteady Freestream," PhD Thesis, Wichita State Univ., 2007.

ANNEX A

FIGURES

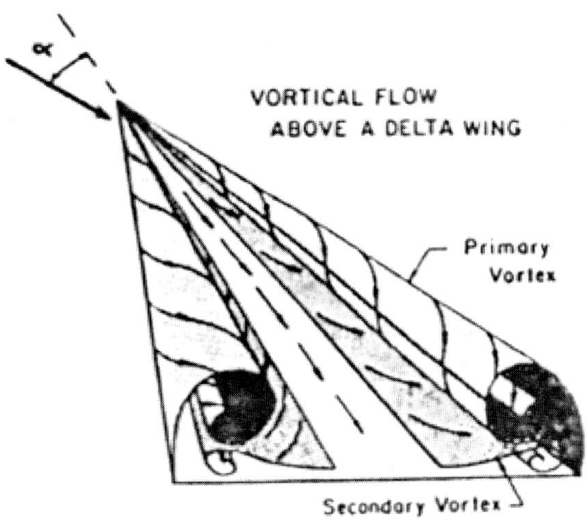

Fig. 1: Top view of vortices over a delta wing [4]

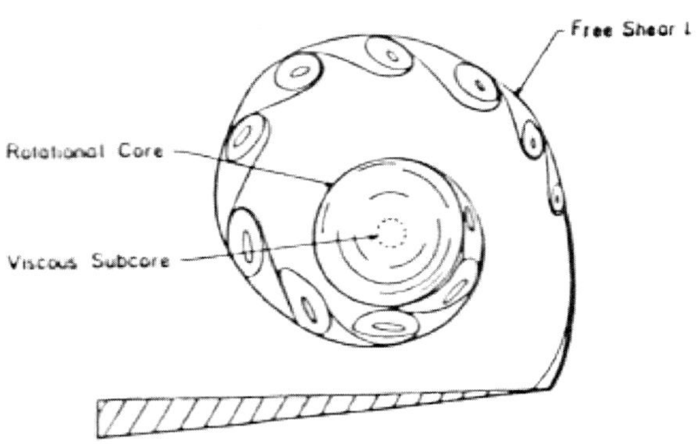

Fig. 2: Schematic description of the vortex regions [4]

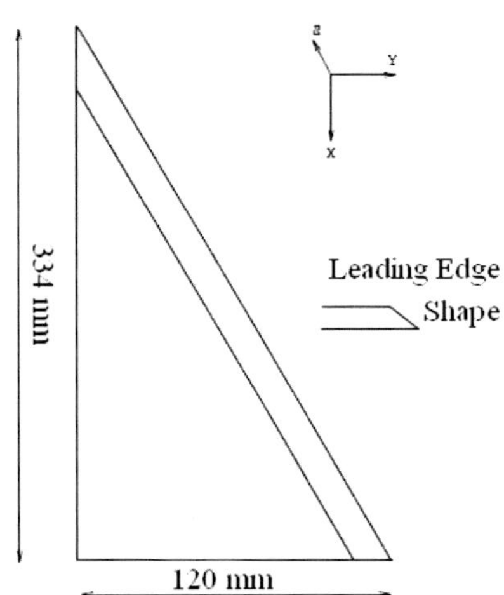

Fig. 3: schematic of the half delta wing model

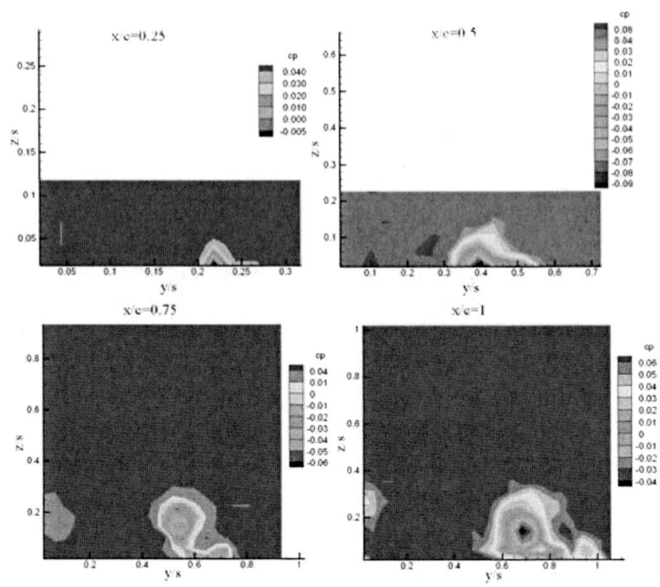

Fig 4: Pressure contour at α=20°

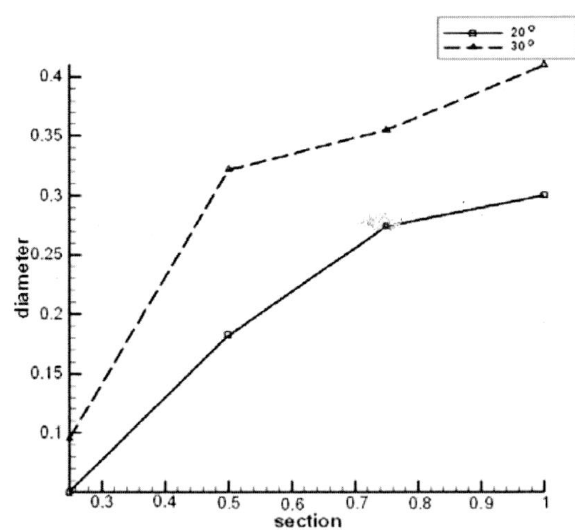

Fig 5: vortex diameter variation at different cross section at constant A.O.A.

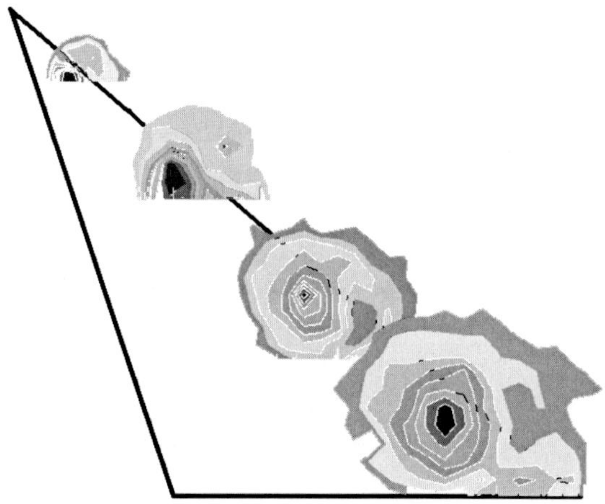

Fig 6: shape of vortex at along the cord at α=30°

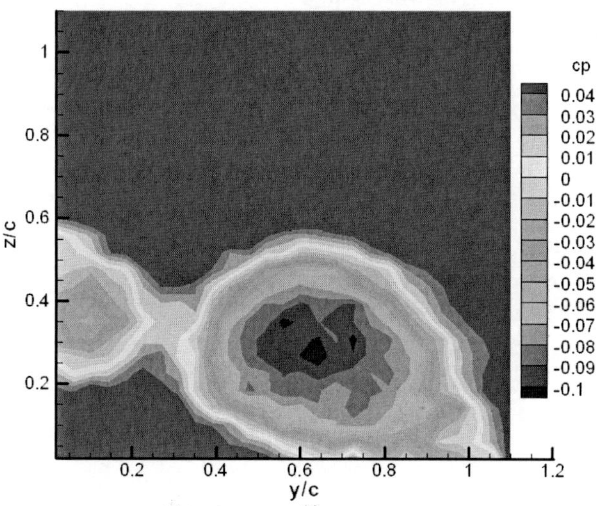

Fig 7: Pressure contour at the L.E. at α=35°

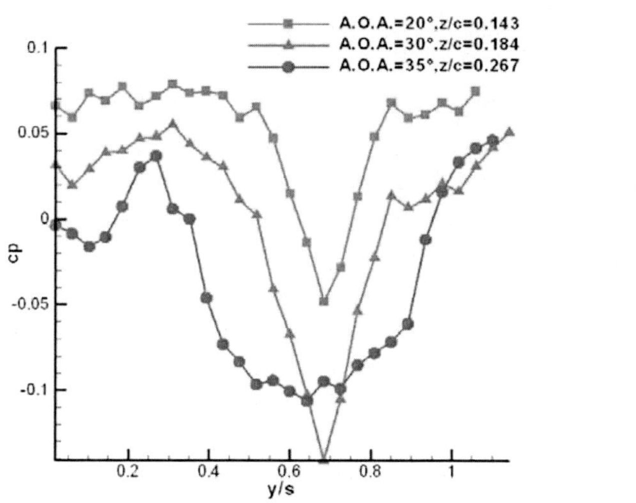

Fig 8: Pressure distribution at leading edge at different A.O.A

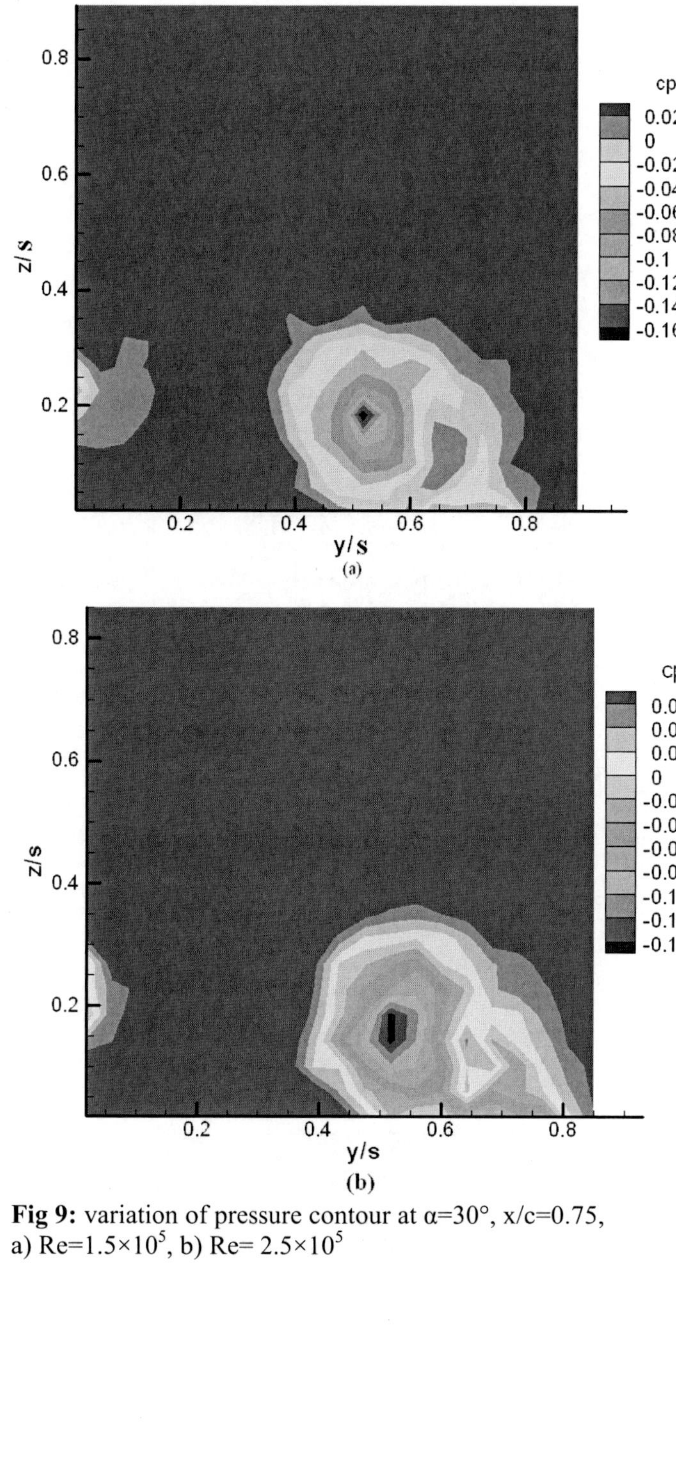

Fig 9: variation of pressure contour at α=30°, x/c=0.75, a) Re=1.5×10^5, b) Re= 2.5×10^5

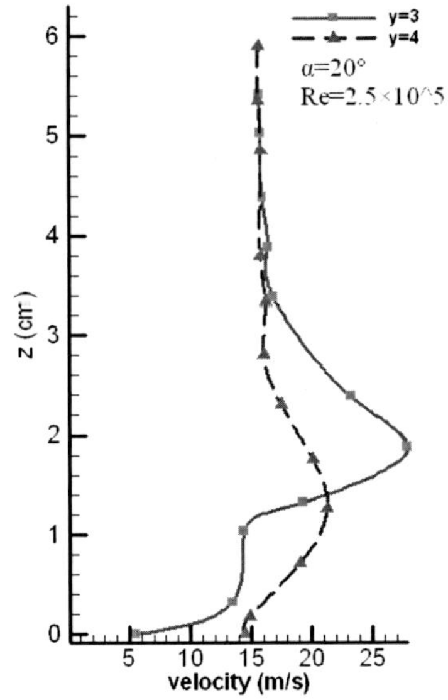

Fig 10: velocity profile at α=20°

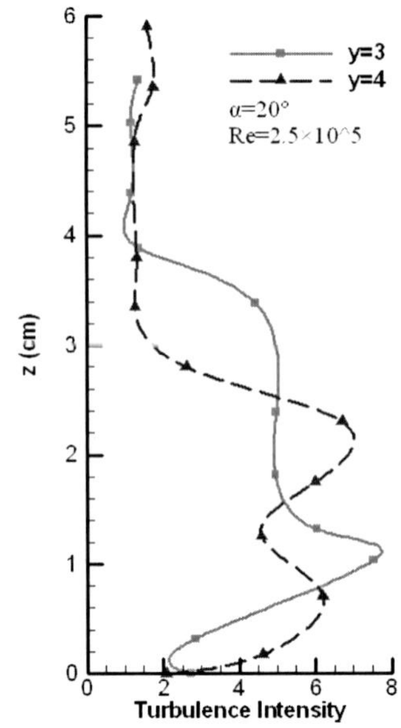

Fig 12: Turbulence intensity profile

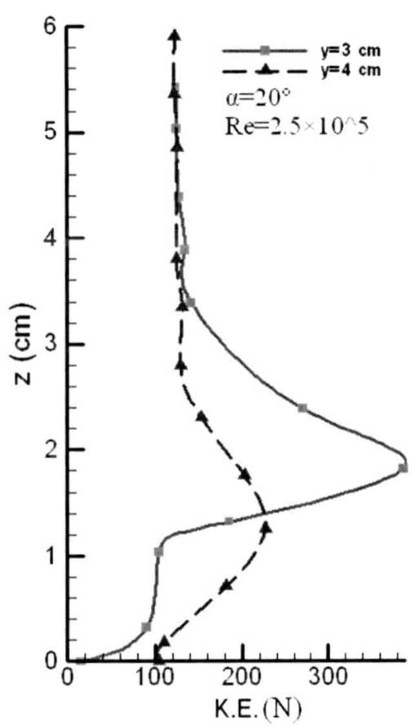

Fig 11: Kinetic energy profile

Proceedings of the ASME 2010 International Mechanical Engineering Congress & Exposition
IMECE2010
November 12-18, 2010, Vancouver, British Columbia, Canada

IMECE2010-37206

CONCEPT DESIGN OF COMPOSITE AIRCRAFT WING

J. S. Rao
Chief Science Officer
Altair Engineering India Pvt Ltd
Bangalore, Karnataka, India

S. Kiran
Formerly Altair Engineering India Pvt Ltd
Autolive India Pvt Ltd.,
Bangalore, Karnataka, India

ABSTRACT

This paper is concerned with an optimal concept design of aircraft wing from an airfoil. The airfoil itself is generated from CFD studies but there is no baseline of the wing structure. Topology optimization is recently applied for weight reduction given an operating baseline structure; here it is demonstrated that this optimization can be used to derive the concept of the wing structure directly. The optimized concept design is realized in to CAD and then a composite free-size optimization is performed to determine material distribution and ply drop regions etc. Finally a composite size and shape optimization is done and the ribs thus realized are presented.

INTRODUCTION

A typical aircraft wing consists of Spars, Ribs, Stringers, Skin and Flaps and Ailerons. Spars are the structural elements that ultimately bear the load carried by the wing. They run through the length of the wing from the point nearest to the fuselage out to the wing tip. Civil aircraft manufactured by Boeing, Airbus and McDonnell Douglas show that the layouts have been extremely stable in the post-World War-II era, with no dramatic variation between manufacturers, and between models of the same manufacturer (1). Wings made by these companies are based on the same few-spars-many-ribs design, with near-constant rib spacing. While most of the designs contain only two spars, a few include an additional mid-spar that goes partway, from fuselage up to wing-mounted engine. An interesting variation is observed among some German aircraft dating back to the pre-World War days; the old Heinkel bomber and the current Dornier aircraft both utilize iso-truss construction in the wing, see (2). The wing structures are thus evolved over several years from experience and tests to provide the light weight structures.

Aircraft structures designed over the years tend to be heavy making the aircraft bulky and inefficient. To reduce the weight two different approaches were evolved; optimize the metallic structures for minimum weight giving the same structural properties of compliance and buckling characteristics as the well proven designs in service and adopt light weight composites. In optimization approach, Topology methods were developed based on Solid Isotropic Material with Penalization (SIMP); see Bendsøe and Kikuchi (3) and Bendsøe and Sigmund (4). With multi-objective functions and several design variables, commercial codes became necessary for the industry to achieve light weight designs from the existing ones. Altair OptiStruct (5) is one of the popular codes available for topology optimization, e.g. Schumacher et al (6) used OptiStruct to assist in the structural design of a new military transport aircraft.

While the topology optimization is generally practiced on an existing structure for mainly weight reduction, the question arises on its application in a new design. As a specific case, if one has derived the airfoil shape needed from CFD analysis to carry the structural loads, is it necessary to develop the base line with the Spars, Ribs, Stringers, Skin and Flaps and Ailerons before applying topology optimization by adopting existing practices? The answer is No. Rao et al (7) adopted topology optimizing in OptiStruct to derive the wing structure starting from the airfoil shape by eliminating the intermediate steps of deriving the base line. The structure derived is metallic with Aluminum as the material.

It is well known that composite structures are lighter in weight and give substantial weight savings compared with metallic structures. For example Poort (8) has shown the optimization of the 787 Dreamliner Horizontal Stabilizer CFRP Composite Main Box. Design and Optimization of Laminated Composite Materials were discussed by Gürdal, Haftka and Hajela (9). Ryoo and Hajela (10) presented a novel approach for handling variable string lengths in GA-based topological design and implemented in representative algebraic problems, truss topology design, and the layout of a stiffened composite panel. Zhou and Fleury (11) discussed the latest capabilities in composite design and optimization available in OptiStruct version 10 (5). They presented Free-Size Topology Optimization, Ply-Bundle Sizing with ply-based FEA modeling and Ply Stacking Sequence Optimization.

This paper illustrates the optimization process to derive light weight composite structure from the metallic wing obtained in [6].

COMPOSITE OPTIMIZATION

Optimization for the composite structure is carried out in three phases:

Phase I - Concept: Free-Size or Topology Optimization

Composite free-sizing optimization is to create design concepts that utilize all the potentials of a composite structure where both structure and material can be designed simultaneously. By varying the thickness of each ply with a particular fiber orientation for every element, the total laminate thickness can change 'continuously' throughout the structure, and at the same time, the optimal composition of the composite laminate at every point (element) is achieved simultaneously. Manufacturing constraints like lower and upper bound thickness on the laminate, individual orientations and thickness balance between two given orientations can be defined.

Phase II – Detailed Design: Ply-Bundle Sizing with ply-based FEA modeling

Sizing optimization is performed to control the thickness of each ply bundle, while considering all design responses and optional manufacturing constraints. Ply thicknesses are directly selected as design variables. Composite plies are shuffled to determine the optimal stacking sequence for the given design optimization problem while also satisfying manufacturing constraints like control on number of successive plies of same orientation, pairing 45° and -45° orientations together etc.

Phase III – Detailed Design: Ply Stacking Sequence Optimization

Composite plies are shuffled to determine the optimal stacking sequence for the given design optimization problem while also satisfying additional manufacturing constraints like,

- The stacking sequence should not contain any section with more than a given number of successive plies of same orientation.
- The 45° and -45° orientations should be paired together
- The cover and/or core sections should follow a predefined stacking sequence

DERIVATION OF COMPOSITE WING

A Solid wing made of Aluminum with Wing Span 13.5 m, Constant chord wing span 4.74 m, Tapered chord wing span 8.76 m shown in Fig. 1 was the starting point in (7). The Chord length at wing root is 2.68 m and at wing tip 1.39 m. Material properties are taken as E = 70000 MPa, Poisson's ratio = 0.33 and Density = 2.60×10^{-09} tonne/mm^3.

The loads taken on the wing include flight load, fuel load and engine load. The airfoil sections, where pressure values are determined by Vortex Lattice method using XFOIL (a public domain code) are given in Fig. 2 along with the pressure values at stations every 500mm apart the wing span.

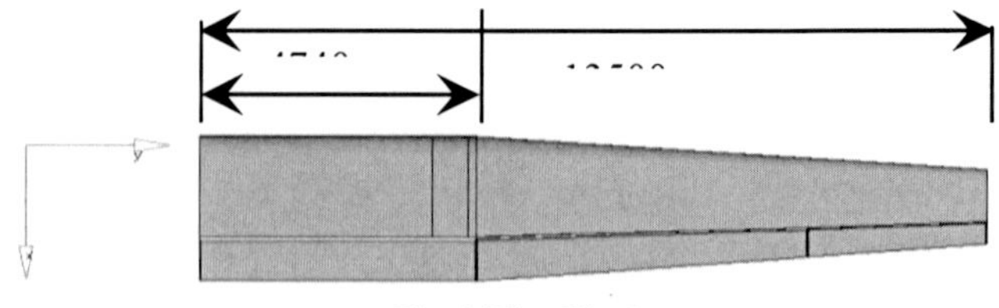

Fig. 1 Wing Planform

Fuel tank starts from 20% and ends at 85% of constant chord wing span, see Fig. 3. Center of Gravity of the fuel tank is at 1.27m from leading edge and 2.5m from wing root. Fuel load is 1730 Kg per wing and is distributed through RBE3 element. The engine is mounted from 85% to 98% of constant chord wing span; see Fig. 4 and its weight is 800 Kg. It is assumed to be mounted on front and rear spars through two ribs and the weight distribution at the four nodes of intersection between the two spars and ribs shown in Fig. 4 is given in Table 1.

Table 1 Engine load

	First Rib	Second Rib
Front Spar	250 Kg	250 Kg
Rear Spar	150 Kg	150 Kg

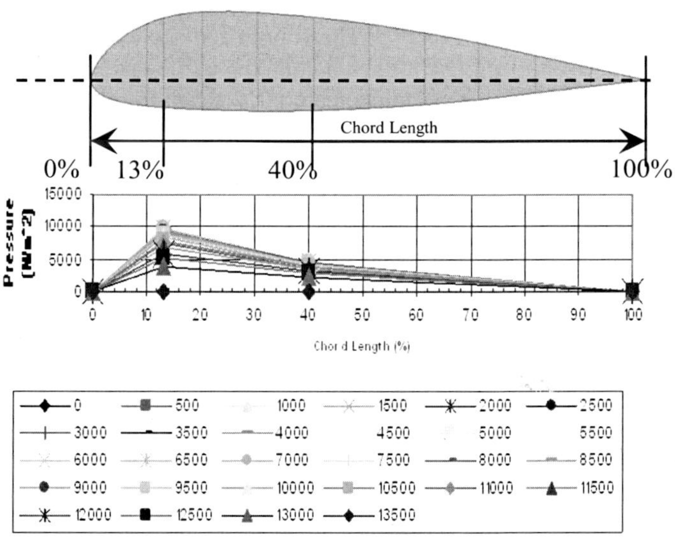

Fig. 2 Airfoil with Pressure Loads

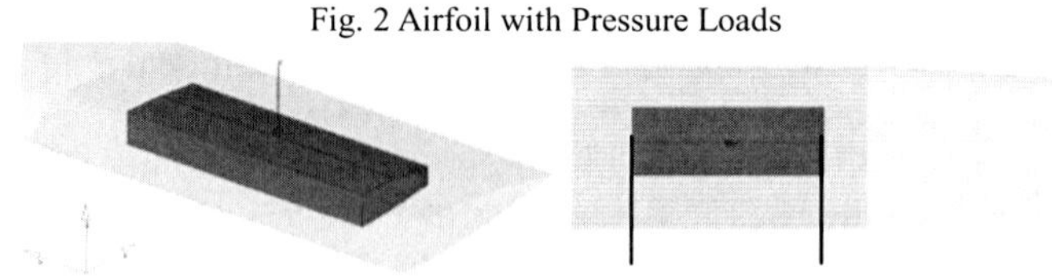

Fig. 3 Fuel tank location and loading in Wing Span

The CAD model of the wing developed in Topology Optimization of Aircraft Wing (7) is shown in Fig. 5 without the skin. The skin is 1.2 mm thick and the total weight is 2100 kg.

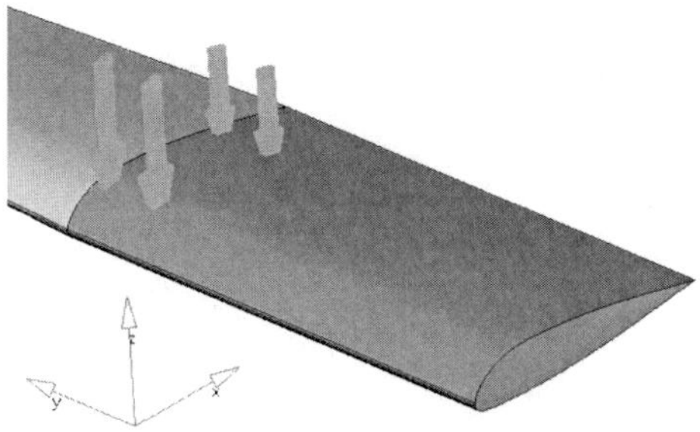

Fig. 4 Engine Load

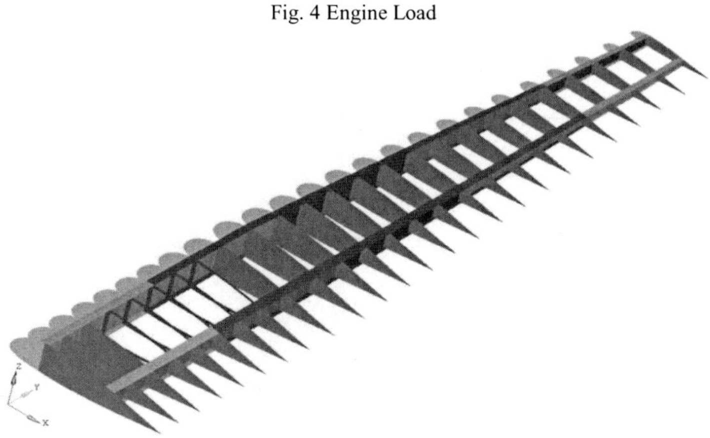

Fig. 5 CAD model realized for the wing

This model is now taken as baseline to develop the composite structure.

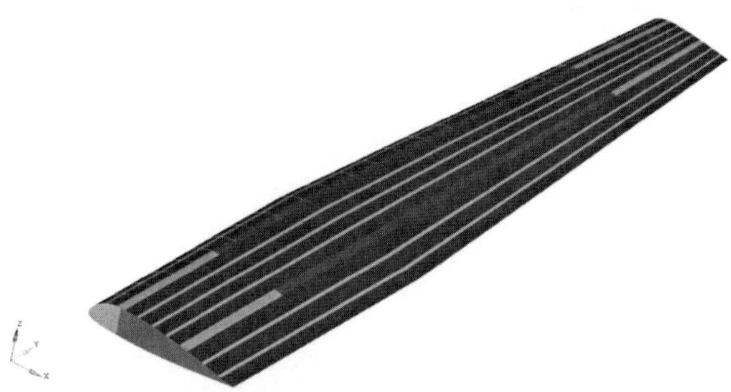

Fig. 6 Regions for Free Sizing

Phase 1: Free-Size or Topology Optimization

Free size optimization was performed to determine the thickness and laminate family within the chosen regions of the CAD model. The wing in Fig. 5 is divided in to regions for size optimization as shown in Fig. 6. Objective functions in this optimization are mass and compliance and minimization of the same. The design space is limited to ribs only in this exercise. The base laminate chosen is symmetric without a core with 10.16 mm thickness for all plies as given in Table 2. Carbon Fiber Reinforced Plastic is chosen for the laminate plies. Table 3 gives the CFRP properties used in the design.

The results of free size optimization for the four plies in all the ribs are given in Fig. 7. We notice that the ribs near the wing root have higher material density or should be thicker and those nearer to the tip do not carry much load and therefore can be lighter.

Table 2 Base Laminate

Review

Ply lay-up: Rib_mm
Total number of plies: 4
Total thickness: 40.64

Ply	Material	Thickness T1	Orientation Degrees
1	CFRP_LA	10.16	0.0
2	CFRP_LA	10.16	45.0
3	CFRP_LA	10.16	-45.0
4	CFRP_LA	10.16	90.0

Table 3 CFRP Properties

	Symbol	Units	Properties
E	E1	N/mm^2	1.52×10^5
E	E2	N/mm^2	9650
G	G12	N/mm^2	5930
n	NU12		0.321
r	Rho	Tonne/mm^3	1.53×10^{-9}

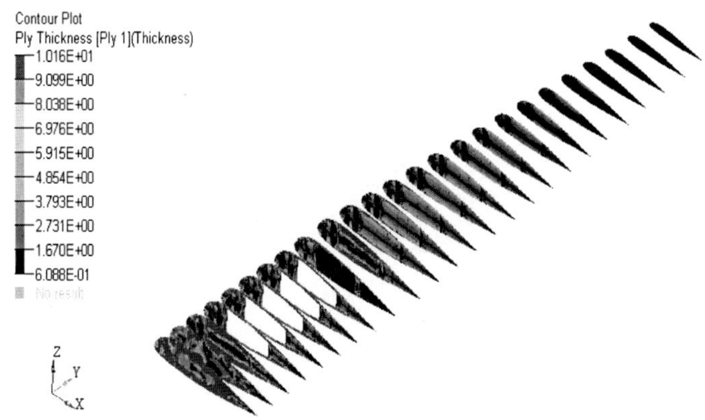

Fig. 7a Free Size Optimization for Super Ply at 0° Orientation of all Ribs

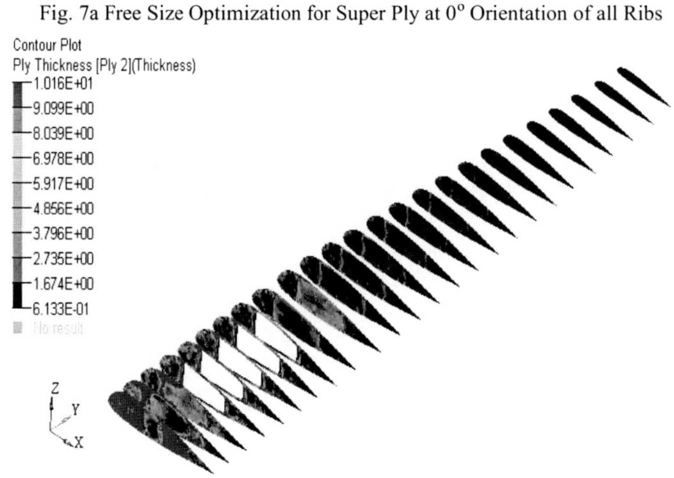

Fig. 7b Free Size Optimization for Super Ply at +45° Orientation of all Ribs

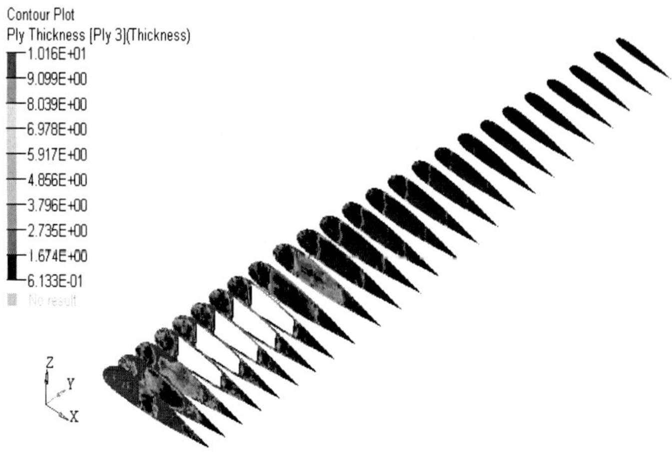

Fig. 7c Free Size Optimization for Super Ply at -45° Orientation of all Ribs

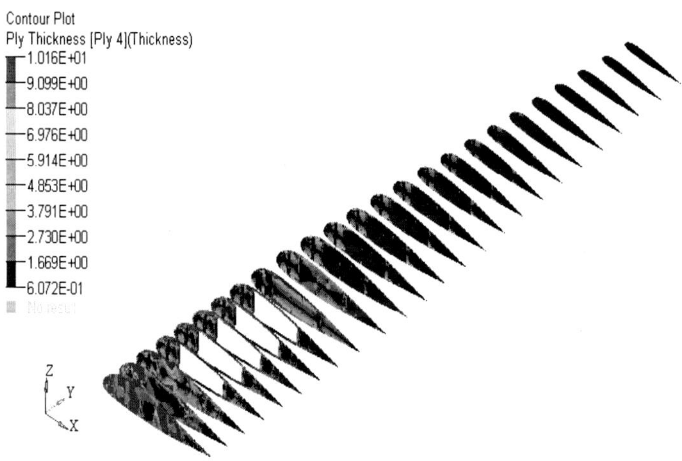

Fig. 7d Free Size Optimization for Super Ply at 90° Orientation of all Ribs

Phase 2: Ply-Bundle Sizing with Ply-Based FEA modeling

The size optimization of the Plies obtained in Phase 1 with Free Size or Topology optimization was next performed to determine the thickness and laminate family within all the Ribs, Skin and Spars. Objective functions are again mass and compliance and minimization of the same. The results obtained are given in Fig. 8.

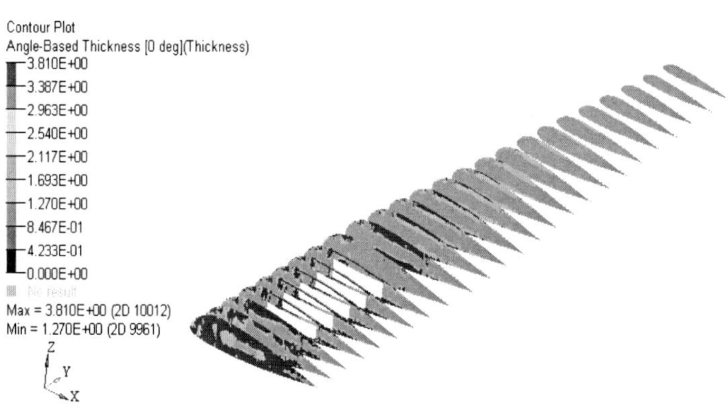

Fig. 8a Size Optimization for Super Ply at 0° Orientation of all Ribs

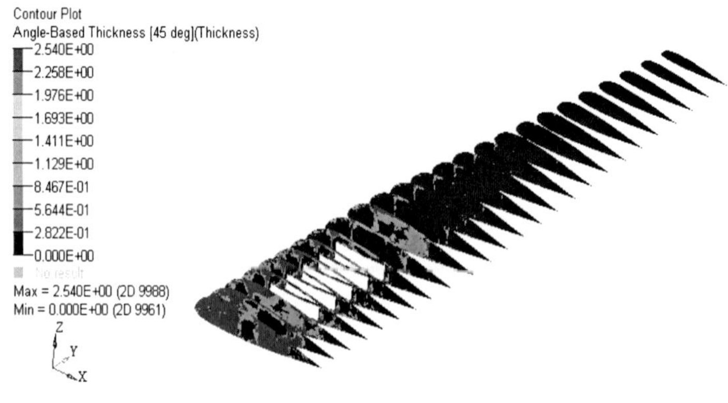

Fig. 8b Size Optimization for Super Ply at +45° Orientation of all Ribs

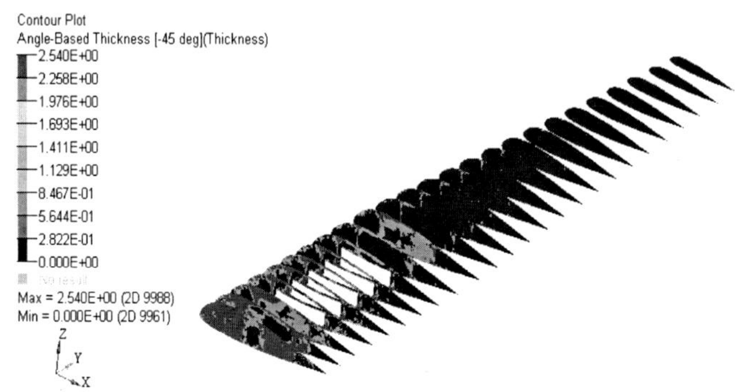

Fig. 8c Size Optimization for Super Ply at -45° Orientation of all Ribs

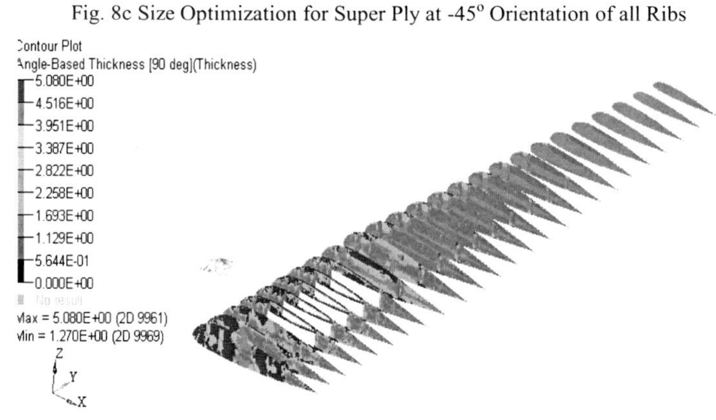

Fig. 8d Size Optimization for Super Ply at 90° Orientation of all Ribs

A review of the design process up to now reveals that we established the optimum ply shape and patch locations in phase 1 (free size optimization) and subsequently optimized the ply bundle thicknesses in phase 2 (ply bundle sizing optimization), allowing us to determine the required number of plies. These ply bundles represent the Optimal Ply Shapes (Coverage Zones).

Figs 9a to 9d show the optimal ply shapes thickness contour of ply bundles of the 90° Super Ply.

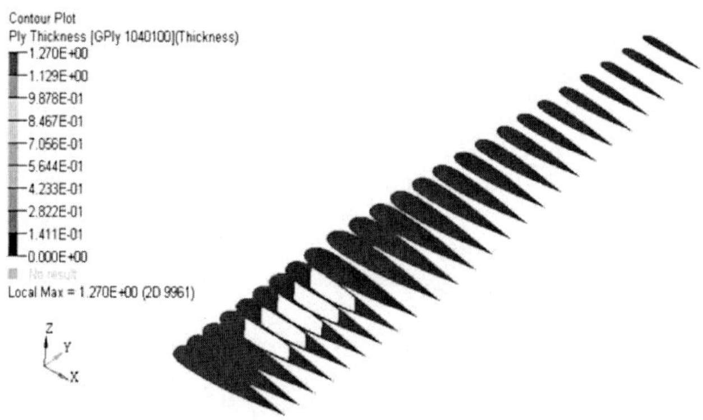

Fig. 9a Optimal Ply Shapes thickness contour of ply bundle 1 of 90° Super Ply

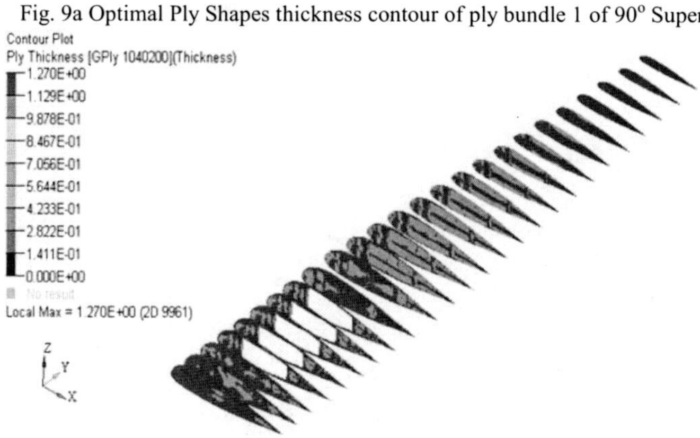

Fig. 9b Optimal Ply Shapes thickness contour of ply bundle 2 of 90° Super Ply

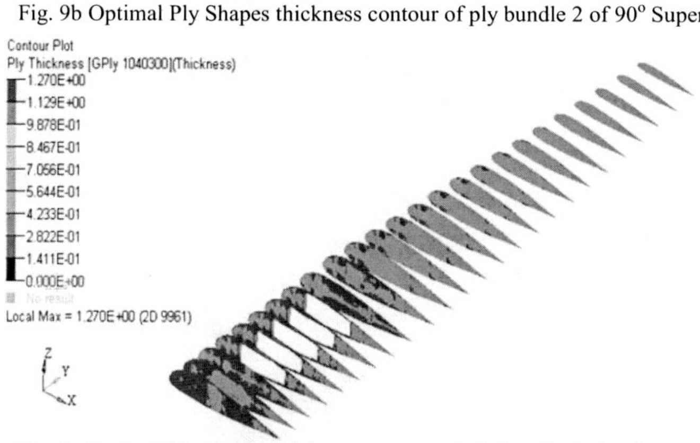

Fig. 9c Optimal Ply Shapes thickness contour of ply bundle 3 of 90° Super Ply

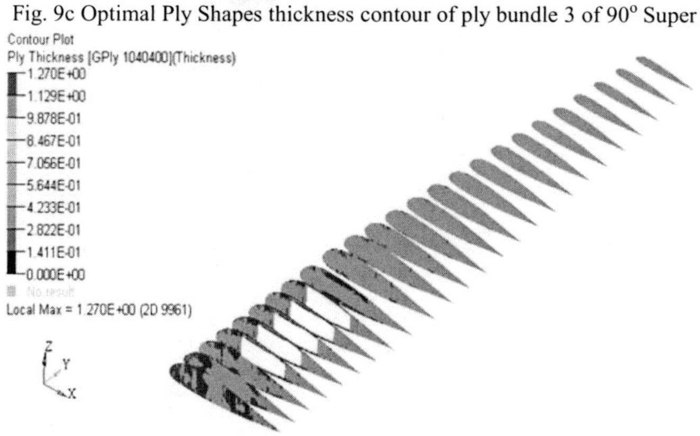

Fig. 9d Optimal Ply Shapes thickness contour of ply bundle 4 of 90° Super Ply

Similarly the Optimal Ply Shapes thickness contours of ply bundles for the 0° +45° and -45° Super Plies are obtained (not given here).

After completing the ply bundle sizing optimization, the optimized thickness contour of all ply bundles are post processed. The total size optimization of the ribs is shown in Fig 10.

Fig. 10 Size Optimization Results of the Total ribs

Phase 3 – Detailed Design – Ply Stacking Sequence Optimization

The algorithm HyperShuffle is more aimed at providing a 'global view' of what the optimal stacking sequence could be. While shuffling the stacking sequence it is important that behavioral and design constraints are preserved. Further, it is required that certain ply book rules be applied to guide the stacking of plies based on specific requirements. Some of the ply book rules that control the stacking sequence are:

- Maximum number of successive plies of a particular fiber orientation
- Stack is balanced, pairing of the plies.
- Identifying a sequence for the core and cover regions
- Symmetric stack required
- Number of plies in any one direction placed sequentially in the stack is limited
- Minimize the number of occurrences of the 0° to 90° (or 90° to 0°) change in any two adjacent plies.
- Minimize the number of occurrences of 45° to -45° (or -45° to 45°) change inside the stack by putting one 0° or 90° ply between them.

The ply bundle thus obtained is used for final laminate stacking. An algorithm HyperShuffle converts the dimensional ply bundle obtained above into local ply configuration or the final laminate. Fig. 10 shows the final laminate stacking sequence.

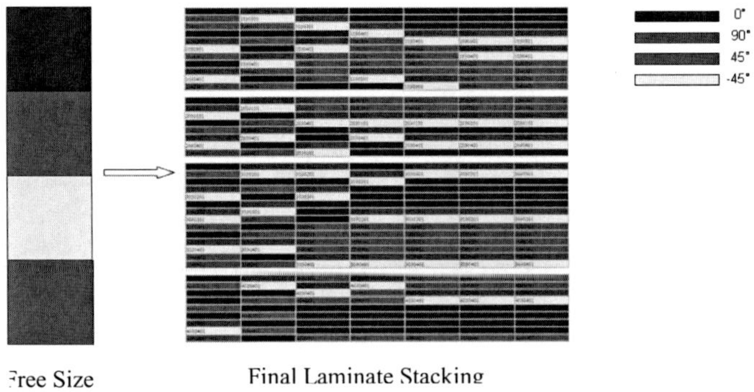

Fig 11 Final Laminate Stacking

Optimized thickness for one of the patch on the rib marked in Fig. 11 is given in Table 2. The plies realized after shuffling the layers is given in Fig. 12.

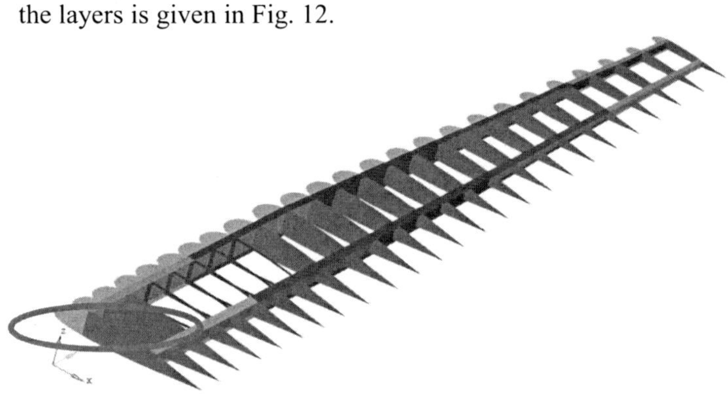
Fig. 12 Plies realized after shuffling the layers

Table 2 Ply for one of the patches in the rib marked on Fig. 11

Review

Ply lay-up: Rib_mm
Total number of plies: 11
Total thickness: 13.97

Ply	Material	Thickness T1	Orientation Degrees
1	CFRP_LA	1.27	0.0
2	CFRP_LA	1.27	0.0
3	CFRP_LA	1.27	0.0
4	CFRP_LA	1.27	45
5	CFRP_LA	1.27	-45
6	CFRP_LA	1.27	45
7	CFRP_LA	1.27	-45
8	CFRP_LA	1.27	90
9	CFRP_LA	1.27	90
10	CFRP_LA	1.27	90
11	CFRP_LA	1.27	90

Total ply thickness in the ribs varies from 2.54 to 13.97 mm.

Similar optimization process is adopted for the wings and the Spars. Free size optimization results in Phase 1 for the four plies of the front and rear Spars are given in Figs. 13 a-d. We notice that the near the Spar root material density is higher or should be thicker and nearer to the tip do not carry much load and therefore can be lighter.

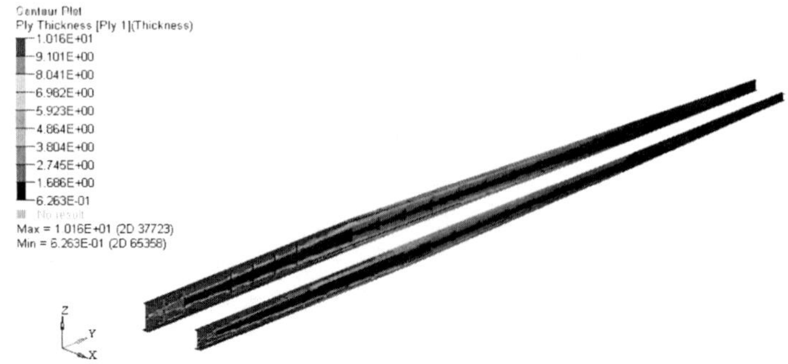

Fig. 13a Free Size Optimization for Super Ply at 0° Orientation of Spars

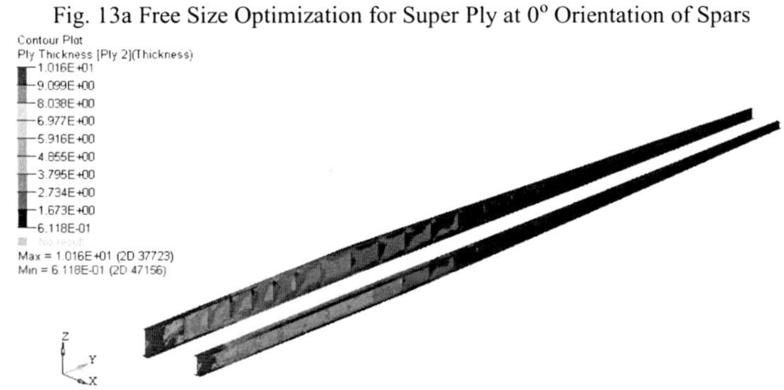

Fig. 13b Free Size Optimization for Super Ply at +45° Orientation of Spars

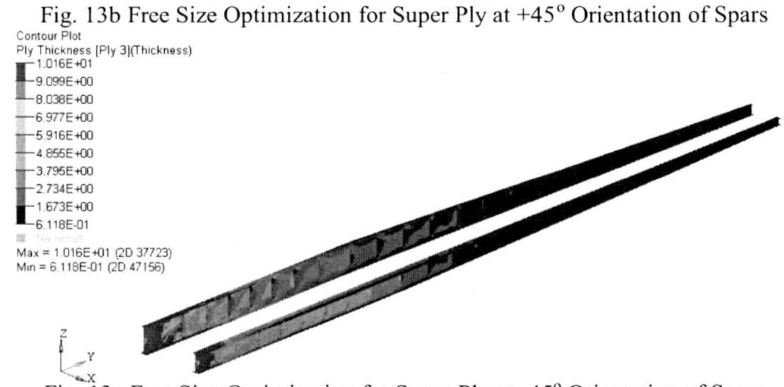

Fig. 13c Free Size Optimization for Super Ply at -45° Orientation of Spars

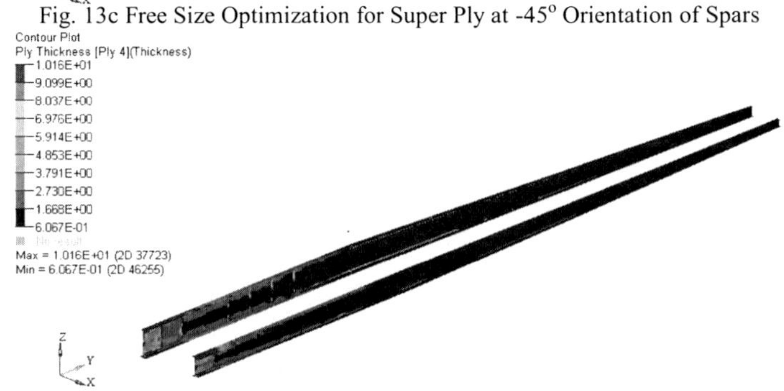

Fig. 13d Free Size Optimization for Super Ply at 90° Orientation of Spars

The size optimization results of the Plies obtained in Phase 2 are given in the following Fig 14 a-d.

Fig. 14a Size Optimization for Super Ply at 0° Orientation of Spars

Fig. 14b Size Optimization for Super Ply at +45° Orientation of Spars

Fig. 14c Size Optimization for Super Ply at -45° Orientation of Spars

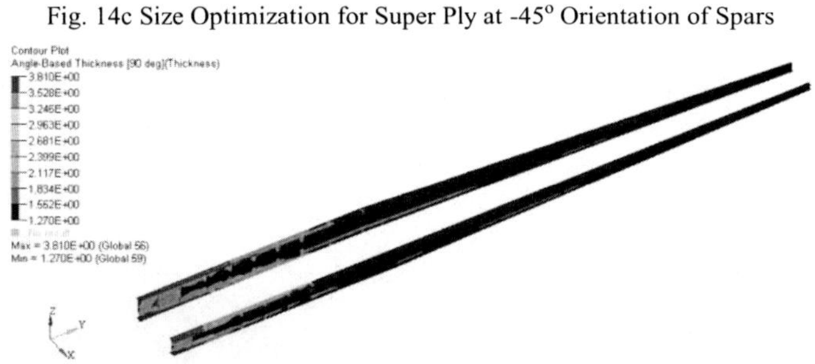

Fig. 14d Size Optimization for Super Ply at 90° Orientation of Spars

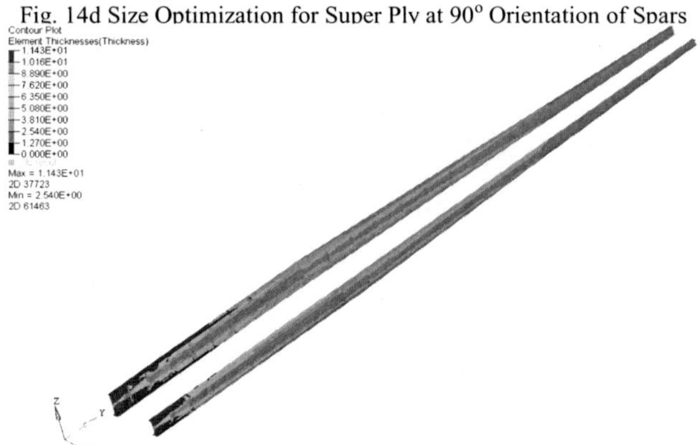

Fig. 15 Total Size Optimization Results of the Spars

The total size of the spars is given in Fig. 15.

The results of free size optimization for the four plies of the Skin with integrated Stringers are given in Figs. 16a-d. We notice that material density is higher near the wing root or it should be thicker and nearer the tip the density is lower and therefore can be lighter.

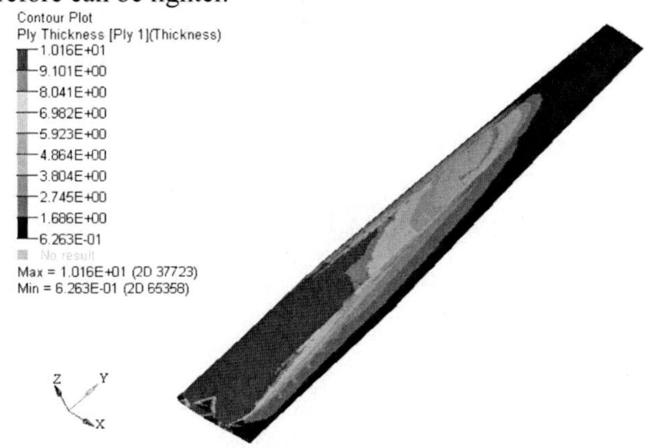

Fig. 16a Free Size Optimization for Super Ply at 0° Orientation of Skin with integrated Stringers.

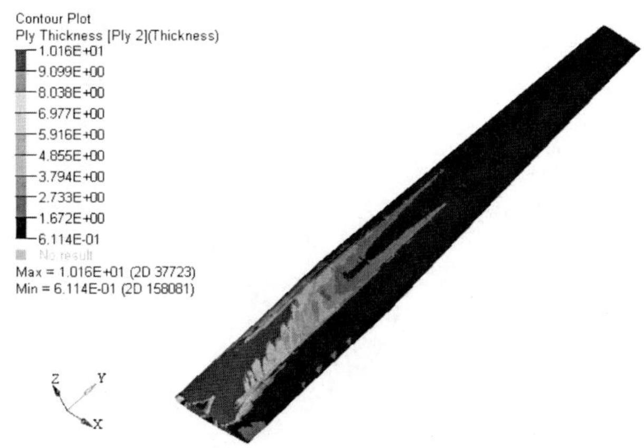

Fig. 16b Free Size Optimization for Super Ply at +45° Orientation of Skin with integrated Stringers.

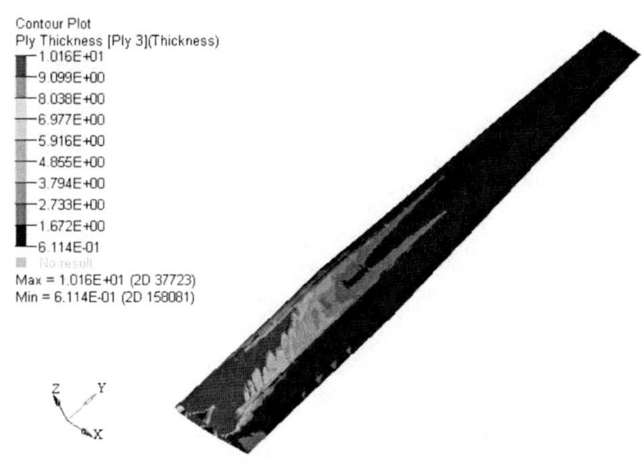

Fig. 16c Free Size Optimization for Super Ply at -45° Orientation of Skin with integrated Stringers.

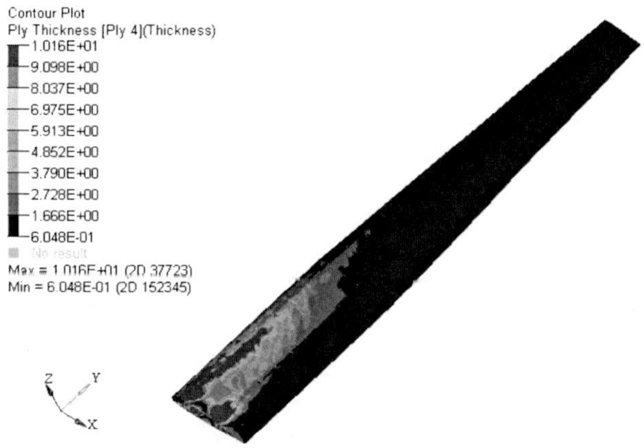

Fig. 16d Free Size Optimization for Super Ply at 90° Orientation of Skin with integrated Stringers.

The size optimization results of the Plies obtained in Phase 2 for the skin integrated with stringers are given in the following Figs 17a-d

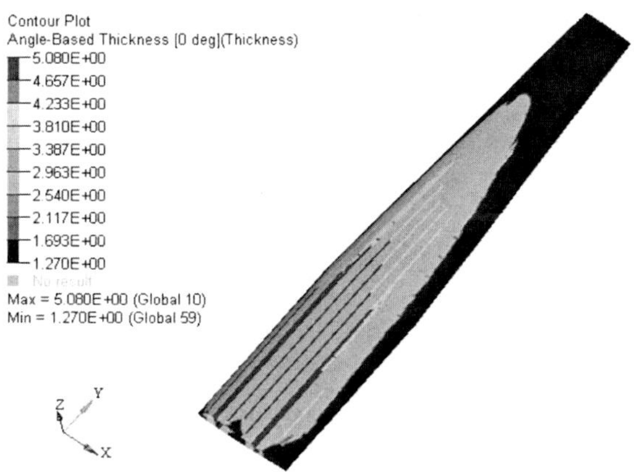

Fig. 17a Size Optimization for Super Ply at 0° Orientation of Skin with integrated Stringers.

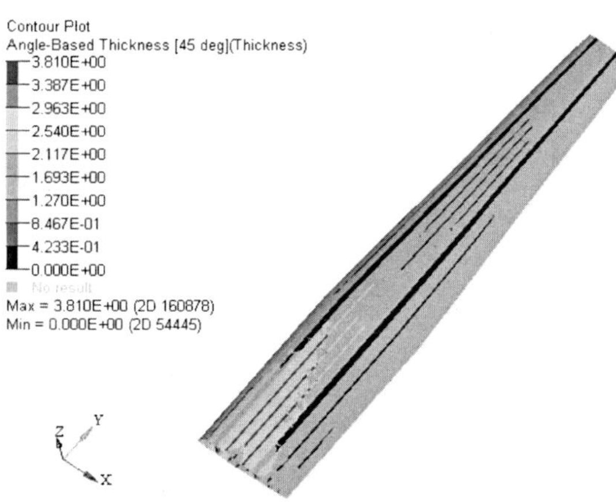

Fig. 17b Size Optimization for Super Ply at +45° Orientation of Skin with integrated Stringers.

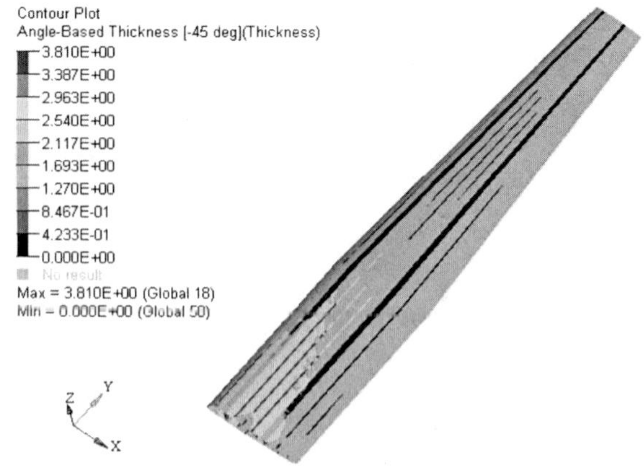

Fig. 17c Size Optimization for Super Ply at -45° Orientation of Skin with integrated Stringers.

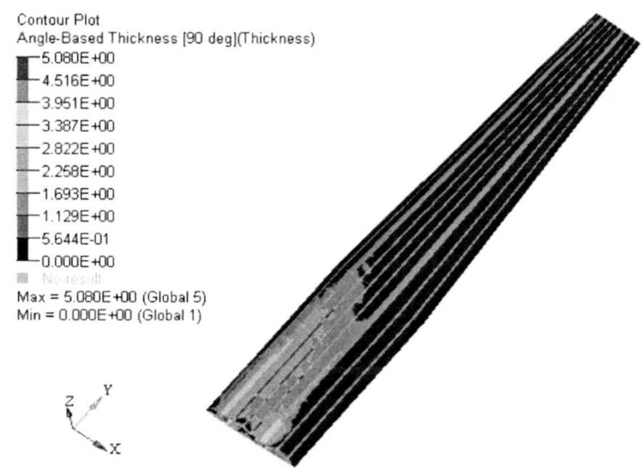

Fig. 17d Size Optimization for Super Ply at 90° Orientation of Skin with integrated Stringers.

The total size of the skin with integrated stringers is given in Fig. 15.

Fig. 18 Size Optimization Results of the Skin with integrated stringers.

The composite wing thus obtained is 1350 kg as against 2100 kg of metallic wing. The mass saving obtained from the optimized CAD model to composite model in the wing is thus around 35.7%.

CONCLUSIONS

The importance of topology optimization in obtaining concept design and the capability of composite optimization is demonstrated. So far topology optimization has been applied to realize weight savings from proven aircraft by keeping the same structural requirements. Now for the first time it has been shown that topology optimization can be used directly to realize the concept of a new design. It is also shown how this concept design can be extended right away for a composite realization. This design process saves considerable time and effort and results in minimum optimum weight for a new aircraft structure design.

REFERENCES

1. Sensmeier, M. D. and Samareh, J. A., A Study of Vehicle Structural Layouts in Post-WWII Aircraft, AIAA 2004-1624.
2. Technical Review of 428JET Wing, Dornier Luftfahrt, 1999.
3. Bendsøe, M.; Kikuchi, N. 1988: Generating optimal topologies in optimal design using a homogenization method. Comp. Meth. Appl. Mech. Eng. 71, 197–224.
4. Bendsøe, M. P and Sigmund, O., Topology Optimization– Theory, Methods and Applications, Springer Verlag, Berlin/Heidelberg, 2003
5. Altair OptiStruct, Users Manual V10.0, 2009, Altair Engineering Inc., Troy, Michigan.
6. Schumacher, G., Stettner, M., Zotemantel, R., O'Leary, O., and Wagner, M., Optimization Assisted Structural Design of a New Military Transport Aircraft, AIAA-2004-4641, Proceedings of the 10th AIAA MDO Conference, Albany, NY, 2004.
7. Rao, J. S., Kiran, S., Kamesh, J. V., Madhusudan A. Padmanabhan and Satish Chandra, Topology Optimization of Aircraft Wing, Journal of Aerospace Sciences & Technologies, vol. 61, No. 3, August 2009, p. 402
8. Poort, J., Optimization of the 787 Dreamliner Horizontal Stabilizer CFRP Composite Main Box, Proceedings of North American HyperWorks Technology Conference, October 2-3, 2007, Troy, Michigan.
9. Gürdal, Z., Haftka, R. T. and Hajela, P., Design and Optimization of Laminated Composite Materials, John Wiley, 1999
10. Ryoo, J. and Hajela, P., Handling Variable String Lengths in GA-based Structural Topology Optimization, Structural Multidiscipline Optimization vol. 26, 2004, p. 318
11. Zhou, M. and Fleury, R., Composites Design Optimization – from Concept to Ply-Book Details, Structure and the first multi-disciplinary seminar on Optimization Theory and Applications, Dalian, China, 3-4 September 2009

Proceedings of the ASME 2010 International Mechanical Engineering Congress & Exposition
IMECE2010
November 12-18, 2010, Vancouver, British Columbia, Canada

IMECE2010-37509

A CFD INVESTIGATION OF WINDOW COOLING FOR AN OPTICALLY ACCESSIBLE COMBUSTION RIG

S. CHEN
Institute for Aerospace Research
National Research Council Canada
Ottawa, Ontario, Canada, K1A 0R6
Suzhen.Chen@nrc-cnrc.gc.ca

ABSTRACT

Numerical simulation was performed for the window cooling system of a new facility, the High Pressure Optically Accessible Combustion Rig, at Institute for Aerospace Research, National Research Council Canada. The Shear-Stress Transport k-ω two-equation model in FLUENT was used to solve the complex flows. The numerical results revealed that a cooling film was formed along the window glass. However, because of the presence of the bridge blocks the cooling film was not quite uniform. The bridge blocks for the cooling slots induced complex flow patterns and thus reduced the cooling effectiveness for the window. The maximum surface temperature was found to be at the trailing edge of the inner window surface. The cooling flow exiting from the downstream end cooling slot formed a large and strong vortex, which entrained hot air from the combustion air flow and caused the described high surface temperature. At the current flow condition, the maximum value of the temperature on the window surface was slightly higher than the allowable temperature for the designed window material.

INTRODUCTION

To support key research topics on sustainable energy and the environment, the Institute for Aerospace Research (IAR), National Research Council Canada (NRC) has advanced its facilities by building a high-pressure Optically Accessible Combustion Rig (OACR) as shown in Fig. 1.

The OACR has a modular construction, which allows it to be configured for a variety of fuel injectors and combustor liner geometries. The rig can accommodate combustor liners with diameters up to 45 cm and lengths up to 120 cm. The five-way optical accessibility provides complete access to laser based diagnostics such as Particle Doppler Anemometry (PDA), Laser Doppler Velocimetry (LDV), Particle Image Velocimetry (PIV) and Laser Sheet Imaging (LSI). These features allow the rig to measure and characterize combustion processes in a detailed and non-intrusive manner under realistic gas turbine operating conditions.

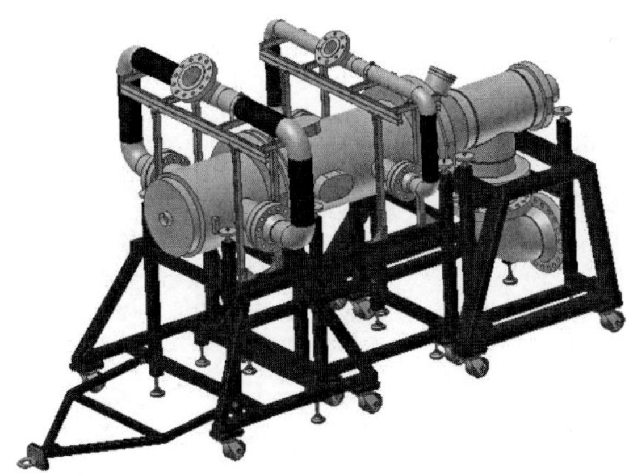

Fig. 1. High-pressure optically accessible combustion rig (OACR)

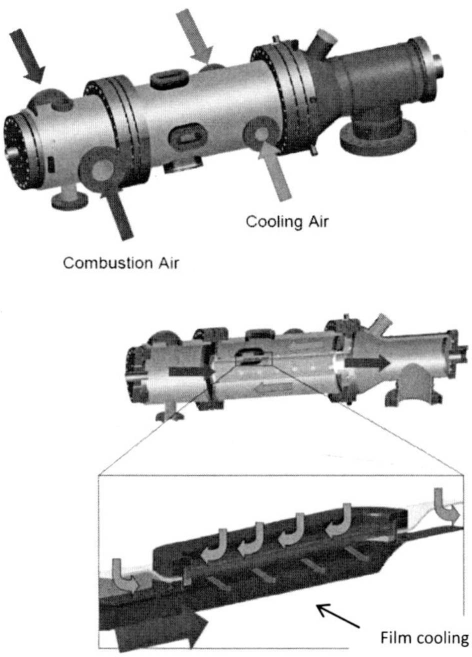

Fig. 2. OACR flow path

The cooling flow distribution for the combustion chamber is one of the key design features. Figure 2 shows the combustion and the window cooling flow paths for this test rig. The cooling air is injected into the annular section through one of both of the two flanges on the pressure vessel. Part of the cooling air goes through the impingement holes on the shroud shell to cool down the combustor liner, and then exits the channel between the shroud and the liner to mix with the combustion flow at the exit of the combustor; the rest of the air goes through the cooling holes around the five optical windows, and is expected to form a cooling film over the window glass, providing protection from the hot gases.

The design of the cooling holes/slots was based on a one-dimensional analysis model. However, CFD techniques are able to provide the details of the flow features using more advanced numerical methods. This report presents the work using CFD method to verify the efficacy of the inner window cooling design.

The CFD technique has been used to study cooling flow performance for engine systems by many researchers, such as the recent works in references [1-3]. Because of the tedious details of film cooling geometries, most of the CFD calculations were performed for a typical section of the geometry to save the computational expense, rather than simulating the whole combustor or turbine film cooling system. For example, Jeromin, et al. [2] used CFD to investigate the effusion-cooled wall of a combustor containing just one single row of cooling holes. The advantage of this approach is that the local flow field can be resolved with very fine mesh and the CFD results were able to reveal the details of flow mechanics or flow patterns and heat transfer processes around the cooling holes. However, the flow interactions induced by the different distributions of these cooling holes, and the overall performance of the cooling system for specific geometries cannot be fully resolved.

The current CFD study was performed for a complete window geometry including the cooling air through 45 holes and its interaction with the high-temperature main flow within the combustion chamber. The objective was to determine if the cooling film can be formed effectively along the surface of the window glass to protect its material from the very high temperature gas in the combustion chamber. The results will provide designers and test engineers with a reference database and can be applied to future improvement for this newly-built unique facility in its capability.

PROBLEM DEFINITION

There are five pairs of access windows for this OACR, four of them are side windows located where the combustion processes take place. They are not distributed uniformly in the radial direction due to optical instrumentation requirement. The fifth is located at the end of the rig looking upstream. Since numerically it would be too expensive to simulate the geometry with all windows, the present numerical study focused on a section containing one side window to study the cooling film formation along the inner surface of the window glass.

There are 45 cooling holes around each window. Part of the annular cooling air enters into the combustion chamber through these holes under the pressure difference between the annular section and the combustor chamber. The arrangement of these 45 holes is shown in Fig. 3. There are 30 holes with a diameter of 0.01969D (referred to as large cooling holes thereafter) on one side of the window, and 15 with a diameter of 0.00984D (referred to as small cooling holes thereafter) on the other side, where D is the diameter of the combustion chamber. The path of the cooling air is sketched in Fig. 4 for a cross section of the OACR internal window. The cooling air is colored in blue, while the window glass is colored in yellow, and high temperature air in red. The cooling air is admitted at the inner surface of the window via six cooling slots located on the side of the window with large cooling holes. Figure 5 shows the cooling air volume. The six cooling slots are shown in the above mentioned figure, and they are bounded by five small walls defined as bridge blocks for structural purposes.

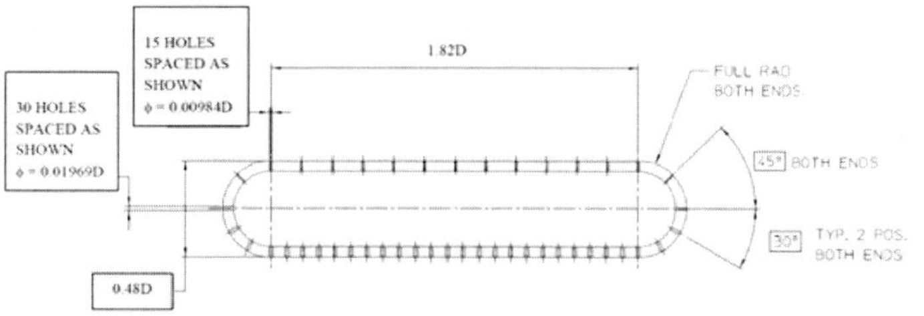

Fig. 3. Distribution of cooling holes for the internal window (D is the diameter of the combustion chamber)

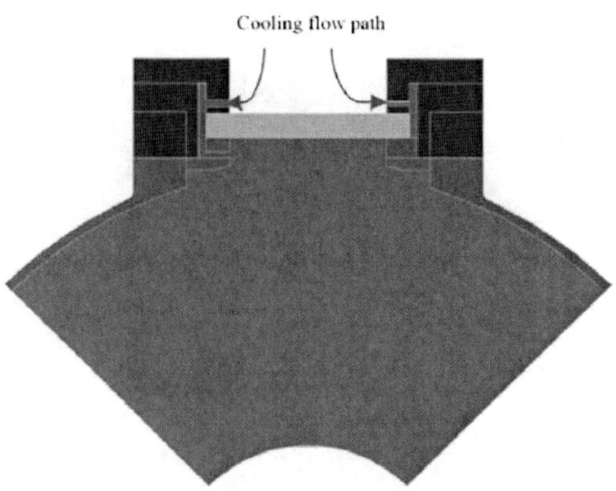

Fig. 4. Cross section of the OACR internal window

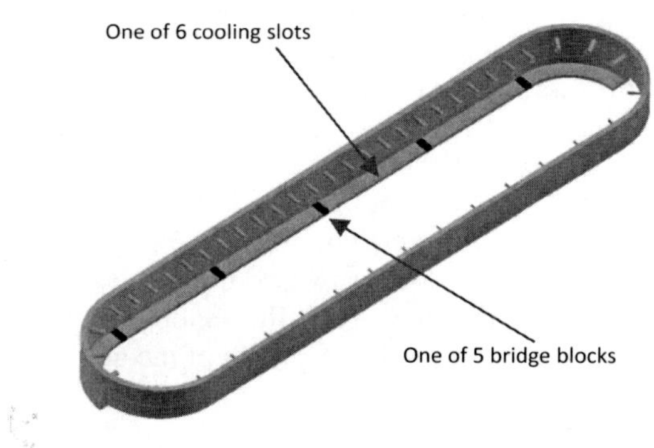

Fig. 5. Cooling air volume and bridge blocks

The energy equations for the solid materials of window glass, frames, and combustor liner were needed to be resolved for current study because of a large temperature range going from 350 K to 2099 K.

The coupled heat transfer calculation between fluid and solid presents a severe challenge for numerical simulations. Care must be taken in defining the numerical models and generating quality meshes close to the interface between the two media.

It is expected that the stainless steel frame wrapping around the cooling air for the outer sections of the window plays little role in the heat transfer calculation of the glass window. As a result, in order to reduce the computational requirement, the solid materials highlighted in purple in Fig. 4 were not simulated.

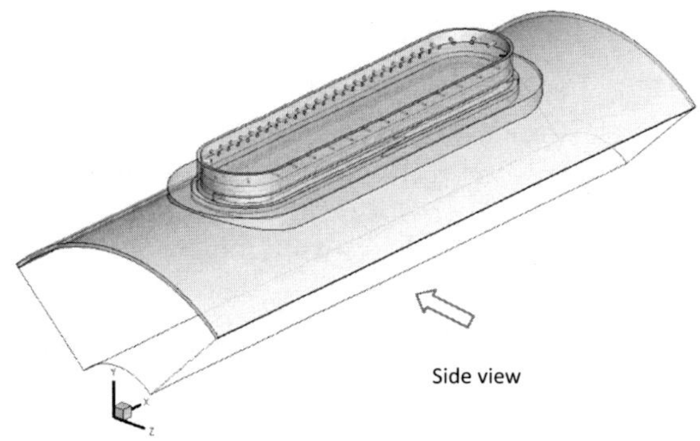

Fig. 6. Computational domain for the window cooling study

The overall computational domain is shown in Fig. 6. The coordinate system in the figure is defined as: x-axial is along the main flow direction; x-y plane goes through the middle section of the window from a side view. The origin of the coordinate is on the inlet boundary, where the main flow enters, and is located at the center of the combustor chamber. Since the combustion process was not the primary concern of the current investigation, the computational domain was simplified by using 2/3 of the combustor chamber in the radial direction. Slip wall boundary was used for the bottom surface of the domain.

The annular section of the combustor was not included. However, the flow conditions for the inlets of the 45 cooling holes were calculated based on a pressure drop of 5% across the window, which was provided by the designers. The pressure in the annular section was assumed to be 2,060,000 Pa, and cooling flow temperature was 353.15 K. Uniform combustion temperatures were assumed inside the combustion chamber and the design condition of 2099.15 K was used at the inlet of the domain, and the mass flow rate of 4.5 kg/s was used to calculate its inlet flow conditions.

The flow was assumed to enter the combustion chamber in the x-axial direction without swirl for the current numerical studies. Steady-state flow conditions were assumed.

NUMERICAL PROCEDURE AND GRID GENERATION

The steady flow numerical solutions were obtained using the commercial CFD code FLUENT 6.3 [4]. It is a multiblock-unstructured code that uses the finite volume method to discretize the governing Navier-Stokes equations in space. The density-based solver in Fluent was selected to solve the compressible flow features for the current geometry since in some areas the flow Mach number was greater than 0.3.

For the studied geometry, cooling air through these small holes formed jet flow; the main flow in the combustion chamber interacting with the flow exiting from cooling slots presented the features of cross flows, and the back-step flow past the window frame and bridge blocks would induce flow recirculation and separation. These complex flow features need to be resolved using higher order turbulence models.

The SST k-ω model developed by Menter blended the robust and accurate formulation of the k-ω model in the near-wall region with the free-stream independence of the k-ε model in the far field. Although different researchers have different preferences in choosing their own turbulence models, the SST k-ω model has been often used in resolving separation flows. Jeromin, et al. [2] used the model to predict the heat transfer coefficients for the effusion-cooled wall for a gas turbine combustor, Ceccherini, et al. [3] applied the model to study the effects of slot injection, effusion array and dilution holes on the cooling performance of a combustor liner. Esch and Menter [5, 6] found that the SST k-ω model was able to provide much better heat transfer predictions than most of the other common two-equation turbulence models. Durbin and Shih [7] came to the same conclusion. Recent work of Zu, et al. [8] validated various turbulence models for a single-jet impingement flow and heat transfer with a set of benchmark experimental data, and concluded that the SST k-ω model was the best compromise between computational cost and accuracy. Therefore, the SST k-ω model was used to resolve the boundary layer right down to the wall for the current study.

The second-order upwind discretization scheme was used for all flow variables to ensure accuracy of the solutions.

Conjugate heat transfer calculations were performed to couple the heat fluxes between the fluid regions and the solid walls. The fluids for both of the cooling and high-temperature flows were assumed to be air and they were modeled using the ideal gas law. The specific heat of air was defined as a function of the temperature. Its thermal conductivity and molecular viscosity were calculated using kinetic theory [4]. In the solid regions of the glass and liner, the energy equation was also solved.

The multi-blocked meshes were produced using Gambit 2.4 [9]. Hexahedral meshes were generated in the regions around the solid walls to ensure more accurate numerical resolution in the boundary layers. Figure 7 shows the surface mesh around the window and the cross section of the window. Fine grids were used to resolve the boundary layers above the viscous walls. The dimensionless normal distance of the first grid points from the glass wall, y^+, was in the range between 0.25 ~ 3.5.

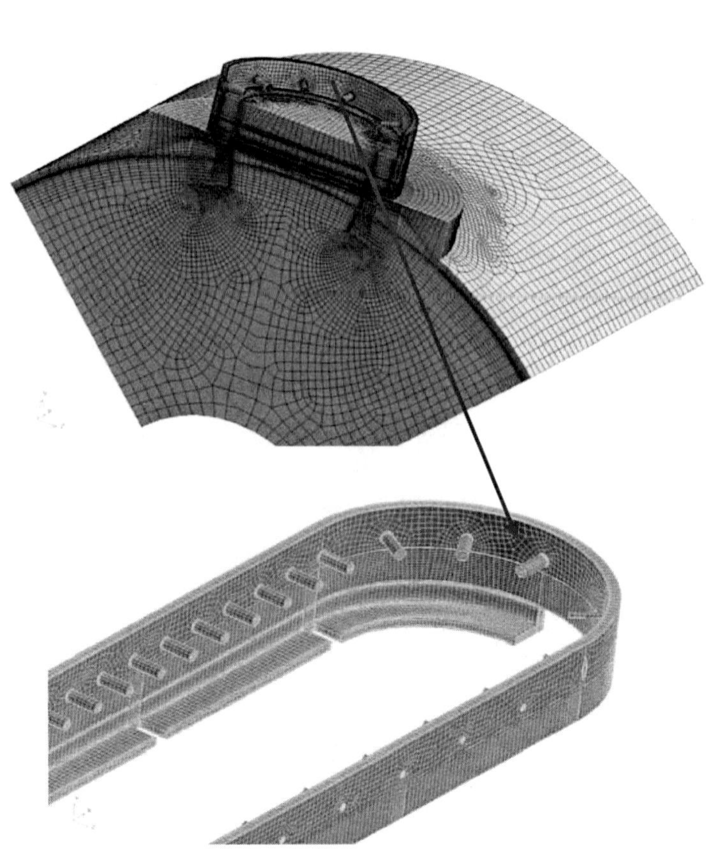

Fig. 7. Surface mesh for a section of the geometry

The 45 very small cooling holes and their non-symmetric distribution created many challenges for mesh generation and control of grid points for this window geometry. Since the primary objective of the current study was to predict the temperature features around the window glass, relatively coarse meshes were used for the cooling fluid volumes (the enlargement in Fig. 7) to reduce the total number of grid points. For the cooling fluid volume, the wall y^+ values were mostly less than 200. Mesh independent studies were performed for a slice of the volume containing two cooling holes. It concluded that the mesh distributions were fine enough for turbulence models to predict the main flow features of the cooling air, such as the mass flow rate.

The total number of grid points was about 5 million for the entire computational domain.

All computations were run on a Linux cluster with Quad-Core AMD Opteron™ Processors 2384 at NRC-IAR. A parallel technique was used to speed up the computations. To stabilize the computation process, the Standard k-ε model with wall function was initially used and then its solution was interpolated as the initial solution for the calculations using SST k-ω model. All variables converged and their residuals were reduced by approximately four orders of magnitude. The convergence of the computations was also monitored with temperature integration on the surface of the glass. At the end of the calculation, the area-weighted average temperature on the glass surface converged to 547 K.

The mass flow flux was examined for all of the inlets and outlets. The mass flow imbalance was less than 1% of the smallest value of the inlet flows. Therefore, the mass flow can be considered to be balanced for the current computation.

RESULTS AND DISCUSSIONS

Figure 8 shows the streamlines around the window for the cooling and main flow fluids colored by temperature. The cooling air exiting the six cooling slots is injected into the high-temperature air of the main cross flow in the combustor chamber. These two sets of air interacted with each other and exchanged their thermal energies. As the result, the cooling air temperature was increased gradually along its streamlines, while some of the main flows were disturbed, cooled, and entered the area close to the window glass.

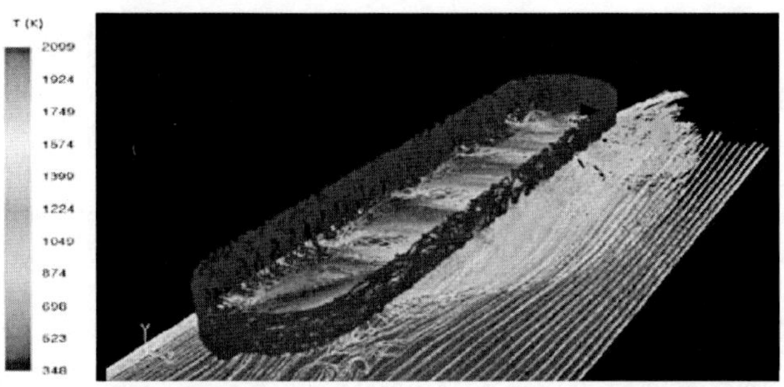

Fig. 8. Streamlines coloured by temperature around the window

Furthermore, the cross flow induced vortices formed by the cooling flows. These vortices stretched downstream into the core region of the main flow. As the results show, the main flow temperature was decreased in a large portion in the downstream region. Figure 9 shows the temperature contours at several cross sections along the main flow direction. The development of the vortices is presented in the figure. The vortices were formed on the opposite side of the cooling slot openings. At the window trailing edge (Section A in Fig. 9), the induced low temperature core occupied almost 1/3 of the whole combustion flow. This large low-temperature region may affect the combustion process and reduce the combustion efficiency.

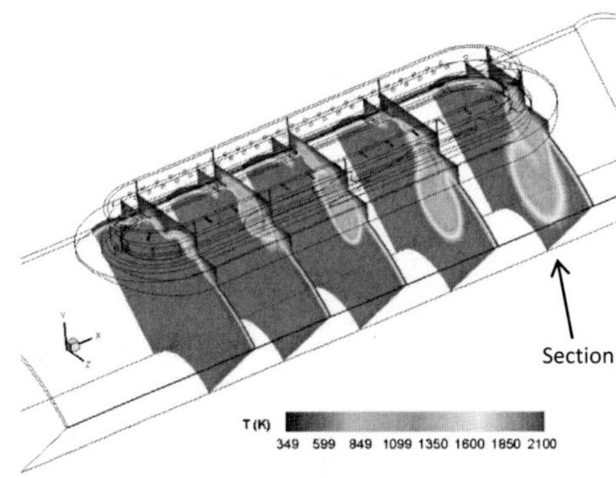

Fig. 9. Temperature contours at several cross sections

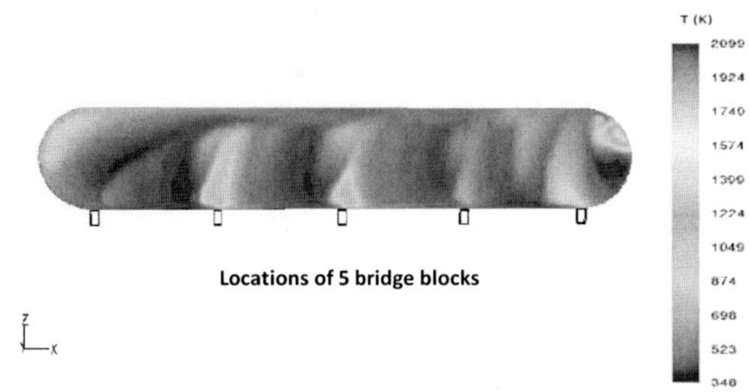

Fig. 10. Window glass surface temperature distribution

The temperature contours on the glass surface are shown in Fig. 10. At the two ends of the window, high temperature areas were observed. They were both located on the side opposite of the cooling slots. However, the temperature at the trailing edge of the glass was higher than those at the leading edge.

Figure 11 shows the temperature contours at the cross section through the middle of the whole domain from a side view of the combustor. Recirculation flows were formed on the leading and trailing edges because of the steps of the window geometry. However, the one at the trailing edge was larger and stronger.

Since the cooling flow at the leading edge of the window exited in the direction of the main flow, it caused only minor turbulence on mixing. However, at the trailing edge where the cooling flow was opposite to the main flow, strong vortices and turbulence were created. These large and strong vortices entrained the hot air downstream of the window into the window corner and increased the surface temperature of the glass in that area.

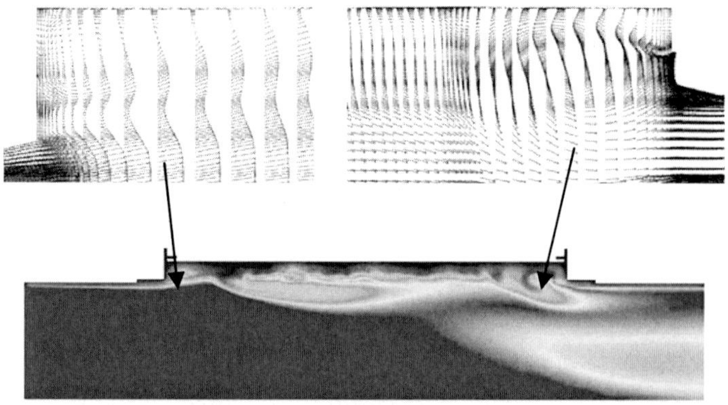

Fig. 11. Temperature contours at the middle cross section

High temperatures were also found on the glass surface above the areas where the five bridge blocks were located (Fig. 12). These blocks induced vortices above them in the window spanwise direction as shown in the figure. Because of the influence of the main flow, these vortices were in count-clockwise direction. The velocity vectors in the figure were above one of these blocks at the cross section at the quarter width of the window (Section A-A). These vortices brought the high temperature air into the cove region of the window and might be responsible for the high temperature at the glass surface. Figure 13 compares the glass surface temperature for the geometries with and without these blocks. The cooling effectiveness for the window was greatly reduced because of the presence of these blocks.

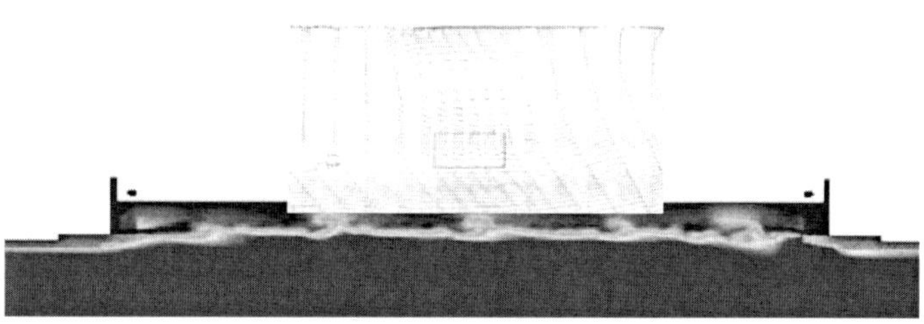

Section A-A

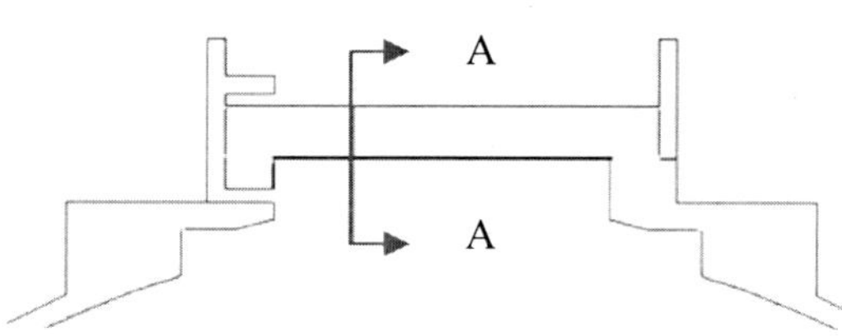

Fig. 12. Temperature contours at the cross section at the quarter width of the window

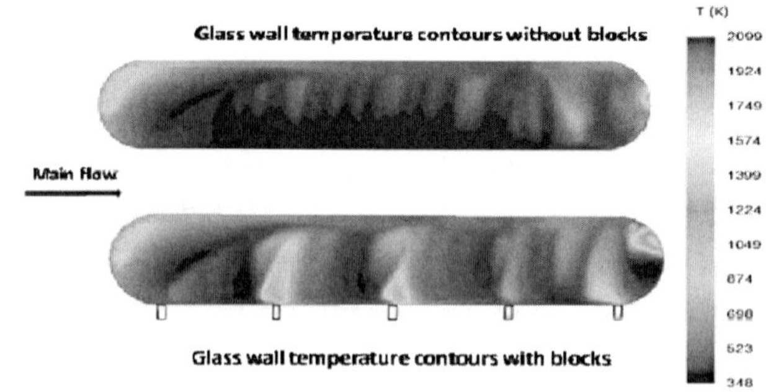

Fig. 13. Comparison of the glass surface temperature between the geometries with and without bridge blocks

The supplier of the glass material recommended a maximum temperature of 1050 °C (1323 K) for the glass in order to prevent devitrification (crystallisation) and embrittlement of vitreous silica. However, the maximum temperature of the current numerical predictions was 1360.5 K on the glass surface, which was about 40 K marginally over the allowable temperature. Note that, for current flow conditions, the flow was assumed to enter the combustion chamber without any swirl. When a swirling flow condition is imposed, the glass surface temperature may be lower than the allowable value of 1323 K.

CONCLUSIONS AND FUTURE WORK

Numerical simulation was performed for the window cooling system of the CNRC's new facility, the High Pressure OACR, using the SST k-ω two-equation turbulence model in FLUENT.

The numerical results revealed that a cooling film was formed over the window glass. However, because of the presence of the bridge blocks, the cooling film was not completely uniform. The bridge blocks for the cooling slots induced complex flow patterns and thus reduced the cooling effectiveness for the window. For the present flow conditions, with uniform main flow entering the combustor, there may be room to improve the cooling effectiveness with more consideration in designing the size, shape, angle and locations of the cooling slots. However, a more realistic swirling flow condition might improve the cooling effectiveness. Further numerical studies in this aspect will provide valuable information for future improvement of the facility.

The maximum surface temperature was located at the trailing edge of the window. The flow exiting from the end cooling slot formed a large vortex because of its interaction with the main flow, which flowed in an opposite direction as that of the cooling flow. The strong vortex entrained some hot air from downstream of the window and might be responsible for the high surface temperature in the corner area. The predicted maximum value of the temperature on the glass surface was 1360.5K for current flow conditions. This temperature is 40 K higher than the allowable temperature of 1323 K for Spectrosil 2000 material.

A deep penetration of the cooling flow into the main flow was also observed. The number and distribution of the 45-cooling holes may need to be optimized in order to reduce the disturbing effects on the combustion process caused by the optical window cooling system.

Currently the computations for swirling flows at the inlet of the combustor are being investigated. The results will provide a comprehensive understanding for this window cooling design.

For the future work, experimental results are desired to examine some of the important flow features revealed in the CFD studies.

ACKNOWLEDGMENTS

The author would like to acknowledge Mr. M. Capurro, Drs J. Boulanger, I. Campbell, and W. Chishty for their contributions to the OACR CFD project. The author would also like to acknowledge Mr. J. Ross of NRC Design Office, who provided the CAD geometry, and Dr. K. Chen of IAR, who provided advice in selecting the thermal data for solid materials.

REFERENCES

[1] Na, S., Williams, B., Dennis, R. A., Bryden, K. M. and Shih, T. I-P., 2007, "Internal and film cooling of a flat plate with conjugated heat transfer", Proceedings of GT2007, ASME Turbo Expo 2007: Power for Land, Sea and Air, May 14-17, 2007, Montreal, Canada. ASME GT2007-27599.

[2] Jeromin, A., Eichier, C., Noll, B. and Aigner, M., 2008, "Full 3D conjugate heat transfer simulation and heat transfer coefficient prediction for the effusion-cooled wall of a gas turbine combustor", ASME Turbo Expo 2008: Power for Land, Sea and Air, June 9-13, 2008, Berlin, Germany. ASME GT2008-50422.

[3] Ceccherini, A., Facchini, B., Tarchi, L., Toni., L. and Coutandin, D., 2009, "Combined effect of slot injection, effusion, array and dilution hole on the cooling performance of a real combustor liner", ASME Turbo Expo 2009: Power for Land, Sea and Air, June 8-12, 2009, Orlando, Florida, USA. ASME GT2009-60047.

[4] FLUENT 6.3 User's Guide, Fluent Inc., September 2006.

[5] Menter, F. R., and Esch, T., 2001, "Elements of industrial heat transfer predictions", 16th Brazilian Congress of Mechanical Engineering, Vol. 20, p. 117-127.

[6] Esch, T., and Menter, F. R., 2003, "Heat transfer predictions based on two-equation turbulence models with advanced wall treatment", *Turbulence, Heat and Mass Transfer 4*, Begell House Inc., p. 633-640.

[7] Durbin, P. A., and Shih, T. I.-P., 2005, *Modeling and Simulation of Turbulent Heat Transfer*, WIT Press, Southampton, p. 3-31.

[8] Zu, Y. Q., Yan, Y. Y., and Maltson, J. D., 2009, "CFD prediction for multi-Jet impingement heat transfer", ASME Turbo Expo 2009: Power for Land, Sea and Air, June 8-12, 2009, Orlando, Florida, USA. ASME GT2009-59488.

[9] GAMBIT 2.4 User's Guide, Fluent Inc., May 2007.

Proceedings of the ASME 2010 International Mechanical Engineering Congress & Exposition
IMECE2010
November 12-18, 2010, Vancouver, British Columbia, Canada

IMECE2010-37574

DELAY IN FLOW SEPARATION FOR CIRCULATION CONTROLLED CYLINDERS

Byron W. Patterson, Gerald M. Angle[II], Emily D. Pertl, and James E. Smith
West Virginia University
Morgantown, West Virginia, USA

ABSTRACT

Recent circulation control testing at West Virginia University, in a closed loop wind tunnel, has been conducted on models where the trailing edge radius was selected to be smaller than that used in literature, such as Loth and Boasson [1], 1.5 inches and Englar [2], 0.4375 inches. The reduced size was chosen in an attempt to minimize the drag experienced during periods of non-activation of the circulation control, and the smaller size was more compatible to the wind tunnel test section size. However, while the drag is lessened by a smaller trailing edge, the performance of circulation control also appears to be dependent upon a multitude of variables including, but not limited to, the trailing edge radius and jet velocity. Through a modeled experiment, the two attributes that influence the circulation control performance were concurrently manipulated by varying the radius of curvature and the velocity of the blown jet. The combination of these characteristics were experimentally explored to determine the location where the jet leaves the surface of the cylinder, also known as the separation point. The optimum separation point is defined as the farthest angular displacement from the plane of the blown jet exit slot, which corresponds to the greatest increase in the circulation around the cylinder, representing the trailing edge of a circulation control airfoil. From the known radius and jet velocity, an expression that relates the separation point and the mass flow rate velocity quantity are compared. Understanding the blowing coefficient and its impact on the separation point, results in a predictive relationship between these two attributes of circulation control. The results of this two-dimensional cylinder study found that an increase in trailing edge radius decreased the location of the separation point. In addition, an increase in the jet velocity resulted in an increase in the separation point location. The combination of these two quantities produced a relationship similar to each individually, illustrated by the mass flow rate velocity value, which is the blowing coefficient excluding free stream conditions, versus the angle of separation. Data is therein compared to the theory by Newman [3], which predicts a maximum separation point location at 245 degrees beyond the jet exit plane and an increase in the separation point as the radius of curvature increases. The results of this study found a separation point maximized at 231 degrees, and, contrary to Newman [3], a decrease in the separation point was found as the radius of the cylinders increased.

NOMENCLATURE

b = slot height
C_L= coefficient of lift
C_μ= blowing coefficient
c = chord length
M_j= mach number of the jet
$\dot{m}$ = mass flow rate
P_o = plenum pressure
P_{static}= static pressure
P_{total}= total pressure
q= dynamic pressure, $\rho_\infty V_\infty^2/2$
R= universal gas constant, 1716 ft lb/ slug °R
r= radius of curvature
T_j= temperature of jet
t = slot height
V_j, V_{jet}= velocity of jet
V_∞= free stream velocity
γ= ratio of specific heat of air
θ, θ_{sep}= angular displacement, angular separation point
ρ_j= density of the jet
ρ_∞= free stream density

INTRODUCTION

The purpose of this research effort is to experimentally determine the ideal combination of jet velocity and radius of curvature for circulation control applications. This information is useful for a variety of circulation control projects conducted by researchers abroad because it generates trends, leading to the optimal relationship between the two properties, thereby increasing the effectiveness of the application of circulation control. This research effort culminates in an experimental determination of the relationship between radius, jet velocity, and separation location, useful for future design of circulation control applications.

BACKGROUND

This section summarizes some prior research examining circulation control, the blowing jet, and the ideal slot height to radius of curvature relationship. The information gathered from literature was used for determining the parameters of the model during the experimental investigation.

Circulation Control

When studying the applications of circulation control, one will become familiar with the Coanda effect. This phenomenon, initially described by Henri Coanda in 1910, is experienced by the common person on a daily basis [4]. One example is when a liquid is poured from a glass. As the liquid exits the top of the glass, it has a tendency to adhere to the edge and curl around. Another, perhaps friendlier example is when one places his finger, or the back of a spoon, under a thin water stream. The stream attaches and flows around, becoming once again a uniform flow on the other side, at an angle offset from the initial direction.

There are multiple definitions for the Coanda effect. Reba [4] defines the anomaly as the tendency of a fluid, either gaseous or liquid, to cling to a surface that is near an orifice (opening) from which the fluid emerges. Another description states the effect as the deflection of plane jets by adjacent boundaries [3].

Similar to the glass of liquid and stream of water visualizations of the Coanda effect, the air flowing around an airfoil can be influenced by a jet of air blown out of a small slot over a region of the airfoil's surface. As this air travels along the surface and particularly around the trailing edge, the fluid will remain attached for values of θ exceeding 180° from the jet exit [3]. The slot height, b, is the dimension of the slot exit normal to the surface of the cylinder where the blowing jet is expelled at a given velocity, V_{jet}, as shown in Fig. 1. The trailing edge of the airfoil can be modified such that it is cylindrical shaped, in lieu of the conventional airfoil pointed trailing edge. The Coanda effect would not be as advantageous for the common airfoil design. For the conventional design with a blown jet, the fluid adheres to the surface until the sharp point, at which time it becomes a jet flap. The jet flap concept can enhance the lift performance of the airfoil but typically at a higher energy cost than a Coanda jet to achieve the same lift coefficient.

However, it is important to note that while the airfoil design employed to exploit the Coanda effect increases the lift performance, there is also an increase in adverse effects, such as profile drag, due to the geometry of the cylindrical trailing edge, as well as the induced drag due to increased lift. Even though drag increases, Fisher [5] found that an isolated circular cylinder consistently produced greater Lift/Drag (L/D) values than the 50% ellipse airfoil, whose results were found to be analogous to a conventional airfoil. Thus, the drag on the circular cylinder was minimized as the lift was optimized and a circulation control airfoil outperforms a conventional airfoil in regards to L/D.

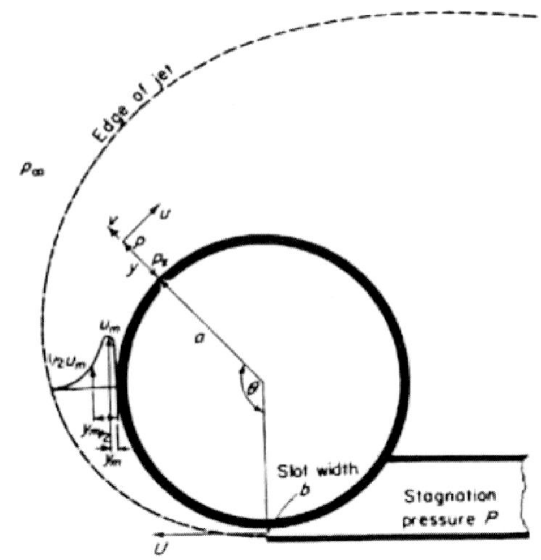

FIGURE 1 FLUID FLOW AROUND A CYLINDER

As the fluid leaves the slot, the jet attaches to the surface, due to the imbalance of centrifugal and suction forces [4]. According to Henri Coanda, as accounted by Reba [4], the pressure on the surface of the cylinder is less than the atmospheric pressure, creating a partial vacuum or suction force pulling the jet toward the surface. As the air is blown out of the slot, the vacuum accelerates the flow around the surface to the point of separation. The flow separation occurs when the pressure force between the jet and the free stream pressure decreases to the point where it is equalized by the inertia (centrifugal) force of the jet, causing a balanced, or zero, net force. During Reba's [4] experiments, it was found that in order to maintain this suction force, the stream of air must hug the surface without actually adhering to it, otherwise, the flow would dissipate. In addition to the suction force, the lower pressure causes the entrainment of fluid outside the flow, drawing more air into the jet. Coanda found that as the jet flowed around the edge it entrained up to 20 times the amount of air, relative to the original jet [4]. By incorporating a slot at the trailing edge of the airfoil, considerable amounts of lift is generated. The significance of the impact circulation control has on the coefficient of lift is demonstrated by the work of Cagle and Jones [6], who compared pulsating blowing coefficients to the traditional circulation method of steady flow. Illustrated in Fig. 2, as the velocity of the jet of air is increased, there is an increase in the coefficient of lift. Cagle and Jones [6] use the non-dimensional blowing coefficient, C_μ, as defined in Eq. (1), to represent jet velocity variations. In the case of a blowing coefficient of 0.060, the coefficient of lift increased about three times that of an airfoil without any energized air, depicted as the $C_\mu = 0$ trend in the chart.

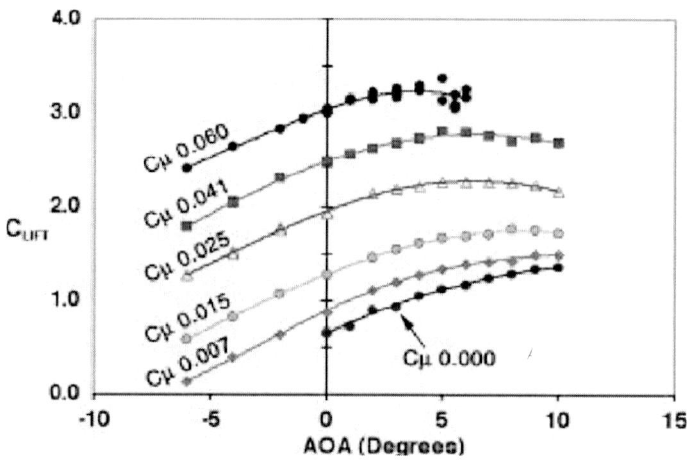

FIGURE 2 EFFECT ON CL BY CM [6]

Blowing Jet

One method to lengthen the attachment of air on a circulation controlled airfoil is to alter the velocity of the jet exiting the jet slot. Ambrosiani [7] states, that the velocity of the jet has a direct effect on the separation point, where the jet stream and the free stream pressures are equal. The definition of the blowing coefficient by Ambrosiani [7] is shown in Eq. (1). Englar [2] offers an alternate definition of the blowing coefficient per unit span in Eq. (2).

$$C_\mu = \frac{t}{c}\frac{2\gamma M_j^2}{\rho_\infty V_\infty^2} P_o \left[1 + \frac{\gamma - 1}{2} M_j^2\right]^{\frac{-\gamma}{\gamma-1}} \tag{1}$$

$$C_\mu = \frac{\dot{m} V_j}{qc} \tag{2}$$

Slot Height and Radius of Curvature Relationship

As outlined by Reba [4], there are ways to influence the stability of the Coanda effect. One of the possible adjustments to the geometry that influences the stability involves the height of the slot. If the slot height is too large, the risk of stream detachment is prevalent. Conversely, if the slot height is too narrow, the stream will stick to the surface, which, as explained in the circulation control section, causes the dissipation of air [4]. Determining a slot height that avoids both extremes is imperative to the design of applications involving the Coanda effect.

Due to the pressure gradient, there is a point where the stream of air separates from the airfoil, the separation point. For the purposes of optimizing the benefits of circulation control, the idea is to have the separation point as far forward (closer to the leading edge) as possible on the bottom surface of the airfoil. This will allow for greater generation of lift, enhancing the airfoil performance. Newman [3] defined the angular separation point in Eq. (3); his efforts also theorized that a maximum separation point occurs at 245 degrees.

$$\theta_{sep} = 245 - \frac{391\,\frac{b}{r}}{\left(1 + 1.125\frac{b}{r}\right)} \tag{3}$$

It is important to note that this relationship does not account for the power required by the blown jet, which leads to a trade-off of performance and power required. In addition, this equation does not consider the impact of varying jet velocities.

In Englar's [2] investigation of circulation controlled wing sections, he analyzed the relationship of the slot height and the radius of curvature in regards to their impact on the circulation controlled wing. The investigation shows that a strong Coanda flow is maintained when the slot height, h, to radius of curvature, r, ratio is in the range of $0.01 \leq \frac{h}{r} \leq 0.05$. Englar [2] also investigated the relationship between the dynamic pressure and the slot height on the blowing coefficient. Englar's [2] work demonstrates how the slot height has a direct effect on the velocity of the jet while both in turn have an effect on the separation point of the jet, and his efforts found that there is a linear relationship between the plenum pressure and the blowing coefficient.

EXPERIMENT

Investigation

The investigation experimentally explores the impact the radius of curvature and the jet velocity have on the delay in flow separation, to achieve the optimal configuration of the two quantities so as to maximize the angle of separation. This was done by changing the value of the radius of curvature and the jet velocity throughout the study.

The variables considered in this investigation are: slot height, radius of curvature, jet stream exit velocity, mass flow rate, and the separation point.

Model

The setup, for testing of the delay in flow separation, involves an apparatus, see Fig. 3, that has a cylinder (representing the trailing edge) with a jet slot blowing tangent to the cylinder's surface. The setup was constructed so that during testing there would be an ease in interchanging the cylinders of varying radii. It is necessary to include this function because alteration of the cylinders represents the changing radius of curvature, and consequently the blowing slot height to radius ratio. The threaded hole of the cylinders were constructed so that a piece of 3/8-16 thread would control the rotation of the cylinder. Embedded in the span wise center of the cylinders is a pressure tap, which was connected to a Heise digital pressure gauge and used for locating the separation point by measuring the surface pressure on the cylinder. With the simple movement of the attached handle, the user was able to locate the measurement point (pressure tap) at any degree of choice.

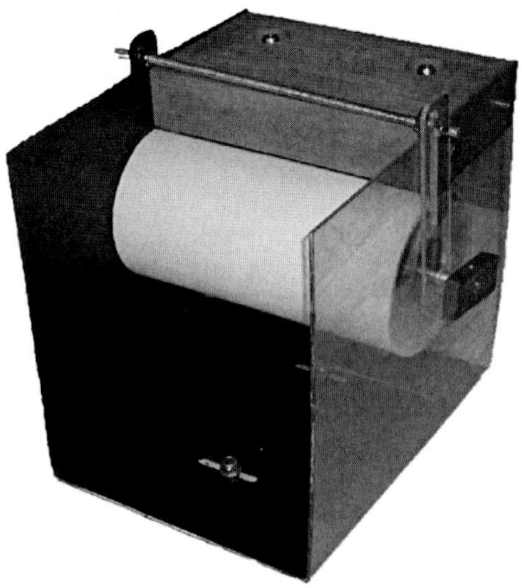

FIGURE 3 MODEL

FIGURE 4 VARYING SIZED CYLINDERS

In addition, the radius of the cylinders were chosen based on the relationship found by Englar [2], where a strong Coanda flow was maintained when the slot height to radius of curvature relationship was in the range of $0.01 \leq \frac{h}{r} \leq 0.05$. Since the slot height is fixed at an average height of 0.030 in., the cylinders where chosen to maintain Engler's [2] findings. Furthermore, the specific cylinders were chosen based on the materials present, which resulted in the application of PVC and Conduit pipes. Subsequently, the radius of cylinders were 0.52 in., 0.659 in., 0.943 in., 1.191 in., 1.622 in., 1.75 in., and 2.18 in., shown in Fig. 4. While the use of piping seems primitive for experimentation, they had a naturally formed cross-sectional circular shape, absolutely necessary for this study.

On the upper portion of the device, Fig. 3, is the plenum, which distributes the incoming air from two hose connecting fittings into a uniform jet. It is necessary to note that while most of the air flow is undisturbed by any obstructions, at approximately one third and two thirds of the chamber span wise are bolts, which support and maintain the consistency of the height at the exit of the slot. While a structural collapse may not have been a probable case, the bolts ensured that the slot exit remained constant throughout the testing.

Prior to considering the slot height to radius ratio, the initial concept was to use a slot height of 0.01 in., the same as used in previous studies. However, when considering the relationship presented by Englar [2], and using the 0.01 in. slot height, the resulting radii would be in the range of $0.2 \text{ in.} \leq r \leq 1 \text{ in.}$. While this range looks simple on paper, building cylinders of this magnitude, especially considering the need for a surface pressure tap, is difficult to fabricate. Thus, the height was chosen based on the cylinder radii, which were a known commodity. These considerations resulted in a larger slot height with an average value of 0.030 in.

Another important characteristic of the model is the presence of the walls on each side of the cylinder. While they are useful for a physical testing perspective, they also serve an aerodynamic purpose. The plates discourage any external environmental influences from impeding the airflow, creating a strictly two-dimensional flow at the centerline of the model.

Model Construction

The first elements of the model that was constructed were the cylinders. To create the circular shaped inserts for the ends of the cylinders, Lexan and plywood was cut-out using a computerized rotor table, as shown in Fig. 5. One end of the cylinder was filled with half inch thick plywood, while the other is half inch Lexan composite. Lexan was needed to maintain threads to hold the 3/8-16 rod used to control the rotation of the cylinder. A bead of weld was applied near the end of the rod as a thread lock when tightened against the Lexan, enabling the cylinder to be controllably rotated to vary the location of the pressure tap. For the testing process, a large metal disc with degree marks was attached to the handle of the threaded rod, easing the task of measuring the location of the pressure tap. Lastly, the plywood inset was press-fit into the cylinder creating a bearing with the centrally located hole, whereas the Lexan side was bolted to provide a secure connection between the angle setting rod and the cylinder.

The second major element constructed in the machine shop was the plenum, formed from a hard plastic. The plenum chamber was created using a mill machine, as depicted in Fig. 6. The height of the slot was manufactured to 0.030 in., with the variation in slot height measured at five locations across the slot in Fig. 7. It is essential to state that while the variations in Fig. 7 seem drastic, the dips are 7% of the plenums height, a little bigger than a human hair.

The second significant part of the plenum construction is the presence of the brass fittings. Visible as the two dots on the upper edge of the plenum in Fig. 3 are 0.5 in. NPT brass hose fittings, this is the connection between the model and the test apparatus.

FIGURE 5 ROTOR TABLE CUT-OUT OF LEXAN

FIGURE 6 MANUFACTURING OF PLENUM

FIGURE 7 WAVES IN PLENUM CHAMBER

Test Apparatus

Along with the construction of the model, components of the test apparatus also needed to be assembled. As shown in Fig. 8, there is a device that contains a multitude of measuring tools. Along with the flexible piping connecting the air supply system to the plenum, a thermo couple, a pressure regulator, a volumetric flow rate sensor, and a shut-off valve were incorporated. All of these measurement devices provided useful data for the calculation of the jet velocity and the mass flow rate.

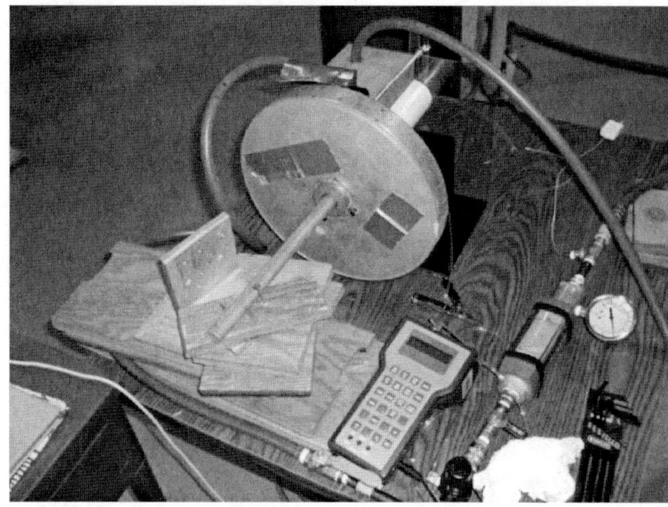

FIGURE 8 TEST APPARTUS

The principle sensor of the test apparatus was the Heise digital pressure gauge. This measurement tool was used to measure the pressure at the regulator, from the pressure tap, and the total and static pressures from the pitot-static probe. The two latter measurements were conducted separately of the former two with the pitot-static tube used in determining the velocity of the air leaving the slot.

The source of the air originates from the WVU Blow-down Supersonic Wind Tunnel and connects to the test apparatus by a quick disconnect fitting. The pressure of the hose was maintained above 120 psig to ensure supply of adequate flow rate of air to the test apparatus.

Experiment Testing Method

Over the course of completing the test plan, the same task was repeated twenty-one times, for the seven different cylinders, and the three different mass flow rates. The three rates used were 10, 20, and 30 ft^3/min. Alteration of the volumetric flow rate directly resulted in a variation in the jet velocity. The repetitious task followed the following sequence:

1. Turn air to test apparatus from blow-down supersonic wind tunnel
2. Adjust the mass flow rate

3. Record the mass flow rate, temperature, pressure from tank (white gauge), regulator pressure (from Heise gauge), and the pressure from the pressure tap (at a location far from the logical separation point)
4. Start pressure tap at zero degrees and record the pressure tap measurement from the Heise gauge
5. Rotate disc ten degrees and record from Heise gauge
6. Repeat step five until the ambient pressure (measured in step three) appears
7. At point just below ambient, record pressure tap every degree
8. Repeat steps one through seven but with the remaining two mass flow rates
9. Remove cylinder from experiment apparatus and replace with new one
10. Repeat steps one through nine until all seven cylinders are tested

One important part of the testing method was determining the angular location where the air flow was separating from the cylinder. This point was primitively estimated during step seven of the testing process. The separation point became defined as the point just before ambient pressure, since the surface pressure is related to the atmospheric pressure through the Bernoulli relation, Eq. (4). Fig. 9 shows the measured pressures as a function of the angular distance from the slot exit for the 1.75 inch radius cylinder. As the angular displacement increased, the pressure further approached ambient pressure, until it finally equaled. The region of pressure lower than ambient pressure is the region where the jet is influencing the surface, hence the Coanda effect is present and a pressure gradient is produced holding the jet to the surface.

Another measurement process was the determination of the jet velocity exiting the slot. To measure this value, a pitot-static tube and a Heise gauge were employed. Just as the slot height was measured at five locations across the span of the slot, the velocities were measured at these same locations. From the pitot-static tube, one could record the total pressure and the static pressure. Through Eq. (4) and Eq. (5), a variant of Bernoulli's continuity equation, the velocity of the jet could be determined.

$$P_{total} = \frac{\rho_j V_j^2}{2} + P_{static} \quad (4)$$

$$\rho_j = \frac{P_{static}}{T_j R} \quad (5)$$

RESULTS

General Results

Across the span wise direction of the jet, the model exhibited a range of velocities, as shown in Fig. 10. From this graph, it can be realized that there is a dip in the velocity from the left of the slot, labeled zero, to the center of the slot, indicated as location three. The explanation for this reaction is the usage of the structural bolts discussed. Since the bolts inhibited the flow of the air in the plenum, a lower velocity was produced. At both the zero and six location, the velocities were expected to be lower than the maximum velocity, due to the interference of the side walls of the chamber on the flow.

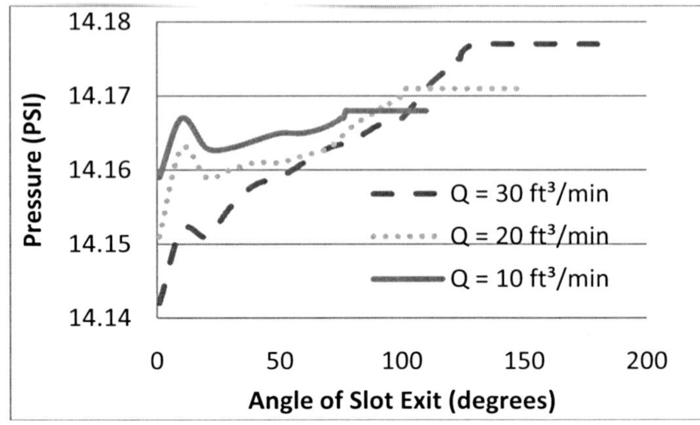

FIGURE 9 PRESSURE TAP MEASUREMENTS FOR 1.75 INCH RADIUS CYLINDERS

The highest velocity in the middle of the curves is a result of the most undisturbed flow at the slot exit. For the purposes of calculations, both the average velocity and the velocity at the center line were used. Fig. 10 is a measurement of the jet velocity; the radius is not an influence in the velocity evaluation, since the plenum exit is upstream the cylinder-airflow interaction.

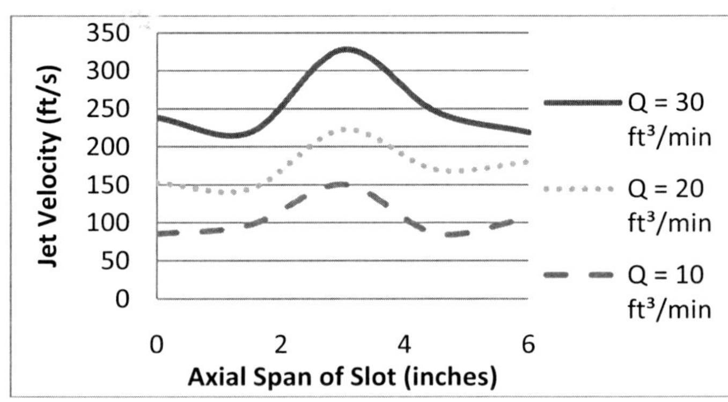

FIGURE 10 JET VELOCITY VERSUS AXIAL SPAN

The next result from this effort was the location of the separation point compared to the mass flow rate velocity quantity. Prior to discussing the findings, it is necessary to focus on the mass flow rate and velocity product in the numerator of Eq. (2), rather than the entire relationship. In the denominator of Eq. (2) is the dynamic pressure, which is undefined in the tested configuration due to the lack of a free-stream velocity. While this effort could model a hypothetical free stream density and velocity, the decision was to focus solely on applying this research to general circulation control applications. Thus, since the denominator of Eq. (2) is contingent on the free stream conditions, the angle of separation will be compared to the numerator of Eq. (2), coined the mass flow rate velocity quantity.

In Fig. 11, the results of the experiment delineate that in fact, the angle of separation is very much affected by the radius of curvature and the velocity of the jet. This study found that as the radius of curvature was increased, there was a subsequent decrease in the adhesion of the flow to the cylinder. Conversely, as the velocity of the jet was increased, the angular displacement increased. Even though the two relationships are evident, it can also be observed that the smallest radius of curvature did not possess the largest angle of separation. This apparent discrepancy can be attributed to the fact that the smallest radius falls below Englar's [2] predictions, which calls for a radius range of 0.6 in. $\leq r \leq$ 3 in., for the particular slot height of 0.030 in.. While not intending to, this event demonstrates Englar's [2] predictions, however, not the upper end of the range.

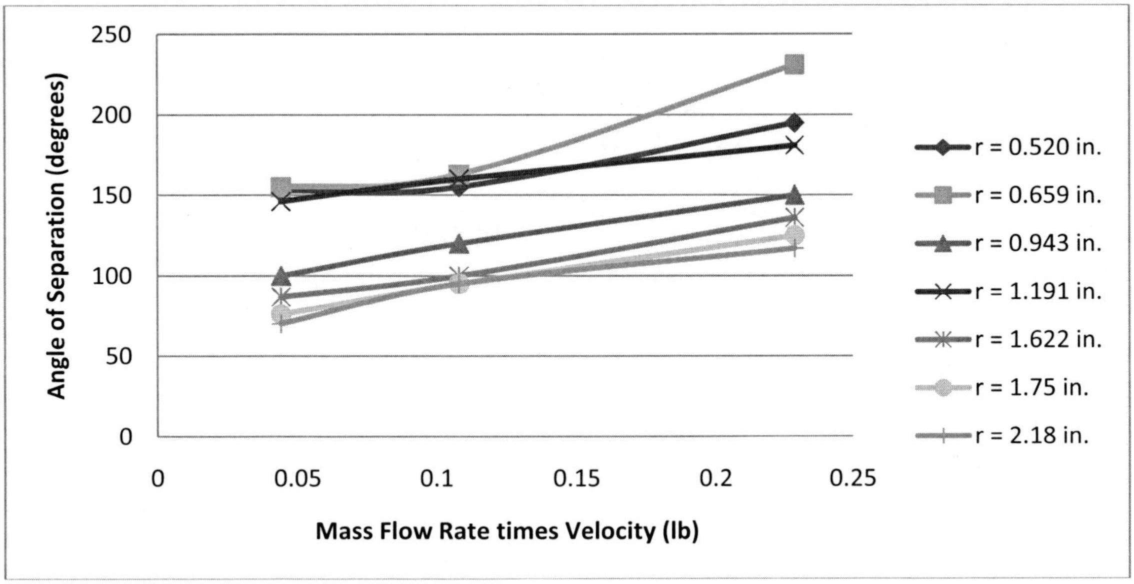

FIGURE 11 ANGLE OF SEPARATION AND MASS FLOW RATE VELOCITY

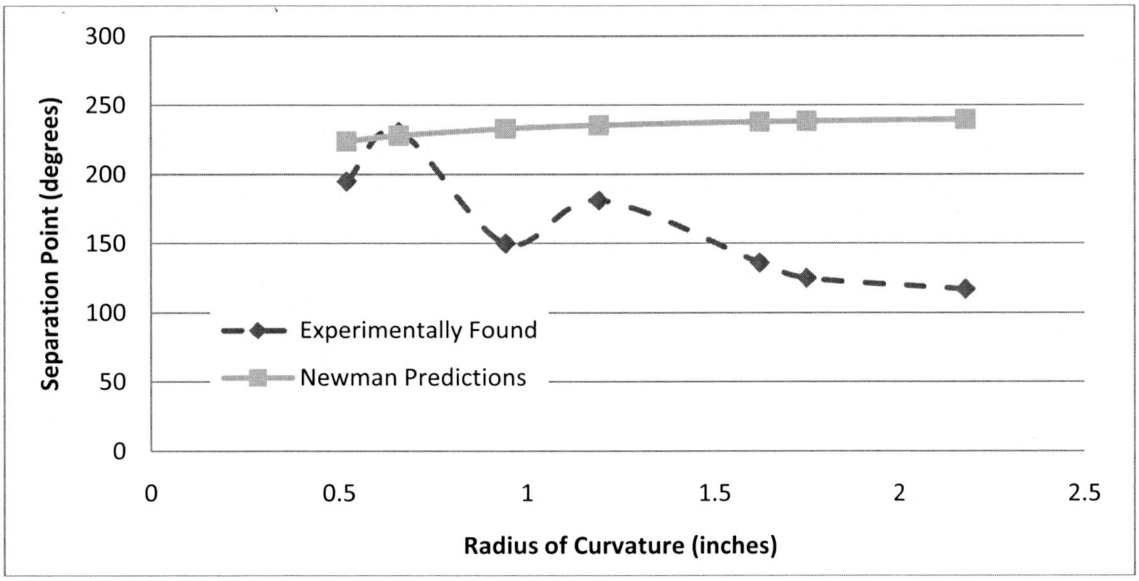

FIGURE 12 EXPERIMENT VERSUS THEORY FOR VOLUMETRIC FLOW RATE OF 30 FT^3/MIN

During the calculation process, it is important to note that both the average velocity and the velocity at the center line were applied. The average velocity was considered in the mass flow rate and the center line velocity was used for the jet velocity, due to its location relative to the surface pressure tap.

Results Compared to Newman [3]

Newman [3] predicts that as the radius of curvature increases, so does the angular displacement of the separation point, for a constant slot height As shown in Fig. 11 and Fig. 12, this is not necessarily the case. A further increase in the radius results in a decrease of the angular displacement. While this proves Newman's [3] theory incorrect for a constant mass flow rate velocity value, this may not be a fair assessment of Newman's [3] work, which focused on locating the maximum point of separation, based on slot height and radius as variables. Newman's [3] research effort identified the maximum turning angle of the jet which was not a consideration in the delay in flow separation study.

CONCLUSION

This effort confirms that the combination of the jet velocity and the radius of curvature have a significant impact on the location of the separation point of a jet blown tangentially to a cylindrical surface. It also shows that Englar's [2] slot height to radius of curvature range for best Coanda effect presence holds true. Subsequently, the radius with the maximum angular displacement fell within the radial limits outlined. This research found that Newman's [3] predictions of an increase in separation point following an increase in the radius of curvature was not true when the product of mass flow rate and jet velocity are held constant. Conversely, an increase in the radius of curvature resulted in a decrease of the angular location of the separation point under the tested conditions.

For future work in this area, there are a few areas of interest and areas of improvement for the experiment model. One topic to look further into is the peculiar result of the 1.191 in. and 0.943 in. radius cylinders. As shown in Fig. 12, there is an unexplainable dip and rise in the experimentally found trend, accredited to the two cylinders, respectively. Cause for error could be human measurement or the geometry of the cylinder, including but not limited to a gap between the cylinder's surface and the jet, or the surface roughness of the model. Further investigation of this quandary would provide further insight into the trend shown in Fig. 12. Another area of improvement would be correcting the variability in the slot height. This process would entail a redesign and reconstruction of the plenum to create the desired slot height, to perhaps the shorter 0.025 in. or to a more uniform 0.030 in., rather than the current average.

ACKNOWLEDGEMENTS

The work of this research project could not have been without the generous help of Dr. Gerald M. Angle[II]. Throughout the entire process, Dr. Angle was willing to help me whenever I came to visit him in his office. In addition, I was able to learn a tremendous amount of information from our verbal discussions and from his doctoral dissertation on blowing slot technology on rotorcraft. Along with Dr. Angle, I would not be able to have this great learning experience without the efforts of Dr. James Smith, Director of the Center for Industrial Research and Applications (CIRA) and Principal Investigator for the Circulation Controlled Vertical Axis Wind Turbine project. Dr. Smith encouraged me as a freshman in his general engineering class to "get involved." His encouragement and willingness to integrate me into CIRA enabled me to gain a once in a lifetime experience.

This work was also made possible through the research funding from the West Virginia NASA Space Grant Consortium.

REFERENCES

[1] John L. Loth and Michael Boasson, "Circulation Control STOL Wing Optimization," *Journal of Aircraft*, pp. 128-134, 1984.

[2] Robert J. Englar, "Subsonic Two-Dimensional Wind Tunnel Investigation of the High Lift Capability of Circulation Control Wing Sections," ASED-274 April 1975.

[3] B.G. Newman, "The Deflexion of Plane Jets by Adjacent Boundaries - Coanda Effect," *Boundary Layer and Flow Control*, p. 232, 1961.

[4] Imantus Reba, "Applications of the Coanda Effect," *Scientific American*, pp. 82-87, 1966.

[5] David T. Fisher, *Wind Tunnel Performance Comparative Test Results of a Circular Cylinder and 50% Ellipse Tailboom for Circulation Control Antitorque Applications*.: Naval Postgraduate School, 1994.

[6] Christopher M. Cage and Gregory S. Jones, "A Wind Tunnel Model to Explore Unsteady Circulation Control for General Aviation Applications," in *AIAA Aerodynamic Measurement Technology and Ground Testing Conference*, St. Louis, MI, 2002.

[7] Jack P. Ambrosiani, *Analysis of a Circulation Controlled Elliptical Airfoil*. Morgantown, WV: West Virginia University, 1971.

Proceedings of the ASME 2010 International Mechanical Engineering Congress & Exposition
IMECE2010
November 12-18, 2010, Vancouver, British Columbia, Canada

IMECE2010-37667

BIRDSTRIKE ANALYSIS ON LEADING EDGE OF AN AIRCRAFT WING USING A SMOOTH PARTICLE HYDRODYNAMICS BIRD MODEL

Vinayak Walvekar
Mechanical Engineering
Wichita State University
Wichita, KS, USA

Chandrashekhar K. Thorbole
Occupant Protection
The Engineering Institute
Farmington, AR,USA

Prasanna Bhonge
Mechanical Engineering
Wichita State University
Wichita, KS,USA

Hamid M. Lankarani
Mechanical Engineering
Wichita State University
Wichita, KS,USA

ABSTRACT

With the increase in air travel, the recent occurrences of birdstrikes on aircraft pose a major threat to human life; hence, there is a need to develop aircraft structures with a high resistance to such occurrences. According to the Federal Aviation Regulation (FAR 25.571) on *Damage-Tolerance and Fatigue Evaluation of Structure* (Amdt. 25-96), an airplane must be capable of successfully completing a flight during which likely structural damage might occur as a result of impact with a four-pound (1.8 kg) bird at sea-level cruise velocity or 0.85 percent of cruise velocity at 8,000 feet (2,400 m). Since the actual physical testing of a birdstrike is expensive, time-consuming, and cumbersome, this paper presents a methodology, based on the use of analytical finite element modeling and analysis, to certify an aircraft for a birdstrike. In actual physical testing for birdstrikes the mass of the bird might not be accurate and hence for certification purpose the computational modelling technique is more accurate and standardizes the certification procedure. The modeling and simulations are carried out as follows: the bird is modeled using the smooth particle hydrodynamics (SPH) technique in the LS-Dyna nonlinear finite element code. To validate this model, birdstrikes are carried out on rigid and deformable plates. The results, including displacement, Von-Mises stresses, forces, impulse, squash time and rise time, are obtained from the simulation, and non-dimensional values are plotted and compared with results from the test data.

The detailed CAD geometry of the leading edge of an aircraft is modeled in CATIA V5. Meshing, connections, and material properties are then defined in the Altair Hypermesh 9.0 program. The results obtained from the birdstrike simulations on this leading edge are compared to data from the experiments, and the process is validated. Parametric studies are carried out by designing the aircraft leading edge for different values of nose radius and by assigning appropriate thickness values for leading-edge components and impacting the SPH-modeled bird at different velocities. The methodology and results obtained from simulation can be utilized in the initial design stages as well as for "certification by analysis" of an aircraft for birdstrike requirements as per federal regulations.

INTRODUCTION

The collision between an aircraft and a bird is a potentially dangerous situation, and the damage caused by such a collision can be catastrophic. As a result, efforts are being taken to manufacture structures that have a high resistance against birdstrikes. Birdstrikes and other wildlife strikes on aircraft have caused an estimated annual damage of $600 million [1] for the U.S. aviation industry and $3 billion for the worldwide aviation sector. It is also estimated that more than 30,000 birdstrikes [1] to civilian aircrafts occur annually. The birds [1] that are generally involved in these strikes are waterfowl (31%), gulls (25%), raptors (18%), and pigeons/doves (7%).These birds represent nearly 81% of the reported birdstrikes that caused damage to U.S. civilian aircraft between 1990 and 2008; about 1,500 Canadian geese strikes were also reported during this period. Table 1 shows the certification requirements, specified with section numbers of FAR, for a bird's mass relative to its impact on different parts of the of an aircraft [2].

Table 1: Birdstrike requirements [2].

Aircraft Part	Bird Mass	FAR
Leading edge wing	4 lb (1.81 kg)	25.571
Leading edge empennage	8 lb (3.62 kg)	25.631
Windshield and Windows	4 lb (1.81 kg)	25.775
Engine	4 lb (1.81 kg)	25.571

One example of a catastrophe caused by a birdstrike is the recent bird impact with an aircraft engine of a U.S. Airways Airbus A320 in January of 2009. The aircraft was ditched in New York's Hudson River after experiencing the loss of both turbine engines due to an impact with a flock of geese.

METHODOLOGY

The bird is modeled in LS Prepost using the smooth particle hydrodynamics (SPH) technique with a pitch distance of 4 mm [3]. The validation of the computational bird model is conducted using published test results of a bird impact with rigid and deformable barriers [4, 5]. The simulation results include displacement, Von-Mises stresses, forces, impulse, rise time, and squash time. Then the non-dimensional values of force, impulse, rise time, and squash time are calculated by carrying out certain mathematical operations, and the results are plotted against the test data and compared.

The effect of pitch distance on birdstrike analysis is observed by varying the pitch distance of the SPH bird model from 4 mm to 10 mm and plotting the results.

The validated SPH bird model is impacted on the leading edge, and the results obtained from the simulation are compared with the test data. The leading edge is modeled in CATIA V5. Meshing and connections are defined in Altair Hypermesh 9.0. The parametric study of the leading edge is carried out by varying its nose radius and the thickness of its main skin, leading edge skin, main rib, and leading edge rib. The bird is impacted with different velocities, and results are tabulated and compared.

SPH Bird Model

The bird characteristics, such as diameter, cylindrical length, and density, are obtained using empirical formulae from the literature [3]. The front view and profile view of the bird model are shown in Figure 1.

Bird mass = 4 lb = 1.81 kg

$$Density\,(plucked) = 959 - 63 \cdot \log_{10}(Mass) = 942.7 kg/m^3$$

$$Diameter = 0.0804 \cdot Mass^{0.335} = 0.098\,m$$

For a cylindrical bird model with hemispherical ends:

$$CylinderLength = 4 \cdot \left(\frac{Mass}{\pi \cdot Density \cdot Diameter^2} - \frac{Diameter}{6} \right) = 0.189m$$

$$TotalLength = 0.189 + 0.098 = 0.287m$$

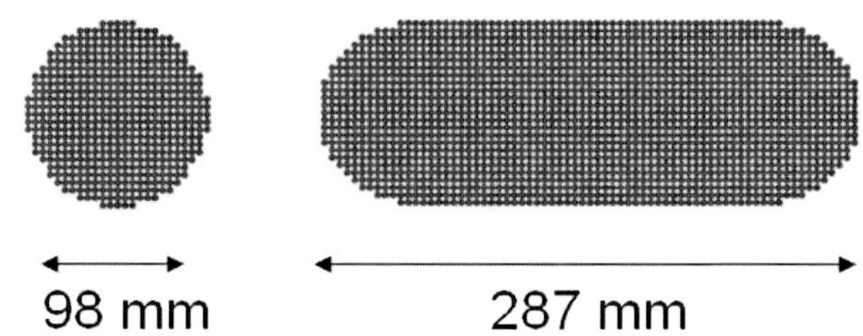

Fig. 1: Dimensions of four-pound bird.

Validation of SPH Bird Model for Impact with Aluminum 7075-T6 Plate

The SPH bird model is impacted on an aluminum 7075-T6 plate, and the simulation results are compared with those in the literature [4]. The bird strike setup consists of an Al 7075-T6 plate with length of 914.4 mm and breadth of 609.5 mm. The SPH bird model with an Equation of state defined [6] is impacted at the center of the plate. This setup is shown in Figure 2.

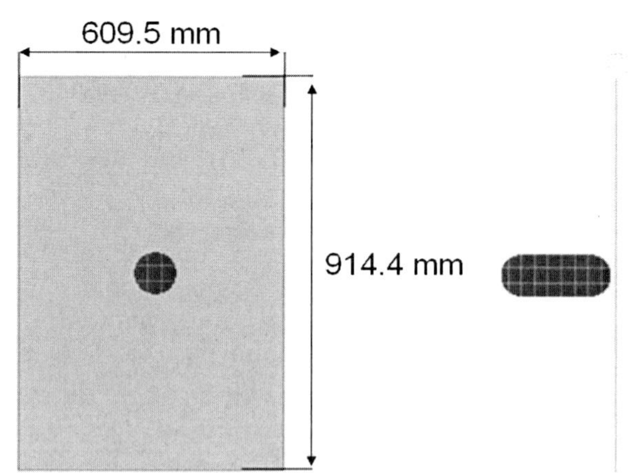

Fig. 2: Birdstrike setup with aluminum plate.

The aluminum plate is constrained in the X, Y, and Z directions. The thicknesses of the aluminum plate are 0.25" (6.35 mm), 0.16" (4.06 mm), and 0.1" (2.54 mm), and the bird is impacted with 265 knots, 269 knots, and

275 knots, respectively. Figure 3 shows the disintegration of the bird on a 0.25 inch plate at different instances.

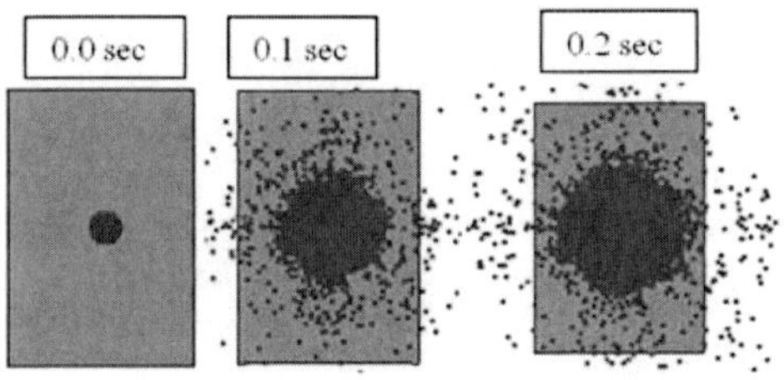

Fig. 3: Disintegration of SPH bird.

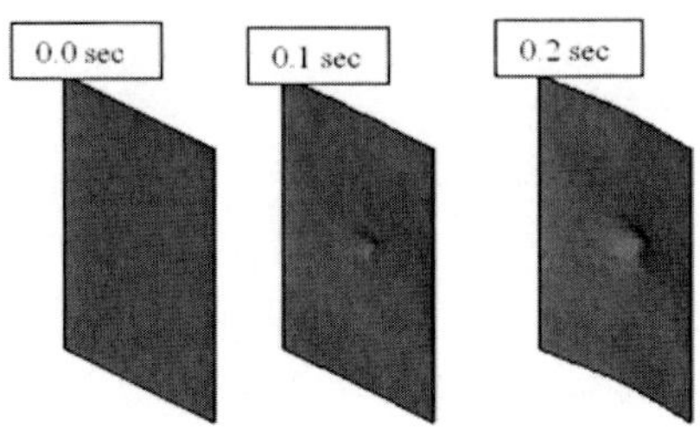

Fig. 4: Deformation of aluminum plate impacted at 265 knots.

Figure 4 shows the simulation sequence for the deformation of an aluminum plate when it is impacted at 265 knots at varying times.

Table 2: Comparison between experimental and simulation results.

Plate Thickness		Test Results	Simulation Results with Lagrangian Model	Simulation Results with SPH Model
0.25 inch	Deformation	1.0" (25.4mm)	1.3" (33.02mm)	1.2" (30.4mm)
	Max Von Mises Stress	NA	71.05 Kpsi (489 MPa)	73.24 Kpsi (504 MPa)
0.16 inch	Deformation	1.5" (38.1mm)	1.3" (33.02mm)	1.8" (45.72mm)
	Max Von Mises Stress	NA	73.2 Kpsi (504 MPa)	73.1 Kpsi (504 MPa)
0.10 inch	Deformation	Material Failure	Material Failure	Material Failure

In Table 2 the comparison between the laboratory test and simulation results with Lagrangian and SPH bird model is tabulated and results show that the variation in test data and SPH simulation is within 20%, which can be considered an acceptable range. When compared with the analysis time of Lagrangian model [4], the SPH model is very less and thus can be said as time efficient technique when compared with Lagrangian model.

Validation of SPH Bird Model for Impact with Steel Plate

In this section the bird model of 600g is validated by comparing the non dimensional parameters as obtained from the simulation to the actual test data [5]. The non dimensional parameters such as, Non dimensional impulse, non dimensional impact duration, and non dimensional rise time are calculated and compared. The SPH bird model is impacted on a steel plate in which the edges are constrained along the X, Y, and Z directions. Figure 5 shows the simulation set up for this condition. The bird is impacted at different impact velocities, and the forces are obtained.

A bird with a mass of 600 g and a pitch distance of 4 mm is impacted on a steel plate at an angle of 90 degrees.

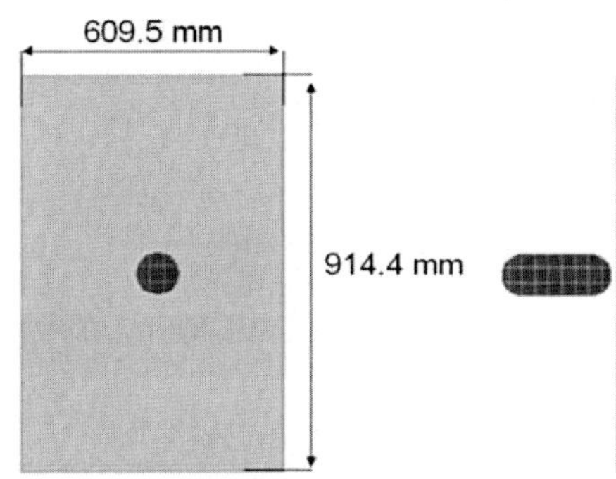

Fig. 5: Birdstrike setup with steel plate.

The bird impact is carried out at 192 m/s, 219 m/s, 249 m/s, and 269 m/s. Figure 6 (a) shows the force vs. time and impulse vs. time plots for impact velocity of 192 m/s, and similarly, Figures 6 (b), 6 (c), and 6 (d), show the plots for 219 m/s, 249 m/s, and 269 m/s, respectively.

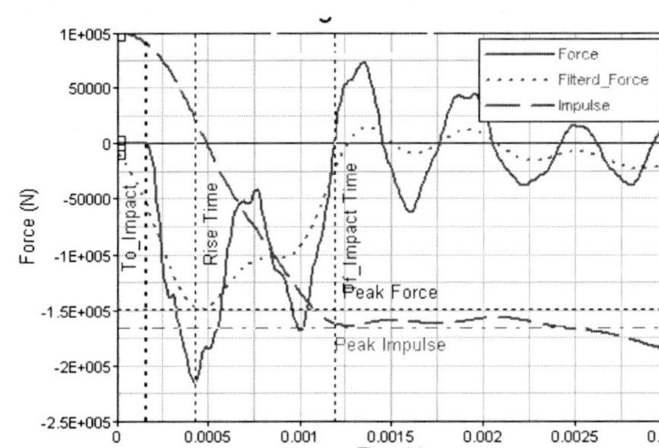

(a) Impact velocity of 192 m/s

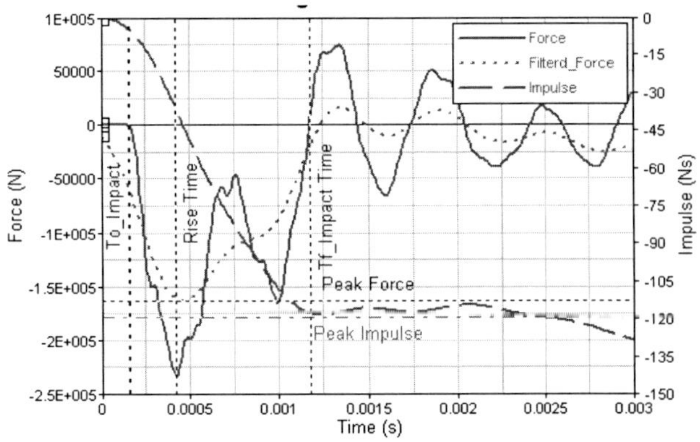

(b) Impact velocity of 219 m/s

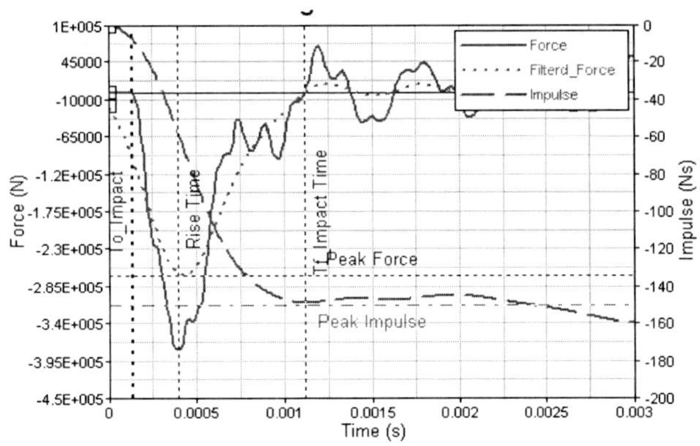

(c) Impact velocity of 249 m/s

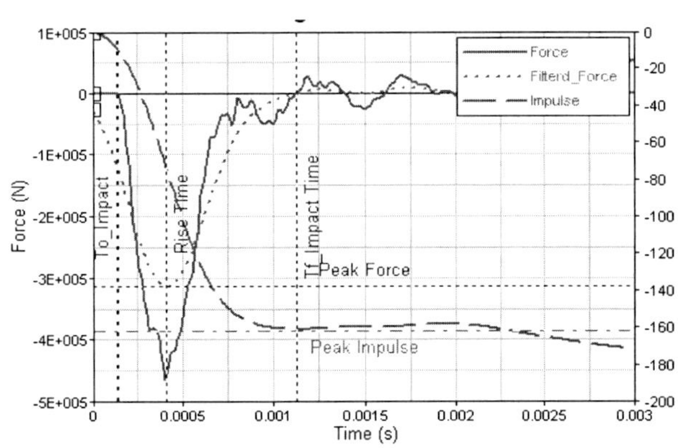

(d) Impact velocity of 269 m/s

Fig. 6: Force vs time and Impulse vs time plots.

Values such as rise time, squash time, peak force, and peak impulse are obtained from these plots.

Rise time is calculated at the instant when peak force is reached on the force vs time plot. Squash time is calculated from the time the impact begins, that is, when the leading edge of the bird initially comes in contact with the rigid plate, and continues until the trailing edge of the bird reaches the target and there is no more bird material approaching at the target. For the force curve, an SAE J211 filter [7] is applied to get a smooth curve and eliminate the high frequency signal due to possible vibration of the plate due to impulsive nature of impact. The impulse vs time plot is obtained by integrating the force curve.

Figure 6 clearly indicates the proportional relation between the bird impact velocity and the peak force/impulse.

Calculation of Non-Dimensional Impulse

The non-dimensional impulse value for the testing is obtained from the literature [5].

$$\text{Non-Dimensional Impulse} = \frac{\int F dt}{m v \sin\theta}$$

Where:

$\int F dt$ = impulse
v = impact velocity
θ = impact angle

For an impact velocity of 192 m/s,

$$\text{Non-Dimensional Impulse} = \frac{108}{0.6 * 192 * Sin 90}$$

$$= 0.937$$

Table 3: Comparison of test and simulation result.

Velocity	Non-Dimensional Impulse	
m/s	Testing	Simulation
192	0.798	0.937
219	0.819	0.951
249	0.889	0.997
269	0.950	1.009

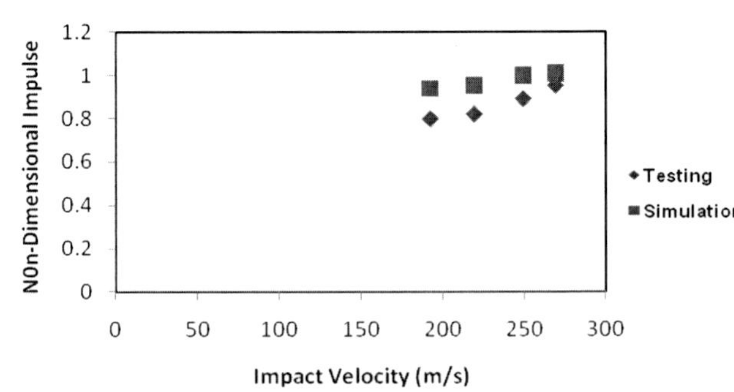

Fig. 7: Non-dimensional impulse vs impact velocity for bird mass of 600 g.

Table 3 shows the comparison of the non dimensional impulse as obtained from simulation to the test result. Figure 7 illustrates that the correlation between the testing and simulation results for non-dimensional impulse parameter is reasonable.

Calculation of Non-Dimensional Impact Duration

$$\text{Non-Dimensional Impact Duration} = \frac{T_s v}{l_{eff}}$$

Where:

$l_{eff} = l + d \tan\theta$ = effective length of bird

l = length of bird

d = diameter of bird

Ts = squash time

Squash time is calculated from the time of initial impact when the leading edge of the bird comes in contact with the rigid plate and continues until the trailing edge of the bird reaches the target. Squash time obtained from the simulation is used for this calculation.

For impact velocity of 192 m/s,

$$\text{Non-Dimensional Impact Duration} = \frac{0.00103 * 192}{192}$$

$$= 1.03$$

Table 4 shows the comparison of the non dimensional impact duration as obtained from simulation to the test result.

Table 4: Comparison of test and simulation result.

Velocity	Non-Dimensional Impact Duration	
m/s	Testing	Simulation
192	0.786	1.03
219	1.041	1.105
249	1.115	1.270
269	0.864	1.457

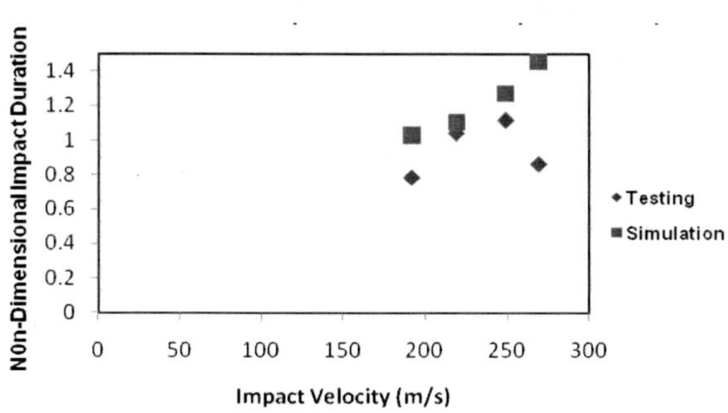

Fig. 8: Non-dimensional impact duration vs impact velocity for bird mass of 600 g.

Figure 8 shows the correlation between the testing and simulation result for the no dimensional impact duration. The non dimensional impact duration is propotional to the impact velocity. As the figure demonstrates that at higher impact velocity the test result is dropped compared to the simulation result. One explaination for this test result data anomaly could be the limitation of the pressure transducers available during that period [8].

Calculation of Non-Dimensional Rise Time

$$\text{Non-Dimensional Rise Time} = \frac{v * T_R}{l_{eff}}$$

Where:

$l_{eff} = l + d \tan\theta$ = effective length of bird

l = length of bird

v = impact velocity

T_R= rise time

The rise time is determined from the instant the bird comes into contact with the target until the peak force is attained in the force vs time plot as shown by Figure 6 previously. The current calculations are done for the impact velocity of 192m/s. The corresponding rise time of the force is obtained from Figure 6 (a).

For an impact velocity of 192 m/s,

$$\text{Non-Dimensional Rise Time} = \frac{192 * 0.00027}{0.192}$$

$$= 0.27$$

Table 5 shows the comparison of the non dimensional rise time as obtained from simulation to the test result.

Table 5: Comparison of test and simulation result.

Velocity	Non-Dimensional Rise Time	
m/s	Testing	Simulation
192	0.206	0.270
219	0.199	0.285
249	0.238	0.337
269	0.457	0.371

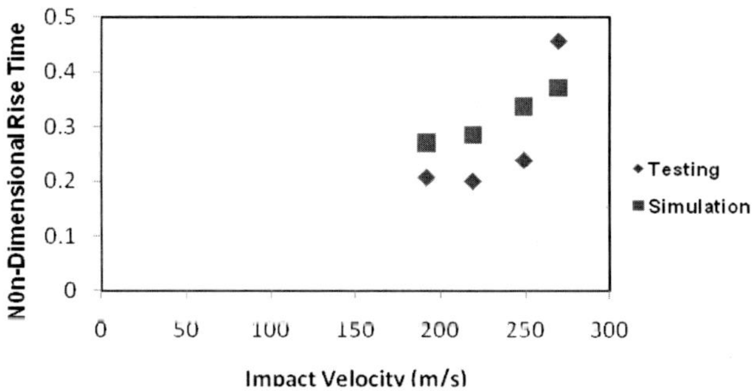

Fig. 9: Non-dimensional rise time vs impact velocity for bird mass of 600 g.

Figure 9 shows that the correlation between the testing and simulation results for non-dimensional rise time parameter is reasonable. Overall, from the above results, it can be understood that the SPH bird model is fairly validated for further simulations with aircraft leading edge.

Bird Models with Different Pitch Distances

The distance between two SPH nodes is referred to as pitch distance. To determine the effect of pitch distance on birdstrike analysis, the bird is modeled for pitch distances of 4 mm and 10 mm, and the simulation results are observed and analyzed. The intention of this analysis is to evaluate the possibility of increasing computational efficiency. In 4 mm bird model each SPH particle weighs 6.02 E-5 kg and in 10 mm bird model each SPH particle weighs 9.82 E-4 kg. Figures 10 show bird models with pitch distances of 4 mm and 10 mm, respectively.

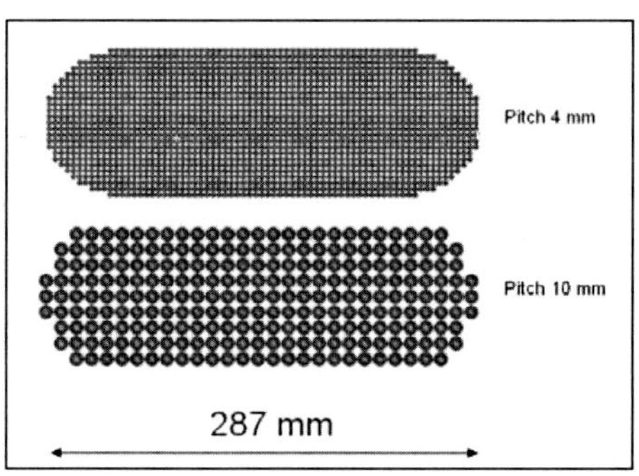

Fig. 10: Bird model with different pitch distances.

The effect of pitch distance on birdstrike analysis is determined by impacting the bird model with pitch of 4 mm and 10 mm on a steel plate at 200 m/s, 250 m/s, and 300 m/s. Parameters such as peak force and impulse are obtained from the force vs. time plots. The non-dimensional impulse, non-dimensional impact duration, and non-dimensional peak force is calculated. These values are plotted against impact velocity for 4 mm pitch and 10 mm pitch bird model. The plots are overlaid to compare the difference.

Figures 11, 12 and 13 show the plot of non-dimensional impulse, non-dimensional impact duration, and non-dimensional peak force vs. impact velocity overlaid for each bird model of pitch distance 4mm and 10mm. The comparison clearly indicates no significant difference or very little variations in the values of non dimensional parameters. The SPH bird with 4mm pitch is selected for all the simulation with the leading edge. This is selected based on a rationale that more realistic disintegration of the bird is represented by the bird model with smaller pitch in the simulation. As observed the pitch change does not affect the quantified parameters.

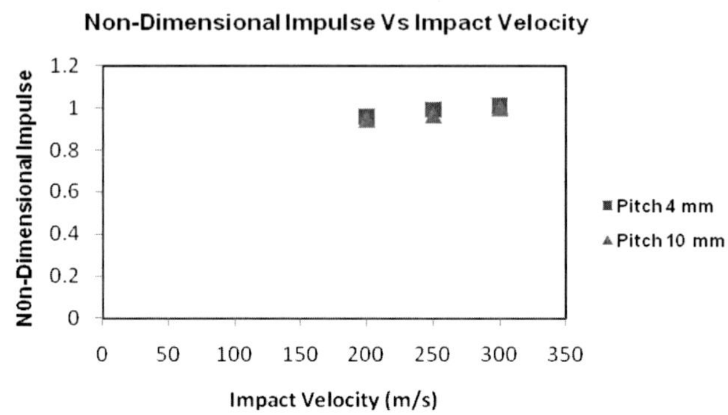

Fig. 11: Comparison of Non-dimensional Impulse.

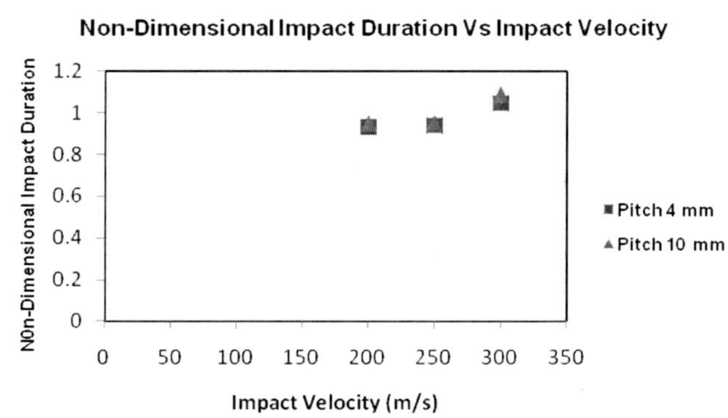

Fig. 12: Comparison of Non-dimensional impact duration.

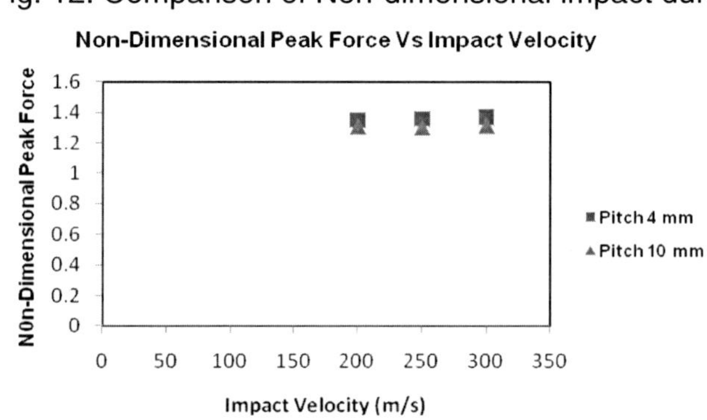

Fig. 13: Comparison of Non-dimensional peak force.

BIRDSTRIKE ON LEADING EDGE OF AN AIRCRAFT WING

The validated SPH bird model is impacted on the leading edge parts with nose radii of 12.7 mm and 62 mm. The solid model of the leading edge is developed using CATIA V5, and meshing and connections are defined in Altair Hypermesh 9.0. The connections between different parts of the aircraft's leading edge are specified using spotwelds and beam elements. The bird is impacted at the leading edge at an angle of 30 degrees. The dimensions of all the components are obtained from the literature [9]. The thicknesses of leading edge parts for different nose radii and the impact velocities are listed in Table 6. The Al-2024-T3 properties [10] are used for the leading edge and they are specified using Johnson Cook material model in LS Dyna [11].

Table 6: Thickness of leading edge parts for different nose radii [9].

Nose Radius (mm)	LE Skin (mm)	Main Skin (mm)	LE Rib (mm)	Main Rib (mm)	Rib Pitch (mm)	Impact Velocity (knots)
12.7	2.03	2.03	1.22	1.22	203.2	261
						427
62	2.03	2.03	1.22	1.22	203.2	146
						294

Figure 14 shows the details of the leading edge and the exploded view.

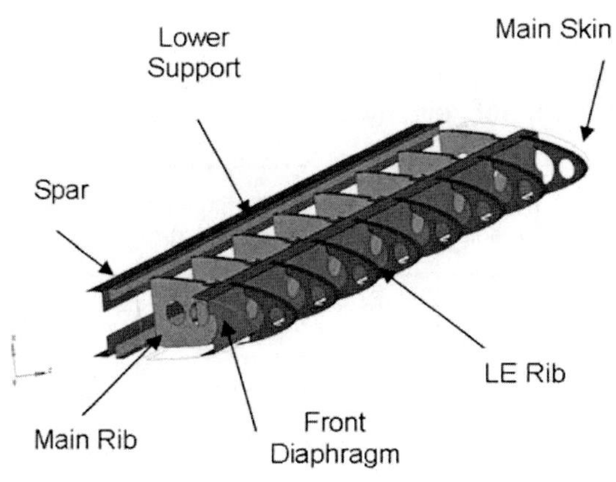

Fig. 14: Exploded view of leading edge.

Birdstrike on Leading Edge with a Small Nose Radius.

A four-pound bird with pitch of 4 mm is impacted with a velocity of 261 knots and 427 knots on the leading edge of a nose radius of 12.7 mm, and the results are analyzed by comparing it with the results from test data [9]. Figure 15 shows the simulation set up for the bird impact on leading edge with small nose radius.

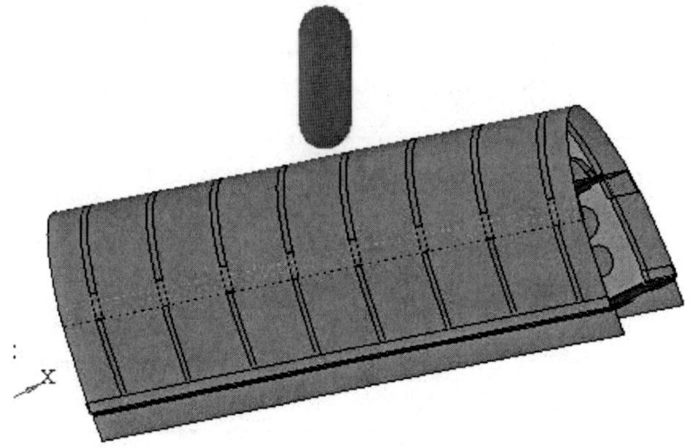

Fig. 15: Simulation set-up for bird impact with 12.7mm edge nose radius.

Case I

The bird strike simulation is carried out by impacting the four-pound bird with the leading edge of nose radius 12.7 mm at an impact velocity 261 knots (134.27 m/sec). The analysis is carried out for 5 milliseconds. Figure 16 shows the simulation sequence of the bird strike analysis with smaller radius with impact velocity of 261 Knots.

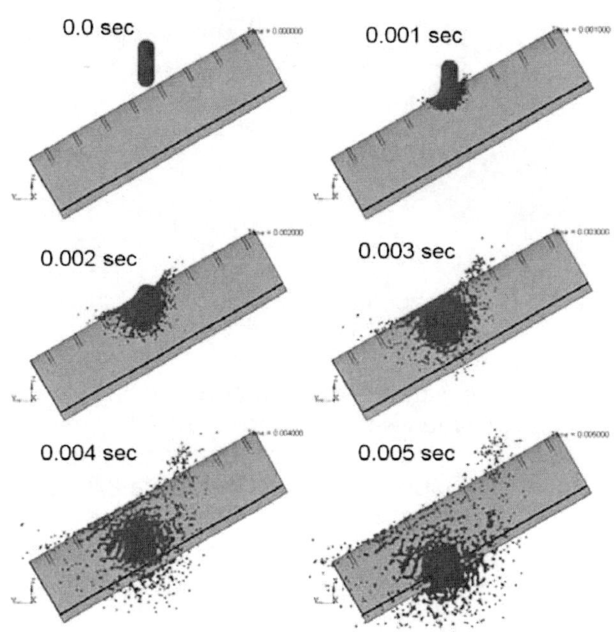

Fig. 16: Simulation sequence for small nose radius and 261 Knots impact velocity.

Table 7 shows the qualitative comparison of the full scale test and the simulation result. This impact shows no damage on the various parts of the leading edge. Thus, there is a good correlation of results between the test and the simulation for the leading edge with nose radius of 12.7 mm at an impact velocity of 261 knots.

Table 7: Comparison of testing and simulation at impact velocity of 261 knots. (134.27 m/s)

Material	Velocity (knots)	Damage on Leading Edge Skin	Damage on Diaphragm	Damage on Spar	Damage on Leading Edge Ribs
Testing 2024-T3	261	None	None	None	None
Simulation 2024-T3	261	None	None	None	None

Figure 17 shows the post impact image of the leading edge of the specimen from the simulation. This clearly indicates no damage with a 12.7 mm nose radius and impact velocity of 261 knots; its parts are highlighted.

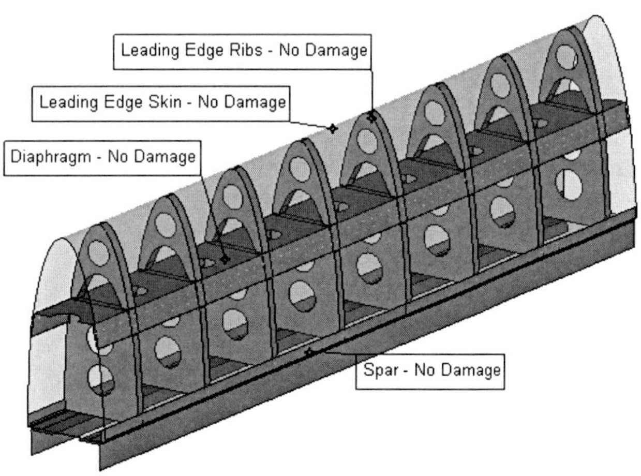

Fig. 17: No damage from the simulation of this impact condition.

Case II

The bird strike simulation is carried out on the leading edge of a specimen with nose radius 12.7 mm and impact velocity of 427 knots (219.66 m/sec) with a bird mass of four pounds. Figure 18 shows the simulation sequence of the bird at each millisecond.

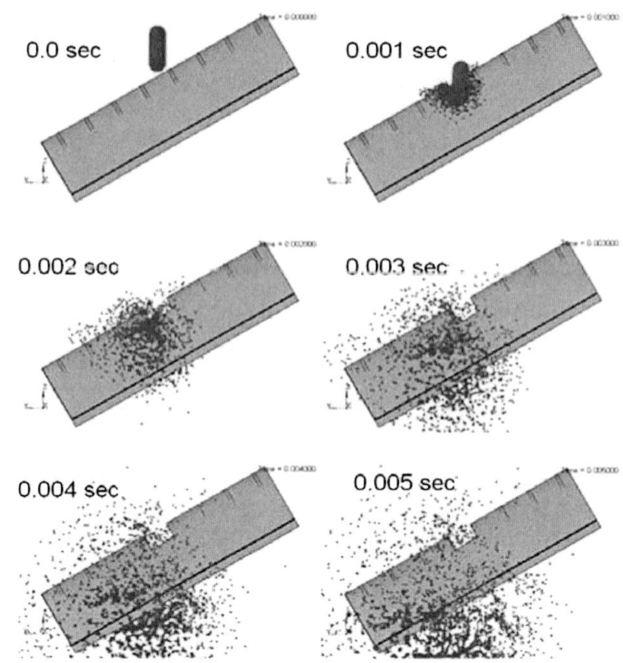

Fig. 18: Simulation sequence for small nose radius and 427 Knots impact velocity.

Table 8 lists the comparison of results between test data and simulation results. As can be seen, the leading edge is torn along the rivet line for both the test and the simulation specimens. In the test, two bays are torn along the diaphragm, and in the simulation, two bays are also torn along the diaphragm. No damage is observed on the spar in both the test and simulation results. In the case of the leading edge ribs, in both the simulation and the test, one is torn off and one is buckled. Thus, it can be observed that there is reasonable correlation between testing and simulation results.

Table 8: Comparison of testing and simulation results at impact velocity of 427 knots. (219.66 m/s)

Material	Velocity (knots)	Damage on Leading Edge Skin	Damage on Diaphragm	Damage on Spar	Damage on Leading Edge Ribs
Test 2024-T3	427	Torn along rivet line	2 bays torn	None	1 torn off 1 buckled
Simulation 2024-T3	427	Torn along rivet line	1 bay torn 1 partially torn	None	1 torn off 1 buckled

Figure 17 shows the structure of the leading edge of the specimen at the end of analysis where there is damage on leading edge skin, leading edge ribs, spar, and diaphragm.

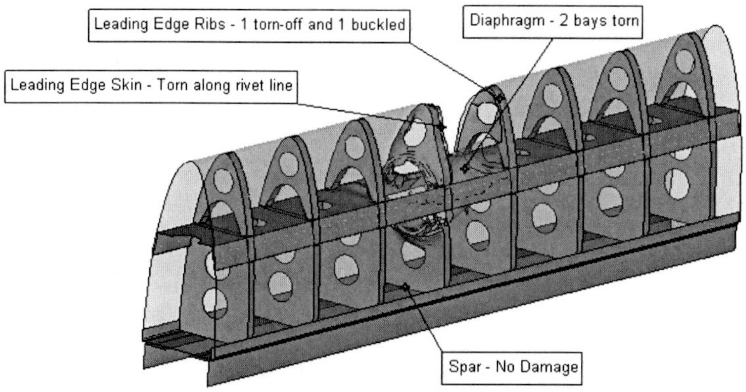

Fig. 17: Post impact damage along leading edge.

Birdstrike on Leading Edge with Large Nose Radius.

A four-pound bird with a pitch of 4 mm is impacted on the leading edge of the 62 mm nose radius of at a velocity of 146 knots and 294 knots, and the results are analyzed by comparing them with results from the test data [9]. Figure 18 shows the simulation set up for the bird impact on leading edge with large nose radius.

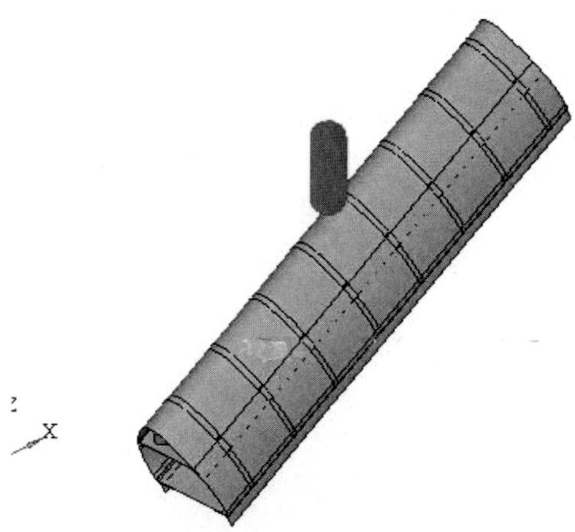

Fig. 18: Simulation set-up for bird impact with 62 mm edge nose radius.

Case I

The bird strike is carried out on the leading edge of the 62 mm nose radius with an impact velocity of 146 knots (75 m/sec) using a four-pound bird. Figure 19 shows the birdstrike on the leading edge at different instants of time for a birdstrike at an impact velocity of 146 knots.

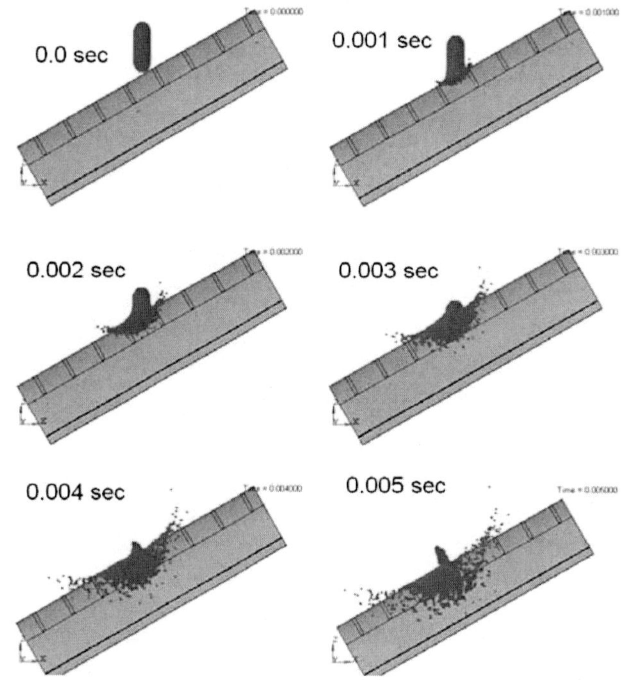

Figure 19. Simulation sequence for large nose radius and 146 Knots impact velocity.

Table 9 shows a comparison of the results of simulation and test data. It clearly indicates that for both the test and simulation, there is no damage on the leading skin. The same is true for the diaphragm, which also shows no damage. And the same result showing no damage is observed in both the test and the simulation for both spar and leading edge ribs. Thus, there is good correlation between the testing and simulation results.

Table 9: Comparison of testing and simulation results at impact velocity 146 knots. (75.10 m/s)

Material	Velocity (knots)	Damage on Leading Edge Skin	Damage on Diaphragm	Damage on Spar	Damage on Leading Edge Ribs
Test 2024-T3	146	None	None	None	None
Simulation 2024-T3	146	None	None	None	None

Figure 20 shows the leading edge structure at the end of the simulation where it can be observed that there is no damage to the physical condition of the different parts of the leading edge, which matches with the results from testing.

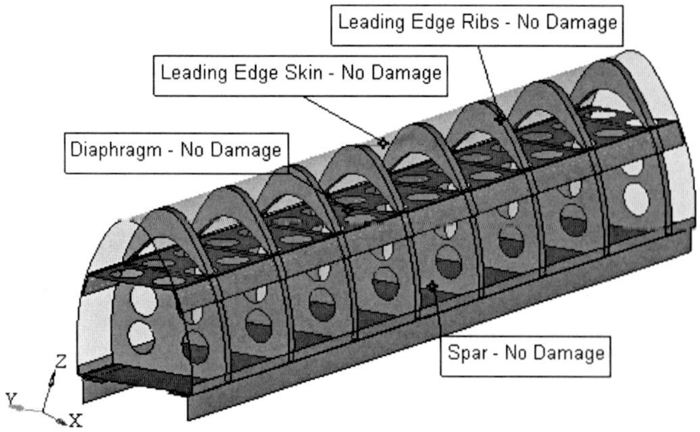

Fig. 20: Post impact condition of the leading edge with no damage.

Case II

The bird strike is carried out on the leading edge of the 62 mm nose radius at an impact velocity of 294 knots (151.24 m/sec) using a four-pound bird. Figure 21 the simulation sequence of the bird strike for this condition.

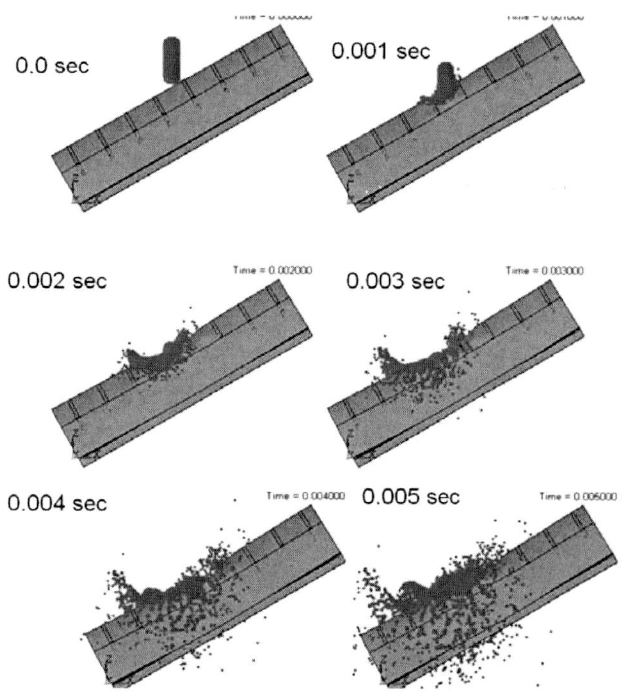

Fig. 21: Simulation sequence for large nose radius and 294 Knots impact velocity.

Table 10 shows the results of the simulation compared to the test data. It indicates that the leading edge of the specimen is torn along the rivet line in the simulation, which is the same as shown in the test data. In the case of the diaphragm, the two bays are torn according to the test but are dented according to the simulation. For the spar, one bay is torn in the test, and three bays are dented in the simulation. For the leading edge ribs, one is torn off and one is dented in the test, and two are torn off in the simulation. There is reasonable correlation between the test and simulation results.

Table 10: Comparison of testing and simulation results at impact velocity of 294 knots. (151.24 m/s)

Material	Velocity (knots)	Damage on Leading Edge Skin	Damage on Diaphragm	Damage on Spar	Damage on Leading Edge Ribs
Test 2024-T3	294	Torn Along Rivet Line	2 bays torn	1 bay torn	1 torn off 1 dented
Simulation 2024-T3	294	Torn Along Rivet Line	2 bays dented	3 bays dented	2 torn off

Figure 22 shows the leading edge structure at the end of the simulation, indicating the physical condition of leading edge ribs, leading edge skin, diaphragm, and spar.

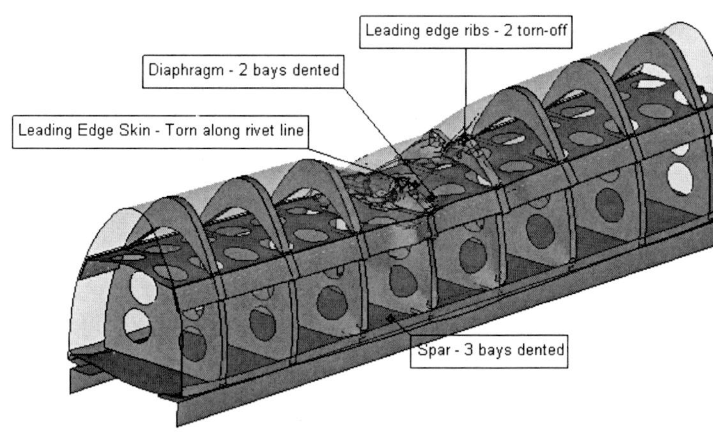

Fig. 22: Post impact damage along leading edge with 294 knots impact velocity.

CONCLUSION

The objective of this study was to devolop a methodology to certify an aircraft for birdstrike using computational technique. To achieve this objective, the bird was modeled using smooth particle hydrodynamics

(SPH) and was validated by two different approaches. In the first case, the SPH bird model was impacted on an aluminum target at different impact velocities for different thicknesses of plate. The results obtained were compared with the test data [4]. The comparison of the test data and the simulation showed a good correlation with the variation of 20%. The SPH technique for the bird model is an efficient technique because of the minimum computation time.

In the second case involving the SPH bird model and a steel target, the results obtained from the simulation were compared with test data from the literature [5]. All plots for the non-dimensional parameters showed good correlation with test results, demonstrating the validity of the bird model.

The effect of pitch distance was observed by varying the bird pitch by 4 mm and 10 mm, and the results show little variation; hence, it was concluded that for birdstrike analysis, the effect of pitch distance between SPH particles is negligible. The validated SPH bird model was impacted on the leading edge of the aircraft wing. The leading edge was modeled in CATIA V5, and dimensions were obtained from the literature [9]. All connections and material properties were defined in Altair Hypermesh 9.0. The connections between different parts of the leading edge were defined using beams and spot welds. The simulation conducted for different impact velocity and the nose radius showed a good correlation with the actual test damage on the leading edge. This clearly demonstrates the feasibility of computational technique for the bird strike certification purpose. The computational bird strike analysis is less time consuming and more economical method than the actual testing of birdstrikes, most importantly, standardizes the method of certification.

REFERENCES

[1] Anon., "Understanding and Reducing Bird and Other Wildlife Hazards to Aircrafts," Bird Strike Committee USA, 2009.

[2] Federal Aviation Administration, http://rgl.faa.gov/

[3] Anon., "The Hazard Posed to Aircraft by birds," Australian Transport Safety Bureau, 2002.

[4] R. D. Furtado, "Finite Element Analysis of Birdstrike on Aircraft Structures," M.S Thesis, Department of Mechanical Engineering, Wichita State University, 2000.

[5] J.P. Barber, H.P. Taylor, and J.S. Wilbeck, "Bird Impact Forces and Pressures on Rigid and Compliant Targets," Air Force Flight Dynamics Laboratory Technical Report, 1978.

[6] C. A. H. Ortecho, "Robust Birdstrike Modeling Using LS Dyna," M.S Thesis, University of Puerto Rico, 2006.

[7] Altair Hyperview Manuals, Version 9.0.

[8] Wilbeck J., Barber J. "Bird Impact Loading," U.S. Naval Research Laboratory, Washington D.C, 1964.

[9] I.I. McNaughtan, "The Design of Leading-Edge and Intake Wall Structure to Resist Bird Impact," Royal Aircraft Establishment, 1972.

[10] G. Kay, "Failure Modeling of Titanium 6Al-4V and Aluminum 2024-T3 with the Johnson-Cook Material Model," Federal Aviation Administration Report DOT/FAA/AR-03/57, 2003.

[11] LS Dyna User Manual V971, LSTC, (Cited 2010).

Proceedings of the ASME 2010 International Mechanical Engineering Congress & Exposition
IMECE2010
November 12-18, 2010, Vancouver, British Columbia, Canada

IMECE2010-37820

ON VISCOELASTICITY IN CNT-REINFORCED POLYMER COMPOSITES

K. Yazdchi
Multi Scale Mechanics, Faculty of Engineering Technology, University of Twente, P.O. Box 217, 7500 AE Enschede, The Netherlands

M. Salehi
Mechanical Engineering Department, Amirkabir University of Technology, Hafez Ave., Tehran 1591634311, Iran

ABSTRACT

The nanocomposites exhibit high electrical conductivity, significant non-linear optical behavior and electroluminescence, while having substantially improved mechanical properties relative to the neat polymer. However, very limited attention has been paid to the viscoelastic behavior of nanotube reinforced polymer composites (NTRPCs). In this paper, the constitutive relation and linear viscoelastic behavior of NTRPC are studied using methods of micromechanics and nanomechanics. First, the effects of volume fraction, aspect ratio and orientation of carbon nanotubes (CNTs), on the overall elastic properties of NTRPC are obtained through a micromechanical technique based on Eshelby's Equivalent Inclusion (EEI) and Mori-Tanaka (MT) method. Secondly, by incorporating the Dynamic Correspondence Principle (DCP), the elastic solution is extended to solve the related linear viscoelastic problem. The results of this study are in good agreement when compared with previous analytical and experimental data.

1. INTRODUCTION

Much research has been carried out on the remarkable carbon-nanotubes (CNTs) mechanical properties in terms of their high strength and stiffness and excellent electronic, optical and transport properties. This interest stems from the fact that CNTs have great potential applications as transistors in microcomputers, tubes for drug delivery, nanoprobes, nanocomposites, etc. A detailed summary of CNTs mechanical properties can be found in [1]. Two main classes of theoretical methods, used for predicting the mechanical properties of CNTs, can be categorized into the atomistic (MD) based and the continuum mechanics based approaches. The former is currently limited by the computational complexity with respect to the number of atoms and timescales (typically several nanoseconds or less). Therefore, researchers have been diligently seeking for more efficient computational methods, such as continuum structural models. A structural mechanics approach (continuum models) was first developed by Natsuki [2] to predict the mechanical properties of SWNTs and MWNTs, with different sizes and structures. This approach was based on carbon bonds beam-like representation and a linkage between the beam sectional stiffness parameters and constants of force field.

The high strength and elastic modulus, fibrous shape and very large aspect ratios of these NTs make them a very promising candidate as the ideal reinforcing fibers for advanced composites with high strength and low density [3]. These polymer based nanocomposites usually show remarkable property improvements in various physical properties even at very low content of CNTs compared to virgin polymer or conventional microcomposites. Lau and Hui [4] and Thostenson et al. [5] give comprehensive reviews of the manufacturing process, the mechanical and electrical properties, and applications of nanotubes and nanotube reinforced composites.

Because the NTs are on the size scale of the polymer chains, they are expected to alter the mobility of the polymer chains and thus change the viscoelastic response of the NRP compared to the pure polymer; this behavior is usually modeled in terms of a reduced mobility of non-bulk interphase region [6, 7]. The schematic of the three phase composite model, including the interphase region, is shown in Figure 1. It is proved that the volume fraction of the interphase region in NTRPCs is much larger than the interphase region in micron-sized composites. In Figure 2, the interphase volume fraction (V_i) as a function of fiber/nanotubes volume fraction for various ratios of interphase thickness t to fiber radius r_f is shown. The value of t/r_f is on the order of 0.05 for carbon fiber composites [8] and 1.0 for nanotube-reinforced polymers [9]. Therefore, also seen in Figure 2, for the case of nanotube-reinforced polymers, a significant portion of the NRP is occupied by the interphase region. In this case, the non-bulk polymer behavior of the interphase region is expected to contribute a lot to the overall viscoelastic response of the material. Recently, Zhang et. al [10] investigate the effect of the thickness of the interphase region on the overall behavior of NTRPC using the method of shear-lag (SL). Since the experimental techniques are so expensive, there is a need to develop accurate analytical/numerical models that can predict the time-dependent viscoelastic behavior of such nanocomposites, by taking into account the reduced mobility of interphase region.

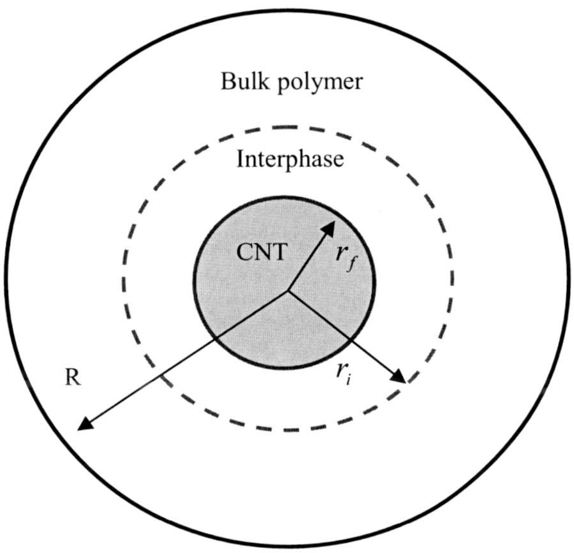

Fig. 1: Three phase model of NTRPCs

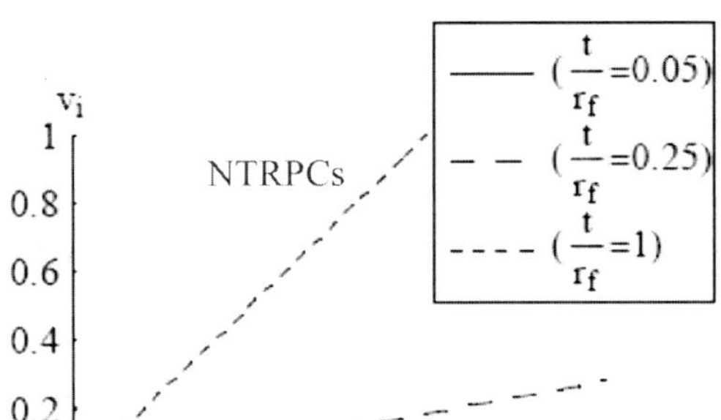

Fig. 2: Interphase volume fraction (Vi) as a function of fiber volume fraction V_f; interphase thickness $t = r_i - r_f$

Suhr et. al [11] carry out direct shear testing of epoxy thin films containing dense packing of multiwalled CNT fillers and report strong viscoelastic behavior with up to 1,400% increase in loss factor (damping ratio) of the baseline epoxy. Using a Molecular Structural Mechanics/Finite Element (MSM/FE) multiscale modeling of CNTs with viscoelastic (VE) polymer matrix, Montazeri et al. [12] showed that the VE polymer matrix, predicts the behavior of real NTRPCs better and thus should be used for more accurate simulations. Li et. al [13] developed a new micromechanical model for describing the linearly viscoelastic behavior of two NTRPC systems: a transversely isotropic one with unidirectionally aligned nanotubes and isotropic one having randomly oriented nanotubes. By introducing a new simplified 3D representative volume element for wavy CNTs, Yazdchi et al. [14] studied the stress transfer in single-walled carbon nanotube-reinforced polymer composites. Based on the pull-out modeling technique, they investigated the effects of waviness, aspect ratio, and Poisson ratio on the axial and interfacial shear stresses. Lu et. al [15] characterized both in-plane and through-thickness linear viscoelastic properties of the single-wall carbon nanotubes (SWNT)/polyelectrolyte multilayer nanocomposite film using nanoindentation. Their results show that the in-plane and through-thickness linear viscoelastic properties are almost identical for a SWNT nanocomposite made by layer-by-layer assembly (LBL) technique, despite the preferred orientation of the SWNT in a nanocomposite film.

In this study the constitutive relation and linear viscoelastic behavior of NTRPC are studied using methods of micromechanics and nanomechanics. First, the effects of volume fraction, aspect ratio and orientation of carbon nanotubes (CNTs), on the overall elastic properties of NTRPC are obtained based on Eshelby's Equivalent Inclusion. Secondly, by incorporating the Dynamic Correspondence Principle (DCP), the elastic solution is extended to solve the related linear viscoelastic problem.

2. MODEL DEVELOPMENT

The first research on viscoelastic behaviour of composites is returned back to the works of Hashin [16] and Schapery [17]. After that, many researches studied the effects of fibre orientation [18] and intermediate phase [19] on the mechanical properties of composites. In the following, the derivation of the constitute relation for viscoelastic behaviour of nanocomposites reinforced with CNTs is presented. The assumptions through out the rest of the paper are:

• The CNTs are straight and the effect of waviness is ignored
• We consider two cases: (a) Random dispersion of NTs which lead to isotropic behaviour and (b) Uniform dispersion of NTs which lead to transverse isotropic behaviour of nanocomposites
• The interphase region is modelled as the elastic and transverse isotropic material. The mechanical properties of this region are determined by the Equivalent Continuum Modelling (ECM) technique [20].
• The overall behaviour of nanocomposites is modelled as linear viscoelastic material
• The mechanical properties of NTs and polymer matrix are independent of temperature

• The micromechanical model is based on Mori-Tanaka (MT) approach [21]

The stress-strain relation for linear viscoelastic material is defined as [13]

$$\sigma(t)=\int_0^t L(t-\tau)\dot{\varepsilon}(\tau)d\tau, \qquad \varepsilon(t)=\int_0^t M(t-\tau)\dot{\sigma}(\tau)d\tau \tag{1}$$

where the dot denotes the differentiation with respect to time t, and $L(t)$ and $M(t)$ are the stress relaxation stiffness and creep compliance tensors, respectively. By applying the Laplace-Carson transformation as:

$$\hat{f}(s)=s\int_0^\infty e^{-s\tau}f(\tau)d\tau \tag{2}$$

Eq. (1) can be rewritten as:

$$\hat{\sigma}(s)=\hat{L}(s)\hat{\varepsilon}(s), \qquad \hat{\varepsilon}(s)=\hat{M}(s)\hat{\sigma}(s) \tag{3}$$

where the hat indicates the transformed function in the Carson domain, and s is the transform variable. In fact, according to the Dynamic Correspondence Principle in viscoelasticity, if a Laplace transformable, analytical solution exists for a problem in linear elasticity, the solution for the corresponding problem in linear viscoelasticity in the Carson (transformed) domain can be directly obtained from the former by replacing stiffness and compliance tensors with its viscoelastic counterpart $\hat{L}(s)$ and $\hat{M}(s)$, respectively. Therefore, the constitutive relation for viscoelastic material can be easily replaced by the elastic relations. In particular, for a transversely isotropic composite material containing unidirectional and aligned identical CNTs, Eq. (3) is expanded as:

$$\begin{cases}\hat{\sigma}_{11}=\hat{L}_{11}\hat{\varepsilon}_{11}+\hat{L}_{12}\hat{\varepsilon}_{22}+\hat{L}_{12}\hat{\varepsilon}_{33},\\ \hat{\sigma}_{22}=\hat{L}_{12}\hat{\varepsilon}_{11}+\hat{L}_{22}\hat{\varepsilon}_{22}+\hat{L}_{23}\hat{\varepsilon}_{33},\\ \hat{\sigma}_{33}=\hat{L}_{12}\hat{\varepsilon}_{11}+\hat{L}_{23}\hat{\varepsilon}_{22}+\hat{L}_{22}\hat{\varepsilon}_{33}, \quad \hat{\sigma}_{12}=2\hat{L}_{44}\hat{\varepsilon}_{12},\\ \hat{\sigma}_{23}=2\hat{L}_{66}\hat{\varepsilon}_{23}=\left(\hat{L}_{22}-\hat{L}_{23}\right)\hat{\varepsilon}_{23}, \quad \hat{\sigma}_{31}=2\hat{L}_{44}\hat{\varepsilon}_{31},\end{cases} \tag{4}$$

where $\hat{L}_{ij}$ are the components of the stiffness tensor. The relation between $\hat{L}_{ij}$ and the normal and transverse Young's modulus, shear modulus and in plane and out of plane Poisson's ration can be determined as:

$$\begin{cases}\hat{E}_{11}=\hat{L}_{11}-\dfrac{2\left(\hat{L}_{12}\right)^2}{\hat{L}_{22}+\hat{L}_{23}},\\ \hat{E}_{22}=\hat{L}_{22}+\dfrac{\left(\hat{L}_{12}\right)^2\left(\hat{L}_{23}-\hat{L}_{22}\right)+\hat{L}_{23}\left[\left(\hat{L}_{12}\right)^2-\hat{L}_{11}\hat{L}_{23}\right]}{\hat{L}_{11}\hat{L}_{22}-\left(\hat{L}_{12}\right)^2},\\ \hat{G}_{12}=\hat{L}_{44}, \quad \hat{\upsilon}_{12}=\dfrac{\left(\hat{L}_{12}\right)}{\hat{L}_{22}+\hat{L}_{23}}, \quad \hat{\upsilon}_{23}=\dfrac{\left[\hat{L}_{11}\hat{L}_{23}-\left(\hat{L}_{12}\right)^2\right]}{\hat{L}_{11}\hat{L}_{22}-\left(\hat{L}_{12}\right)^2},\end{cases} \tag{5}$$

the out of plane shear and bulk modulus are not independent and can be obtained as:

$$\hat{G}_{23}=\frac{\hat{E}_{22}}{2\left(1+\hat{\upsilon}_{23}\right)}=\hat{L}_{66}=\frac{1}{2}\left(\hat{L}_{22}-\hat{L}_{23}\right), \quad \hat{K}_{23}=\frac{1}{2}\left(\hat{L}_{22}+\hat{L}_{23}\right), \tag{6}$$

Therefore, the five independent components of creep compliance tensors are:

$$\hat{M}_{11}=\frac{1}{\hat{E}_{11}}, \quad \hat{M}_{22}=\frac{1}{\hat{E}_{22}}, \quad \hat{M}_{44}=\frac{1}{\hat{G}_{12}}, \quad \hat{M}_{66}=\frac{1}{\hat{G}_{23}}, \quad \hat{M}_{23}=\frac{1}{\hat{K}_{23}}, \tag{7}$$

It should be noted that in case of isotropic composites containing randomly oriented NTs, the above five independent constants are reduced to only two independent constants, namely $\hat{E}=\hat{E}_{11}=\hat{E}_{22}$ and $\hat{G}=\hat{G}_{12}=\hat{G}_{23}$. Easily, by taking the inverse Laplace transformation from $\hat{M}(s)$, which is known from MT method (see next section), we can obtain the time dependent creep compliance tensor $M(t)$ as:

$$M(t)=\frac{1}{2\pi i}\int_{\delta-i\infty}^{\delta+i\infty}\frac{\hat{M}(s)}{s}e^{st}ds, \tag{8}$$

As previously mentioned, it is assumed that the CNT and the nanotube/polymer interface can be modeled as an effective continuum fiber by using an equivalent-continuum modeling (ECM) method [20]. This method consists of two major steps; first, an appropriate representative volume element (RVE) of the nano-structured material is chosen and the interactions of the atoms are described in terms of molecular mechanics force constants. Secondly, an equivalent-continuum model of the RVE is developed in which the total strain energies in the molecular and equivalent continuum models, under the same loading conditions, are set equal. The effective mechanical properties of the equivalent-continuum is then determined from equating strain energies. By applying this procedure and replacing the NT and interphase region with an equivalent transverse isotropic fiber, the effective properties of the fiber can be obtained as:

$$\begin{cases} L_{11}^f = 457.6\ GPa, & \\ L_{12}^f = 8.4\ GPa, & \\ L_{22}^f = 14.3\ GPa, & \Rightarrow G_{23}^f = 4.4\ GPa,\quad K_{23}^f = 9.9\ GPa, \\ L_{23}^f = 5.5\ GPa, & G_{12}^f = 27.0\ GPa,\quad E_{11}^f = 450.4\ GPa, \\ L_{44}^f = 27.0\ GPa, & \end{cases} \tag{9}$$

For the matrix we use power law (PL) model to determine the creep compliance of the polymer. Therefore we have:

$$M_0(t) = D_0 + D_1 t^n \tag{10}$$

where D_0, D_1 and n are the material constants. By taking the Laplace transformation from Eq. (10) we get:

$$\hat{M}_0(s) = D_0 + \frac{D_1 n!}{s^n} \tag{11}$$

In this paper we use LaRC-SI polyimide as a matrix, which its material constants are:

$$D_0 = 0.37\ GPa^{-1},\quad D_1 = 0.05\ GPa^{-1} hour^{-1},\quad n = 0.41,\quad \upsilon_0 = 0.37 \tag{12}$$

Therefore our problem is now reduced to the calculation of overall properties of two phase viscoelastic composite material using MT method.

3. MICROMECHANICAL ANALYSIS - MT METHOD

The Mori-Tanaka (MT) method has been used by a wide range of researchers to model the effective behavior of composites. Zhao and Weng [22] used MT method to study the effect of inclusion shape on composite moduli and Fisher et. al [23] to study the viscoelastic behavior of polymer-matrix composites. The MT method has been successfully extended to cover composites with multiple inclusion phases [24] and random orientations [25] as well.

Assuming that the composite is composed of N phases; the matrix will be denoted as phase 0 with a corresponding stiffness L_0 and volume fraction f_0, while an arbitrary rth inclusion phase (where $r = 1$ to N-1) has a stiffness of L_r and a volume fraction f_r and each phase is linearly elastic and isotropic, and there is perfect bonding between the inclusions and the matrix, the stiffness tensor of the unidirectionally aligned NTs can be expressed as:

$$L_{ij} = \left(f_0 L_0 + \sum_{r=1}^{N-1} f_r L_r A_r \right)\left(f_0 I + \sum_{r=1}^{N-1} f_r A_r \right)^{-1}, \tag{13}$$

and for randomly aligned NTs is:

$$L_{ij} = \left(f_0 L_0 + \sum_{r=1}^{N-1} f_r \langle L_r A_r \rangle \right)\left(f_0 I + \sum_{r=1}^{N-1} f_r \langle A_r \rangle \right)^{-1}, \tag{14}$$

where A_r is the dilute strain-concentration factor of the rth phase and relates the average strain in the rth inclusion to the average strain in the matrix. As it is seen, Eq. (14) is the direct analog of Eq. (13) except that appropriate averaging is used here to account for the inclusion orientation. For more details see [6].

4. NUMERICAL RESULTS

In this section the effects of aspect ratio and volume fraction of CNTs on the overall mechanical properties of NTRPCs, for isotropic and transversely isotropic cases, will be discussed.

Isotropic behavior

In the case of randomly oriented NTs, the overall behavior of composite will be isotropic. In Figure 3 the effect of aspect ratio (ratio of the length to diameter of NTs) on the axial creep compliance (inverse of Young's modulus) is shown. As it is seen with increasing the aspect ratio, the axial creep compliance of composite is reduced or in other words, the nanocomposites became stiffer. In Figure 4 the effect of volume fraction on the shear creep compliance (the inverse of shear modulus) is shown. With increasing the volume fraction, the shear creep compliance is decreased. By comparing the Figure 3 and 4, we can see that the effect of aspect ratio is higher than volume fraction. It should be mentioned that at the longer time, the shear and axial creep compliance are increased.

Transverse isotropic behavior

In the case of uniformly distributed NTs, the overall behavior of composite will be transversely isotropic with five independent material constants. In Figure 5 the effect of aspect ratio on the axial creep compliance M_{11} and axial shear creep compliance M_{44} is shown. As it is seen with increasing the aspect ratio, the creep compliance of composite is reduced. Of course the effect of aspect ratio on M_{11} is clearly higher than M_{44}.

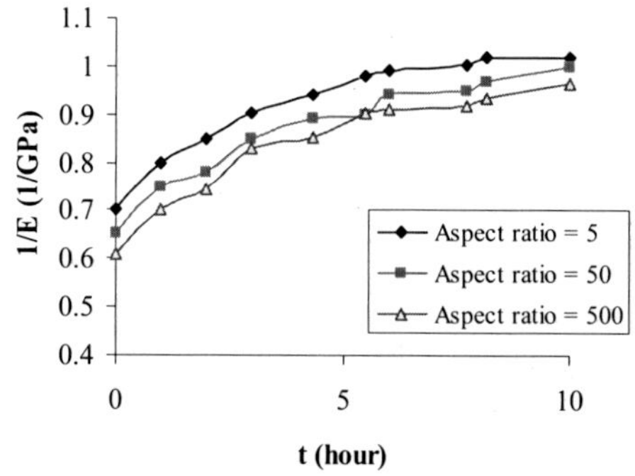

Fig. 3: Variation of axial creep compliance with time at different aspect ratios

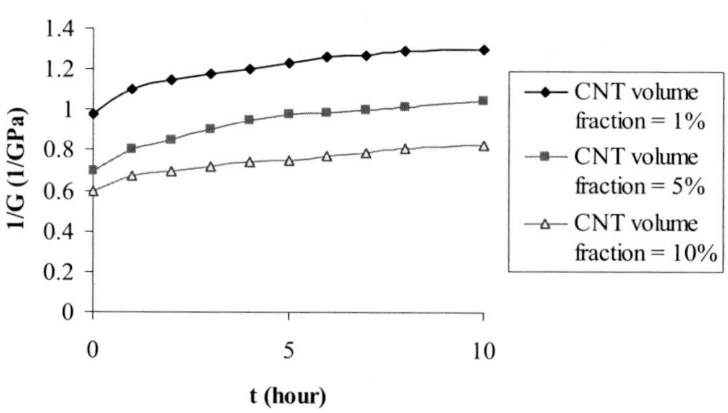

Fig. 4: Variation of shear creep compliance with time at different volume fractions

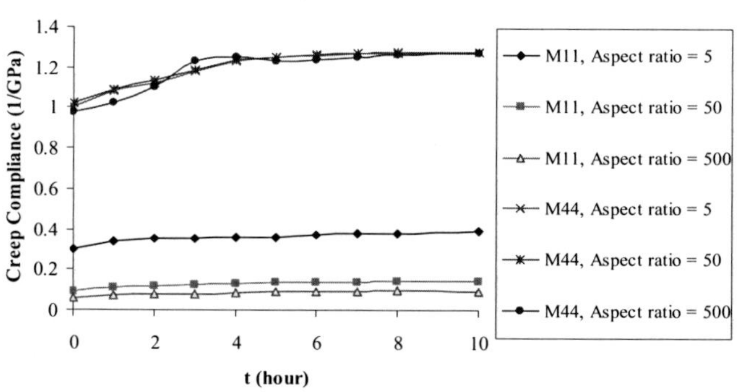

Fig. 5: Variation of axial M_{11} and shear creep compliance M_{44} with time at different aspect ratios

In Figure 6 the effect of aspect ratio on the transverse creep compliance (M_{22}) and the plane strain bulk modulus (k_{23}) is shown. With increasing the aspect ratio, the transverse creep compliance is decreased. However aspect ration has almost no effect on the plane strain bulk modulus. In Figure 7 and 8 the effect of volume fraction on the axial creep compliance (M_{11}), the axial shear creep compliance (M_{44}), transverse creep compliance (M_{22}) and the plane strain bulk modulus (k_{23}) are shown. With increasing the volume fraction, M_{11} and M_{44} are decreased; however k_{23} and M_{22} are almost constant. These trends are in good agreement with previous results [26, 27]. Furthermore, with 0% and 100% volume fraction of NTs we get the mechanical properties of neat polymer and NTs respectively.

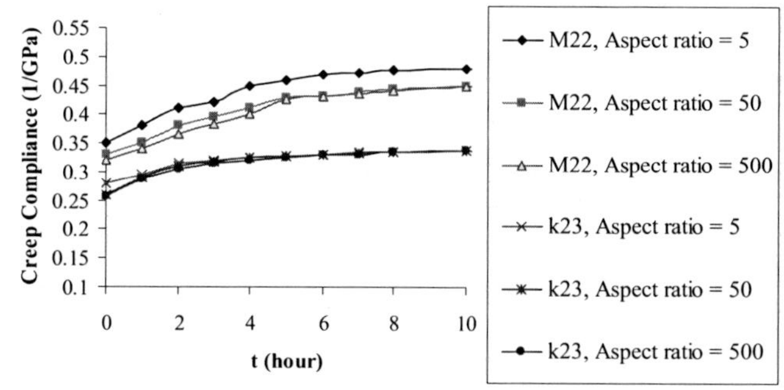

Fig. 6: Variation of transverse creep compliance M_{22} and the plane strain bulk modulus k_{23} with time at different aspect ratios

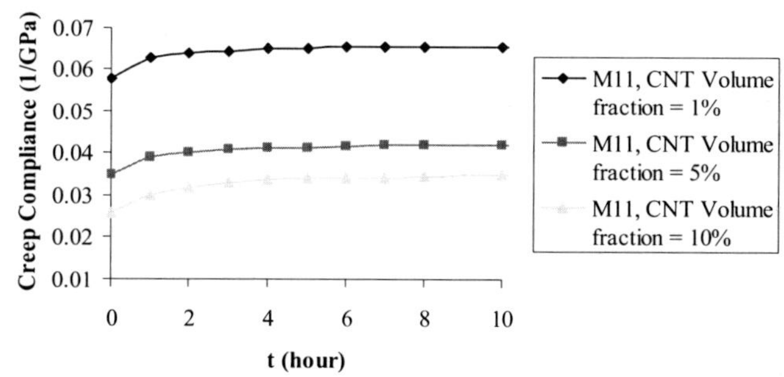

Fig. 7: Variation of the axial creep compliance M_{11} with time at different volume fractions

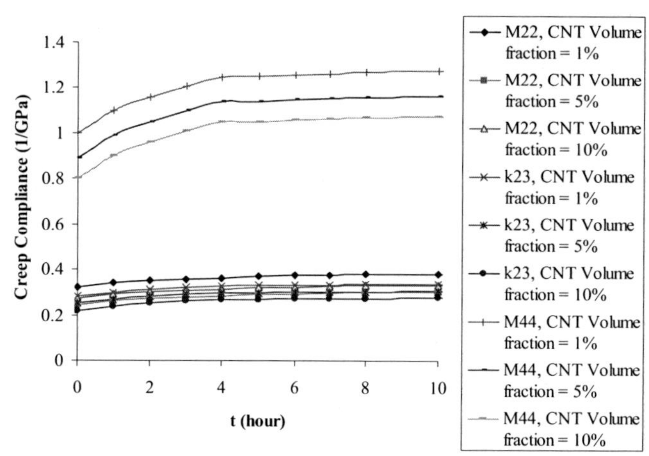

Fig. 8: Variation of transverse creep compliance M_{22}, axial shear creep compliance M_{44} and the plane strain bulk modulus k_{23} with time at different volume fractions

5. CONCLUDING REMARKS

Based on Dynamic Correspondence Principle (DCP) and method of micromechanics, the effects of volume fraction, aspect ratio and orientation of NTs on the viscoelastic behaviour of polymer composites reinforced with NTs are obtained. Some of the main conclusions are: Our numerical results show that for randomly oriented NTs: (a) by increasing volume fraction or aspect ratio the axial creep compliance and axial shear creep compliance are decreased and (b) The effect of volume fraction is higher than aspect ratio. However, for uniformly distributed NTs: (a) the effect of aspect ratio on axial shear creep compliance and the plane strain bulk modulus is insignificant but has a great effect on the axial creep compliance, (b) With increasing the aspect ratio, the axial stiffness of the composite is improved and (c) by increasing the volume fraction the axial creep compliance is decreased however other mechanical properties are almost constant. The results of this study are in good agreement when compared with previous analytical and experimental data. The results of this study can be used in manufacturing of advanced nanocomposites with improved properties for structural applications.

ACKNOWLEDGMENTS

The author would like to thank financial support of the Dutch Technology Foundation STW, which is the applied science division of NWO, and the Technology Programme of the Ministry of Economic Affairs (10120).

REFERENCES

[1] Salvetat JD., Rubio A., Carbon; 40 (2002), 1729-34.

[2] Natsuki T., Tantrakan K., Endo M., Appl Phys A; 79 (2004), 117–124.

[3] Grujicic M., Angstadt D.C., Sun Y.P., Koudela K.L., J Mater Sci.; 42 (2007), 4609– 23.

[4] Lau KT, Hui D., Comp Pt B: Eng; 33 (2002), 263–77.

[5] Thostenson, E.T., Li, C.Y., Chou, T.W., Compos. Sci. Technol.; 65 (3–4) (2005), 491.

[6] Fisher FT., PhD thesis, Northwestern University, Evanston, Illinois, 2002.

[7] Bradshaw R.D., Fisher F.T., Brinson L.C., Compo. Sci. Technol.; 63 (2003), 1705.

[8] Thomason, J. L., Composites; 26 (1995), 487-498.

[9] Wise, K. and J. Hinkley, American Physical Society Spring Meeting, Seattle, WA, 2001.

[10] Zhang J., He C., Acta Mech; 196 (2008), 33-54.

[11] Suhr J., Koratkar N., Keblinski P. and Ajayan P., Nature Materials; 4 (2005), 134 – 137.

[12] Montazeri A., Naghdabadi R., Polymer Composites; 30 (2009), 1545-51.

[13] Li K., Gao X.-L., Roy A. K., Mechanics of Advanced Materials and Structures; 13 (2006), 317–328.

[14] Yazdchi K., Salehi M. and Shokrieh M.M., Mechanics of Composite Materials; 45 (2009), 207-12.

[15] Lu H., Huang G., Wang B., Mamedov A., Gupta S., Thin Solid Films; 500 (2006), 197 – 202.

[16] Hashin, Z., Int. J. Solids and Struc.; 6 (1970), 797-807.

[17] Shapery, R., Journal of Composite Materials; 1 (1967), 228-267.

[18] Brinson, L. and W. Lin, Composite Structures; 41(1998), 353-367.

[19] Li, J. and G. Weng, Composites Part B; 27B (1996), 589-598.

[20] Odegard G.M., Gates T.S., Wise K.E., Park C., Siochi E.J., Composites Science and Technology; 63 (2003), 1671–1687.

[21] Mori T., Tanaka K., Acta Mater.; 21 (1973), 571- 574.

[22] Zhao, Y. H. and G. J. Weng, Journal of Applied Mechanics; 57 (1990), 158-167.

[23] Fisher, F. T. and L. C. Brinson, Composites Science and Technology; 61 (2001), 731-748.

[24] Benveniste, Y., Mechanics of Materials, 6 (1987); 147-157.

[25] Tandon, G. P. and G. J. Weng, Composites Science and Technology; 27 (1986), 111-132.

[26] Xiao K. Q., Zhang L. C., Journal of Materials Science; 39 (2004), 4481 – 4486.

[27] Coleman J. N., Khan U. and Gun'ko Y.K., Adv. Mater.; 18 (2006), 689–706.

Proceedings of the ASME 2010 International Mechanical Engineering Congress & Exposition
IMECE2010
November 12-18, 2010, Vancouver, British Columbia, Canada

IMECE2010-38237

A Multi-Criteria Simulation Framework for Civil Aircraft Trajectory Optimisation

Hasan Zolata, Cesar Celis, Vishal Sethi and Riti Singh
Department of Power and Propulsion, School of Engineering, Cranfield University, Cranfield, Bedfordshire, MK43 0AL, United Kingdom
hzolata@yahoo.it

David Zammit-Mangion
Department of Aerospace Engineering, School of Engineering, Cranfield University, Cranfield, Bedfordshire, MK43 0AL, United Kingdom

ABSTRACT

Over the past few years, great concern has been raised about the impact of commercial aviation on the environment. In a Business As Usual approach, the expected growth in air traffic is going to affect climate change even more unless mitigation policies are devised and implemented. Although there is a tendency to focus on long-term technological solutions and breakthroughs, short-term improvements applicable to existing aircraft/engine configurations are also very important to fully realise the benefits of new technologies. Aircraft trajectory optimisation presents the opportunity to effectively reduce fuel consumption and pollutants emitted providing a feasible short-term strategy to be applied to the existing aircraft fleet. The present study focuses on preliminary results obtained using a multi-disciplinary aircraft trajectory optimisation simulation framework. Three in-house computational models are implemented in the framework to model the aircraft and engine performance, as well as to predict the level of gaseous emissions produced. A commercially available optimiser is integrated within the framework to analyse and optimise single flight path elements (e.g., climb), as well as the entire flight profile. For the purpose of this study, the climb and the whole flight profile are divided in four and eight segments respectively. Trajectory optimisation processes are then carried out in order to minimise three different objective functions: flight time, fuel burned, and mass of oxides of nitrogen (NOx) emitted. The results of the trajectory optimisation processes performed confirm the validity, effectiveness, and flexibility of the methodology proposed. In future, it is expected that these types of approaches are utilised to efficiently compute complete, optimum and „greener' aircraft trajectories, which help to minimise the impact of commercial aviation on the environment. Other computational models that simulate several other aspects such as aircraft and engine noise, weather conditions and contrails formation, among others, need to be also included in the optimisation processes.

NOMENCLATURE

AL	Altitude
CO_2	Carbon Dioxide
EI	Emission index (or indices)
EAS	Equivalent Air Speed
FPA	Flight Path Angle
H_2O	Water vapour
M	Mach Number
NOx	Oxides of Nitrogen
P	Pressure
R	Ground Range
T	Temperature
TAS	True Air Speed
TET	Turbine Entry Temperature
W	Aircraft Weight

Subscripts

i	Initial
f	Final

INTRODUCTION

The commercial aviation sector is expected to develop threefold by 2050 [1] as aircraft will continue to represent the fastest mode of transport available. In order to have a "sustainable" growth it is of paramount importance to take into account the impact of aircraft operations on the environment. Two main issues raising the highest concern of the scientific community can be addressed: gaseous emissions and noise pollution. The recent climate changes suggest to leave ‚laissez-faire' attitudes and to start thinking about possible solutions to mitigate the emissions problem. In response to the high level risk concerning the effects of aircraft operations on climate change, different projects have been undertaken worldwide in the aeronautical field which seek to evaluate and demonstrate the feasibility of possible solutions.

In the United States a project known as PARTNER – Partnership for AiR Transportation Noise and Emissions Reduction – was launched in 2006. PARNER is a leading aviation cooperative research organization collaboratively formed by academia, industry and government, which will make advances in environmental performance, efficiency, safety and security [2].

In Europe a similar project started in 2001 and as well as PARTNER, it is focused on the sustainability of the growing air traffic demand. This sustainability is based not only on the preservation of the flights security and safety but also on the minimisation of the impact of aircraft operations on non-travellers. In 2001 an Advisory Council for Aeronautics Research in Europe (ACARE) was created together with the first European Technology Platform which produced a set of top level objectives to be achieved by 2020 as listed in SRA-1 and SRA-2 [3-4]. The project is structured into six Integrated Technology Demonstrators – ITDs, whose purpose is to develop, integrate and validate through full scale demonstrators the technological solutions studied. Among others, Cranfield University – CU – is actively participating in the European project in several areas, including the ITD known as Systems for Greener Operations – SGO. The two main research fields covered by the SGO ITD are the Management of Aircraft Energy – MAE, and the Management of Trajectory and Mission – MTM. CU contributions relate to the development of computational algorithms for aircraft trajectory optimisation for the MTM area [5].

Technological solutions currently being developed can be classified into long-term solutions, applicable in decades time, and short-term solutions, that will be ready by the next decade. Hence, different approaches apply to solve the common problem of finding more environmental friendly solutions. In the long term it is possible to explore new design solutions for the aircraft, the engine, the combustion chamber, etc. Nevertheless, in the short term the impact of aircraft operations on the environment can be reduced by finding solutions applicable to the existing fleets.

In the last years, long term solutions have seen a radical shift of the imperatives of the air transportation industry future. Twenty years ago during the Concorde-Era three key words were used to describe the guidelines in aircraft design: "Higher, Further, and Faster". Nowadays, the environmental impact of aircraft operations combined to the need of cost-effective solutions resulted into a shift of these guidelines towards: "More Affordable, Safer, Cleaner and Quieter" [6]. To meet these requirements, new design issues have been defined to address the trend inversion in aircraft design: "slower, lower, greener" [7]. Hence, a trade-off analysis should be adopted to take into account all the design priorities, including the economic cost of polluting effects. Quantifying the economic impact of aviation effects on the environment is not easy or unique. As highlighted before, the two environmental issues of highest concern are noise and gaseous emissions. New designs account for a consistent reduction of engine noise and reduced fuel consumption. Even though noise pollution is treated equally to gaseous emissions pollution, the cost of the effects related to the former is not even comparable. Economists refer to these two sources as "externalities" as these effects of air transportation affect third parties (not only passengers). Weighting these externalities is not an easy task and different methodologies can be used. To have an idea of the cost of the two different externalities, the additional cost for a Heathrow passenger ticket has been considered. Assuming a 50-50 short-long haul proportion on the flights leaving Heathrow airport, the aircraft noise adds a cost of 0.40£ to the passenger ticket, while the climate change related effects produce a consistent 30£ extra cost [8]. Summing the externalities over a year time, the external cost of civil aviation for the UK is about 25m£ for the noise and 1,400m£ for the climate change [9]. In addition to this, a noise efficient aircraft design is not so fuel-efficient yet. That means that a noise-optimised design consumes more fuel and produces more gaseous emissions than a fuel-efficient design [7]. This is why it is really important to address properly the influence of the various requirements to be met.

In order to guarantee the key role of aviation in the transport sector for the forthcoming years, it is fundamental to act quickly in order to mitigate its environmental damaging effects. Only by this way, it will be possible to elect the aviation as a „sustainable' or greener mode of transport. This is why short-term improvements applicable to existing aircraft/engine configurations are very important to fully realise the benefits of new technologies. Different mitigation solutions can apply to minimise the negative impacts of emissions on climate change [10]:

1. Introduction of taxation policies based on contraction and convergence, although against actual ICAO rules.
2. Application of trading schemes for the aviation sector as currently applied to other energy sectors.
3. Shift from air travel to other „greener' transportation solutions to reduce the growth prediction in the sector.
4. Introduction of fuel efficient technologies involving fuel usage reductions and/or using alternative fuels.
5. Introduction of a more efficient air traffic management in order to optimise flight routings.
6. Introduction of modifications in aircraft operational rules and redefinition of flight procedures.

The last two points represent solutions associated with the introduction of modifications into aircraft operations. Particular attention is given to the last one that represents a feasible option for reducing the impact of aircraft operations on the environment via the redefinition of aircraft trajectories specifically optimised to minimise the polluting effects. Advanced aircraft trajectory analyses enable the identification of "greener" trajectories, i.e. trajectories that provide a minimum environmental impact in terms of noise perceived, fuel consumption and pollutants emitted in the atmosphere along the flight profile.

To successfully perform aircraft trajectory optimisation processes, two different tasks have to be accomplished:

- The creation of a Simulation Framework

The simulation of the aircraft performance requires in general the implementation of different computational models able to simulate specific aircraft operation-related disciplines such as engine performance, airframe performance, atmosphere, noise, gaseous emissions, etc. In order to optimise a given aircraft trajectory, it is necessary that all the different models operate in a common environment in order to exchange data and work jointly. This virtual common environment is called Framework.

- The selection of an Optimisation Approach

Depending on the structure of the problem, it is possible to use different optimisers already available in the public domain or to develop an optimiser specific for that type of problem. Although the second option is considered more reliable, the first option was selected in this work for the optimisation of aircraft greener trajectories.

The first version of the framework deployed in this study has already been tested [11] using as a test case an unconstrained climb optimisation process. The present work focuses on the conception, development, and implementation of a methodology capable of carrying out multi-disciplinary trajectory optimisation processes. For the purpose of this analysis, the framework has been implemented into a commercial optimizer [12-13] with different numerical methods eligible for trajectory optimisation. After a brief description of the selected numerical method, results relative to a constrained climb profile and a whole flight profile are presented and discussed.

FRAMEWORK DESCRIPTION

„Framework' is used in this context to define the virtual environment created to enable three simulation programs to interface with each other and work jointly to provide the data required for the optimizer at each step of the optimisation process. The three CU in-house programs used to simulate the aircraft-engine system performance in this work are: TurboMatch [14] for the engine performance simulation, APM [15] for the aircraft performance simulation, and Hephaestus [16] for the emissions prediction.

The sensitive data required for the optimisation are collected in two files created from the main program: CompModels.inp – which contains the essential input data; and CompModels.out – which contains the essential output data. The input file contains the following information:

- The engine Design Point configuration (DP) in terms of Mach number, pressure ratio split for the compressors, by-pass ratio, TET and inlet mass flow.
- A default configuration for the emission prediction model, e.g., combustor configuration.
- The mission specification including number and length of segments, initial/final altitudes and velocities for each segment, and the speed mode. The latter enables the user to specify the aircraft speed in terms of Mach number, mode 1, Equivalent Air Speed (EAS), mode 2, or True Air Speed (TAS), mode 3 at each flight segment.

The output file contains information provided from the different models:

- The aircraft model gives the net thrust, F_N, the acceleration along the segment and the time required to fly between the initial and final stage. It also provides rate of climbs and speeds in terms of TAS, EAS, etc.
- The engine model gives the fuel flow, specific fuel consumption and TET necessary to match the net thrust computed from the aircraft model.
- The emissions prediction model gives the emissions indices for the carbon dioxide, nitrogen oxides and vapour water. The EI are used to calculate the level of emissions produced, in kg once the time and fuel flow are computed.

It should be noticed that only a part of all the results available in the single output files of the different simulation models appears in the output file. The purpose of this choice is related to the need of integrating the optimiser easily and quickly inside the Framework structure. The input and output files are the only ones accessed from the optimiser thereby minimising the computational time required to access the different files and read the data of interest.

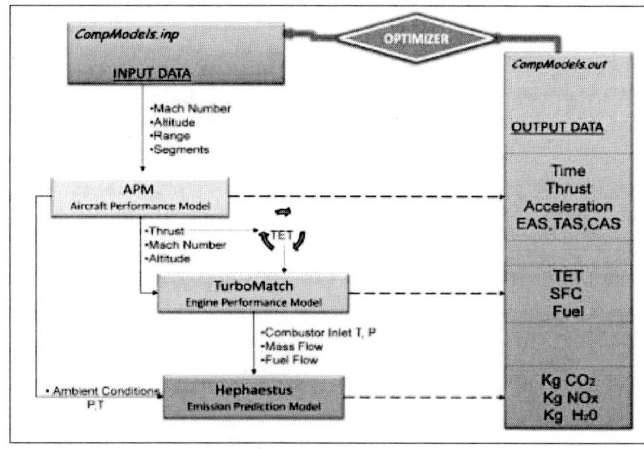

Figure 1: Framework structure

The Framework computational path can be summarised in the main following steps represented in Figure 1:

- The aircraft model reads from the main input file the range, speed and altitude associated with each flight segment.
- The aircraft model computes the rate of climb and net thrust required (F_N) for each segment and provides these values plus the speed and altitude to the engine model for the OD point engine calculation.
- A sub-routine calculates the TET that matches exactly the net thrust required for the given segment: an iterative algorithm runs TurboMatch until convergence of the achieved and required thrust is obtained.
- The engine model runs the off design performance simulation computing all the thermodynamics variables along the engine plus the fuel flow, mass flow, TET and many others.
- The emissions prediction model receives from the engine model all the thermodynamic conditions at the inlet of the combustion chamber and other relevant additional information from the aircraft model, including flight atmospheric conditions.
- The emissions prediction model computes the emission indices (g/kg fuel) for the pollutants of interest.
- The main program elaborates the information about time, fuel flow and EI, to calculate the fuel burned and the emissions produced at each flight segment as well as during the whole flight mission.

Due to the fact that the engine performance model used here (TurboMatch) currently does not take thrust as an input, an iterative process has been designed and implemented with TurboMatch in order to carry out the optimisation processes, as observed in Figure 1. This iterative process basically involves an initial guess of TET, execution of TurboMatch for this guessed TET and comparisons of the calculated thrust with the required one. At the end of the iterative process, the main parameters characterising the operation of the engine, including fuel mass flow rate and combustor inlet conditions, are known. The flight path angle has a fixed limit set to FPA ≤ |7.5°|. This limitation is due to operational constraints more than engine performance issues. Outbound solutions are „excluded' from the optimisation process.

AIRCRAFT PERFORMANCE MODEL

The program used for the aircraft performance simulation uses quasi steady state equations [15] to compute aerodynamic lift and drag, and determine the thrust required for a given segment. The first version of the program was not capable to compute acceleration phases along a given segment; hence only the average Mach number was evaluated [11]. The second version of this program is capable to compute acceleration phases along a given segment thereby solving the discontinuity in the flight profile. Further versions of the aircraft performance model are being developed to take into account additional effects as critical and transonic wave Mach numbers.

The APM program uses the atmosphere conditions of the initial point of each segment to calculate the required thrust to climb and accelerate to the initial point of the next segment. This procedure implies that variations in atmospheric parameters are not evaluated along the segment, but they are kept constant until the beginning of the next segment. In order to have a realistic representation of the real atmospheric conditions it would be necessary to adopt an infinite number of segments along the climb path. By this way, the accuracy of the model would be preserved as the flight path would be represented by an infinite number of small altitude steps.

Five initialisation variables are required to compute the thrust for a given flight segment: aircraft speed, e.g., Mach number (M), flight path angle (γ), initial mass (m), initial altitude (h) and ground range (dx). Given the initial and final altitude, the FPA can be directly computed once the ground range is specified. In case of acceleration, an additional term must be added in the thrust equation that can be obtained from the knowledge of the initial and final velocity.

Table 1: Aircraft model specifications

NUMBER -TYPE OF ENGINES	2 TURBOFAN
AIRCRAFT WEIGHT[1]	72000 KG
TOTAL WING AREA	122.4 m^2
WING ASPECT RATIO	7
WING SPAN	34.09 m
WING ROOT CHORD	6.10 m
WING SWEEPBACK ANGLE	25°
ZERO LIFT DRAG COEFFICENT	0.03
ZERO AOA LIFT COEFFICIENT	0.1
OSWALD SPAN EFFICIENCY	0.8
LIFT CURVE SLOPE	999

The aircraft can be classified as a generic single aisle large transport aircraft powered by two high BPR turbofan with separate exhaust. Table 1 presents the ten parameters that define the aircraft aerodynamic performance: wing reference area, aspect ratio, wing root chord length, wing span, wing sweepback angle, Oswald span efficiency, zero angle of attack lift coefficient drag, lift curve slope and lift induced drag factor. The aircraft speed used inside the algorithm is True Air Speed, TAS, whilst the input file provides the velocities in terms of EAS, M or TAS. The EAS is proportional to the TAS multiplied by the current altitude density as the SL density is a constant. M is proportional to the TAS divided by the speed of sound.

[1] The aircraft weight is fixed only at the beginning of the flight profile while it must be calculated in the complete flight profile analysis

ENGINE PERFORMANCE MODEL

The engine performance was simulated using the in-house CU gas turbine performance code TurboMatch [14], which is capable of modelling the behaviour of gas turbine engines in different operative scenarios and working conditions (design point and off design analyses are easily implemented). The program allows accounting for all ambient conditions as well as simulating degradations of components due to aging, providing a reliable estimate in case of engines already in service. TurboMatch can be extensively used for various Aero and Industrial applications ranging from simple to complex engine configurations.

TurboMatch was used to model a high by-pass ratio two-spool turbofan engine with separate exhausts, HBPRTF. The engine model was created using information available in the public domain and verification processes of the model at design and off design point conditions were carried out using information publicly available. Table 2 shows the information collected from the public domain:

Table 2: Data collected from public domain for a generic high by-pass ratio twin spool turbofan with separate exhausts

MAX T-O THRUST (KN)	120.10
T-O MASS FLOW (KG/S)	406.900
T-O BYPASS RATIO	5.70
TOC OVERALL PRESSURE RATIO	32.6
MAXIMUM CLIMB THRUST	25.040
MAXIMUM CLIMB BYPASS RATIO	5.46
FAN DIAMETER (m)	1.73

The Design Point of the HBPRTF is top of climb, TOC, ISA conditions, altitude of 10660m and Mach number of 0.8. The DP performance was identified after running several simulations to match the performance requirements listed in Table 2. When necessary, educated guesses were made for some characteristic parameters (component efficiencies, bleeding flows, pressure losses, etc.), which were required for the modelling of the engine. A summary of the engine parameters and specifications used in the HBPRTF engine model DP simulation are shown in Table 3:

Table 3: HBPRTF engine model design point at TOC

ENGINE MASS FLOW W (KG/S)	180
ENGINE BPR	5.7
OVERALL PRESSURE RATIO	32.6
PRESSURE RECOVERY	0.99
FAN PRESSURE RATIO	1.80
BOOSTER PRESSURE RATIO	1.811
HP COMPRESSOR PRESSURE RATIO	10
FAN / BOOSTER ISENTROPIC EFFICIENCY	0.9
HPC ISENTROPIC EFFICIENCY	0.885
COMBUSTOR EFFICIENCY	0.99
COMBUSTOR PRESSURE LOSS (%)	0.05
TURBINE ENTRY TEMPERATURE TET (K)	1340
HPT / LPT ISENTROPIC EFFICIENCY	0.91
OPERATING ALTITUDE (m)	10668
FLIGHT MACH NUMBER	0.80

A close matching between the HBPRTF and the data collected from the public domain was achieved even in lack of essential information as TET or components efficiencies. The matching procedure was deemed satisfactory as to small negligible discrepancies. The performance comparison is presented in Table 4 for the TOC phase and TO phase. The specifications obtained from the public domain are enclosed in brackets with the percentage deviation from the HBPRTF.

Table 4: Comparison between data from public domain and TurboMatch results for HBPRTF engine model

TOP OF CLIMB – TOC PHASE			
THRUST	25.110KN	(25.040KN)	0.3 %
BPR	5.7	(5.46)	4.2 %
OVERALL PRESSURE RATIO	32.6	(32.6)	0.0 %
TET	1340K		-
MASS FLOW W	180KG/S		-
POWER OFF TAKE FROM HPT	200KW		-

TAKE OFF – TO PHASE			
THRUST	120.98KN	(120.10KN)	0.7 %
MASS FLOW W	401.390	(406.900)	1.3 %
TET	1600K		-
BPR	5.73	(5.70)	0.5 %
POWER OFF TAKE FROM HPT	200KW		-

EMISSIONS PREDICTION MODEL

The program used for the emissions prediction modelling is called Hephaestus. The particular Hephaestus emissions prediction methodology used adopts a Multi-Stirred Reactors configuration to describe the combustion process [16]. This methodology is a midway approach between two opposite strategies that apply in combustor emissions prediction: empirical correlations and detailed numerical simulations. The stirred reactor approach provides a higher level of accuracy than the correlation based models plus a lower computational time than the detailed numerical simulation based ones. Likewise, it allows computing exactly the time dependent chemistry of pollutant formation without an excessive idealization of the flow turbulence. The reactor configuration is made of three different reactor models: a perfectly stirred reactor model – PSR, a series of PSR – SPRS and a partially stirred reactor model – PaSR.

The PSR aim is to represent the influence of turbulence level on chemical reactions neglecting the spatial non-homogeneities of the velocity field. The two streams entering the PSR - a gas stream and fuel/air stream, are instantaneously mixed to reach the chemical equilibrium. To account for the species in non equilibrium, chemical kinetics calculations are implemented. Two parallel series of PSR are used to model the core and near-wall regions of the Intermediate Zones (IZ) of the combustion chamber increasing the accuracy of the air flow distribution. Each PSR in the series has its own amount of fuel and air. An additional PSRS is added at the end of the two previous series to model the Dilution Zone (DZ).

The partially stirred reactor, PaSR, is used to simulate the core region of the primary zone called Flame Front (FF), while the near-wall region is modelled with a PSR (NW PZ). It accounts for all the parameters that affect the pollutants reaction rates as variations in gas composition, temperature, residence time. Hence, the introduction of this model enables taking into account partial mixing and consequently inhomogeneities in this particular combustor region.

AIRCRAFT TRAJECTORY OPTIMISATION

Two different approaches can be chosen when dealing with aircraft trajectory optimisation. It is possible to use a very fine segmentation of the range yet adopting a coarse modelling of the aircraft-engine systems or use a coarse range segmentation coupled with a high accuracy computational models. The first option enables to account for minimal variations in the atmospheric thermodynamic properties with altitude at the expense of a low level modelling accuracy of, for instance, engine performance and emissions prediction. The second option has been chosen for the present study as more importance has been given to the level of accuracy of the simulation programs: the flight path is then divided into only four segments for the climb phase and into eight segments for the complete flight profile. This obviously results in a non-smooth variation of the atmospheric parameters with altitude. Hence, a coarse horizontal segmentation affects the accuracy of the representation of the actual optimal trajectory. To obtain satisfactory results the flight profile segmentation needs to be consistently increased until the trajectory shape will be independent from the range segmentation. In case of the climb phase optimisation, the optimal number of segments has been identified for the minimum fuel configuration.

Among the optimizers suitable for trajectory optimization, the MatLab Optimization Toolbox Version 4 [12] was used in this study. The toolbox can be used either by a graphical interface or by the command line depending on the information the user requires at the end of the optimization process. Different solvers that can be selected from this commercial package: pattern search methods, genetic algorithms, linear and quadratic programming, least squares and binary integer programming, and many more. The attractiveness of this toolbox is also related to the possibility of implementing multi-objective optimisation processes and Pareto Front curves.

OPTIMISATION METHOD

The word „optimisation' is frequently used and usually implies carrying out a trade off among striving targets. Optimisation processes can be applied to most of the engineering problems though there is not any universal optimisation method, i.e. each particular problem usually requires a specific optimisation method in order to efficiently determine the global optimum. To determine the most suitable method, it is fundamental to identify the characteristics of the problem under analysis. The aircraft trajectory optimisation problem can be classified as multi-objective, constrained, dynamic, deterministic (mostly) and real valued [17]. Accordingly, three main groups of optimisation methods can be identified: Hill Climbing Methods, Random Search Methods and Evolutionary Methods. In the trajectory optimisation processes carried out in this wok, the functions relating input (design variables) and output (target function as time, fuel, emissions etc) are considered as being unknown, highly non-linear, non-smooth and non-differentiable. Hence, Evolutionary Algorithms were selected due to their suitability for this particular type of optimisation problems.

Evolutionary algorithms are based on processes that happen in natural environments according to Darwin's principles of species evolution. These methods try to imitate biological structures and processes in order to solve engineering problems. Among the features reproduced in a mathematical environment, there are: reproduction, mutation and crossover. Adopting these strategies, the algorithm tries to follow the principles of natural selection where the most suitable individuals produce on average more descendants and continue to assure the survival of the species adapting to a mutating environment. Among the most known algorithms in scientific application, there are the Genetic Algorithms (GAs) that are based on the principle of natural selection. They use only the value of the objective function and do not require an initial guess. Another advantage is related to the capability of finding global minima thereby escaping local minima.

The MatLab Optimization Toolbox v4 [12-13] was chosen to perform the optimization process. A careful evaluation of the objective function properties plus the type of constraints applied reduced the search for the most suitable algorithm to GAs. In GAs usually the objective function is called „fitness function' since the purpose of the algorithm is to improve the fitness of the population thereby achieving an optimal point. The particular implementation of GAs utilised does not use the raw fitness score to classify the individuals in the population but scales the scores in order to create a better selection method called „Rank'. Individuals with a lower score than the average are maintained for the next generation as they are more suitable to survive. For this reason, they are called „elite' individuals. The new population is formed from these individuals plus the children created in the previous iteration [13]. In the present case, an initial population of 100 individuals was selected against the default value of just 20 individuals.

CASE STUDIES

The flight mission analysis was performed dividing the mission into three main phases: climb, cruise, and descent. Several simulations were run in order to evaluate the impact of the climb phase on the overall mission. The underlying concept for this choice is related with the thrust settings required during the climb phase. To enable the aircraft to climb from SL up to 35,000ft, if maximum thrust settings are required, then maximum TET values are experienced. The first outcome of these thrust settings is a substantial production of combustion products as NOx whose concentration depends strongly upon TET. Hence, the climb phase analysis is fundamental in order to minimise aircraft emissions along the flight profile. The take-off and approach phases are not considered in this work as their constant contribution can be readily implemented introducing an additional factor to the objective function value (time, fuel or NOx). Hence, the reference altitude for the beginning of the climb or approach phase is fixed at 457m (1,500 ft). The last segment altitude in the climb analysis is fixed as well to the cruise flight level of 10,668m (35,000ft).

Three main objectives are evaluated: flight time, fuel burned and NOx emitted. In order to minimize the impact of aircraft operations on the environment the two objectives of main concern are the amount of fuel burned, which provides a measure of the complete combustion products (CO_2 and H_2O), and nitrogen oxides. In order to produce reasonable trajectories flight time is evaluated as well.

CONSTRAINED CLIMB DESCRIPTION

The constrained climb analysis features a four variables optimisation (two altitudes and two EAS) alongside two set of constraints on the admissible values of altitude and speed. The term constrained is used to highlight the introduction of operational constraints on speed limits below given altitude levels. Additionally, an implicit constraint on the Mach number to be flown from the aircraft in the last flight segment is implemented in the framework.

The baseline trajectory is divided into four segments with a cumulative range of 190 km. The reference climb is defined in terms of initial and final altitude of every segment plus the relative EAS or Mach number at the beginning of the segment as shown in Table 5. In the first segment, the aircraft performs a climb at constant EAS1 (speed mode 1) starting from the minimum fixed altitude $AL1_i$ until reaching $AL2_i$ that is a fixed as well. Then, in the second segment, it accelerates from EAS1 to EAS3 maintaining a constant flight level. In the next segment, the aircraft starts to climb again at constant EAS3 up to an altitude ($AL4_i$) where M 0.8. In the last segment the climb is now performed at constant Mach number (speed mode 2) up to the fixed cruise altitude $AL4_f$. Hence, the variables to be optimised are the altitude at the beginning of the third and fourth segment plus the EAS in the first and third climb segments.

Table 5 : Constrained climb baseline definition

Segment No.	AL_i [m]	AL_f [m]	Speed Mode	EAS_i - M_i [m/s]	R [km]
1	457	3,048	1	109.0	20
2	3,048	3,048	1	109.0	10
3	3,048	7,724	1	164.6	60
4	7,724	10,668	2	0.800	100

The baseline proposed can be regarded as a typical climb schedule as in first part of the climb the aircraft is flown according to EAS constraints, while in the second part according to the Mach number specified as cruise M. Flying at constant EAS in the first part of the climb would allow the Mach number to increase up to a value close to the cruise Mach number. Hereon, the aircraft would continue to climb at constant Mach number up to the cruise altitude, resulting in this way in EAS to decrease.

The reference EAS1=109m/s is an average between the stall speed of 89m/s and the maximum admissible speed of 128.6m/s. This maximum value corresponds to a speed constrain of 250kts below 10,000ft. The $AL3_i$ that in the baseline trajectory is kept constant to $AL2_i$ can reach the maximum admissible of 4,365m, in case the aircraft would climb at maximum FPA. The EAS3 is initially set equal to 164.6m/s to provide M=0.8 in the last segment at an altitude of $AL4_i$=7,724m. This value corresponds to a reference speed of 320kts to be achieved at the beginning of the third segment. EAS3 can vary in the range 260-430kts that correspond to 133.75-221.21m/s. The altitude at the beginning of the fourth segment, $AL4_i$, can vary in the range 3,048-10,668 m.

To preserve the climb schedule, an implicit constraint is set on the Mach number at the beginning of the fourth segment. The EAS3 must be selected in such a way to provide M=0.8 at the beginning of the fourth segment, i.e. at an altitude equal to $AL4_i$. The Mach number is checked at the end of every iteration and a penalisation value is assigned to the fitness of solutions that account for a deviation from Mach 0.8 higher than ±0.5%. Hence, the Mach number must be within the range 0.796-0.804 to avoid the optimisation process to discard the solution.

CONSTRAINED CLIMB RESULTS

Three different objectives were evaluated in the optimisation of the constrained climb profile: flight time, fuel burned and NOx emitted. Figure 2 provides a summary of the optimised flight profiles corresponding to the minimisation of the three different objectives.

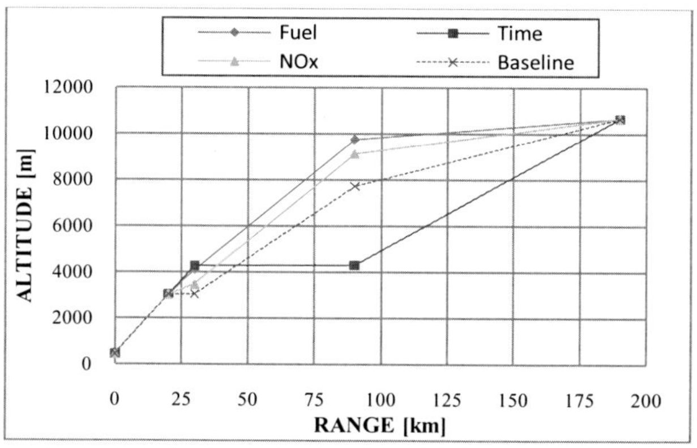

Figure 2: Constrained climb – Optimum trajectories

The optimisation process produced interesting results and provided different solutions for each of the selected objectives, highlighting once again the need for a trade-off among objectives. Table 6 shows percentage variations of the three objectives relative to the baseline trajectory.

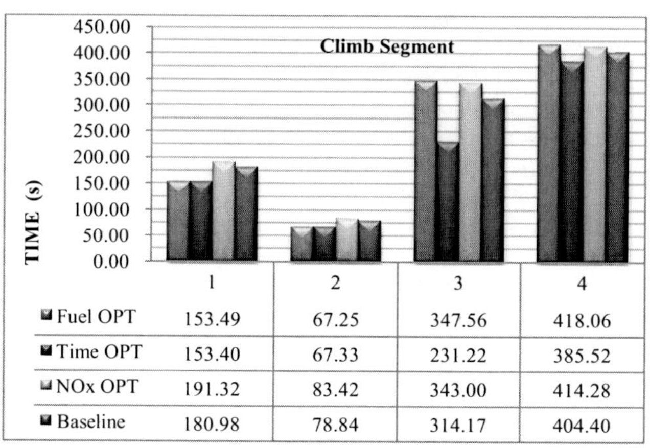

	1	2	3	4
Fuel OPT	153.49	67.25	347.56	418.06
Time OPT	153.40	67.33	231.22	385.52
NOx OPT	191.32	83.42	343.00	414.28
Baseline	180.98	78.84	314.17	404.40

Figure 3: Constrained climb – Flight time per segment

With the adoption of specifically optimised flight profiles it is possible to achieve reductions of about 15% in flight time, almost 7% in fuel burned and 10% in NOx emissions.

In the time minimisation case, the aircraft tries to maximize the TAS (to fly as fast as possible) thereby postponing the second part of the climb to the last segment. As the TAS is proportional to the EAS and inversely proportional to the ambient density, the aircraft strategy is to fly close to the maximum EAS keeping a low flight level for as long as possible before climbing rapidly to the final altitude. Therefore, the EAS1 value matches the maximum value, while the EAS3 is close to the upper bound and corresponds to an initial TAS in the fourth segment of 260m/s. As shown in Figure 5, the last segment gives the maximum contribution to the total amount of NOx production. This is a result of the relatively high thrust setting utilised to fly this particular segment and the large amount of fuel burned. However, the maximum TET>2,000K is reached in the second segment due to the high acceleration occurring in this segment, which in turn results in a huge amount of thrust required, and consequently a relatively high TET value. As this TET value was too high, a control routine was introduced in the framework to limit the value to 1900K in the complete flight profile optimisation.

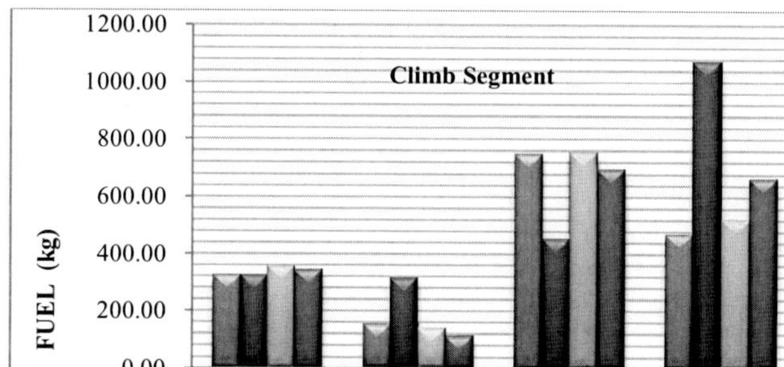

	1	2	3	4
Fuel OPT	321.84	152.24	747.25	469.35
Time OPT	321.77	316.20	452.58	1072.06
NOx OPT	353.40	138.05	752.48	522.09
Baseline	341.95	113.37	694.50	661.98

Figure 4: Constrained climb – Fuel burned per segment

In the fuel minimisation, the aircraft tries to fly at the highest possible levels in order to minimise the aircraft drag, which mainly depends upon altitude (i.e., air density) and speed. The air density is inversely proportional to the altitude while the TAS depends on a combination of EAS and flight level through the density. While the EAS1 is the same as in the minimum time case, the EAS3 is the lowest one. The fuel optimised trajectory is characterised by a quick climb first segment followed by a second segment in which a small acceleration occurs at maximum climb angle. To minimise the fuel consumption in the first segment the optimiser suggest to maximise the EAS as the fuel burned is obtained multiplying the time elapsed by the fuel flow. As the latter depends on the thrust setting, the choice is to couple a small flight time with a thrust setting higher than in the minimum NOx case. The latter case performs a slower climb in the first segment that results in higher fuel consumption because of the longer flight time involved even though the thrust setting is lower.

Table 6: Constrained climb - Objective functions variations as to baseline trajectory

	Flight Time		*Fuel Burned*		*NO_x Emitted*		*DECISION VARIABLES*			
	S	%	KG	%	KG	%	EAS1 (m/s)	$AL3_i$ (m)	EAS3 (m/s)	$AL4_i$ (m)
Baseline	978.39	-	1811.79	-	21.21	-	109.00	3048.00	164.6	7724
MINIMUM TIME	837.47	-14.40	2162.61	19.36	76.04	258.52	128.60	4282.70	209.05	4313.8
MINIMUM FUEL	986.36	0.81	1690.69	-6.68	21.28	0.33	128.60	4124.90	140.86	9766.7
MINIMUM NO_x	1032.03	5.48	1766.02	-2.53	19.16	-9.69	100.70	3481.40	147.64	9146

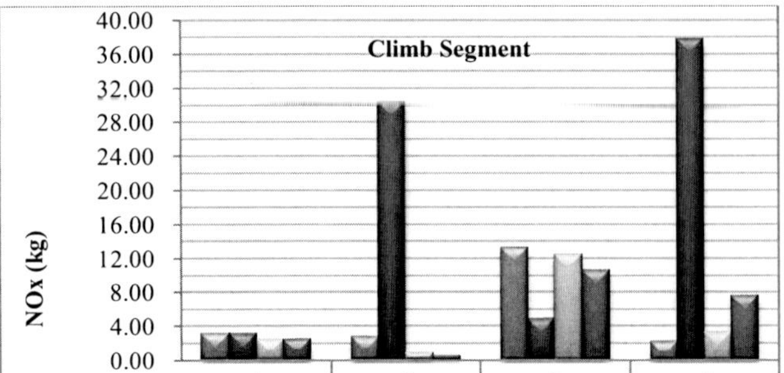

	1	2	3	4
Fuel OPt	3.16	2.76	13.21	2.16
Time OPT	3.16	30.31	4.84	37.73
NOx OPT	2.46	0.91	12.48	3.31
Baseline	2.53	0.51	10.58	7.59

Figure 5: Constrained climb – NOx emitted per segment

To understand properly the strategy adopted in the NOx minimization problem it is important to address the factors that influence the NOx production. The mass (in kg) of NOx emitted are evaluated as the fuel burned multiplied by the Emission Index, EINOx:

$$kg[NO_x] = (FF * TIME) * EINO_x$$

The fuel burned depends upon many variables including flight time and thrust setting, while the EI depends mainly on TET. Values of TET in excess of 1800K result in exponential increases in NOx emissions. Accordingly, the strategy suggested by the optimisation process seeks to maintain a low TET profile, i.e., a low thrust produced by the engine in all the segments. Hence, TET is considered to be the main driver for the NOx production as flight times and fuel burned play a secondary role in NOx optimised trajectories. Consequently, the NOx optimised climb presents a constant TET profile, with the highest peak in the third segment of about 1550K. Temperatures in excess of this value are never reached to avoid entering in the region where NOx production becomes exponential. Due to the complexity of the NOx optimization problem, it is not possible to identify the solution simply summing the contributions of the single segments as the trajectory flown in each segment is affected from the previous segment variable selection. The aircraft flies slower in the first segment while in the third and fourth segment it is slightly faster than the minimum fuel case (Table 6). The optimised trajectory features a climb phase in the second segment that produces more NOx than the baseline but the strategy is to start the third segment climb from a higher flight level and speed. The NOx savings in the third phase definitely compensate the extra amount produced in the second phase as to the baseline trajectory. Nevertheless, the higher EAS3 results in a TAS3=175.7m/s that is close to the fuel optimization value (TAS3=173.4m/s) as in the latter the aircraft performs a higher climb.

COMPLETE FLIGHT PROFILE DESCRIPTION

The simplified flight profile considered in this second optimisation case study is divided in three phases: climb, cruise, descent. Four operational constraints apply to the optimisation process that deals with six variables: three velocities - EAS1, EAS3 and EAS8 - plus three altitudes - $AL3_i$, $AL4_i$, $AL7_i$. Since the overall range involves 880 km, the mission can be classified as a „short range' mission. The FPA in the climb/descent phase is limited to the maximum value of ±7.5°. Likewise the constrained climb analysis, an implicit constraint on the Mach number at the end of the third segment is introduced.

Table 7: Complete flight profile baseline definition

Segment No.	AL_i [m]	AL_f [m]	Speed Mode	EAS_i - M_i [m/s]	R [km]
1	457	3,048	1	109.0	20
2	3,048	3,048	1	109.0	10
3	3,048	7,724	1	164.6	160
4	7,724	7,724	2	0.800	230
5	7,724	7,724	2	0.800	230
6	7,724	3,048	1	164.6	140
7	3,048	3,048	1	109.0	20
8	3,048	457	1	109.0	70

The baseline trajectory is divided in eight segments with a cumulative range of 880km. Table 7 shows the baseline profile defined in terms of initial and final altitude of each segment plus the speed at the beginning of the same. The climb phase features a climb at constant EAS1 from $AL1_i$ to $AL2_i$ in the first segment, an acceleration phase from EAS1 to EAS3 at levelled flight in the second segment, and a constant EAS3 climb up to $AL4_i$ in the third segment. The cruise is made of two segments flown at levelled flight ($AL4_f$=$AL5_f$) at constant Mach number M=0.8. The descent phase that begins with the final cruise altitude features a constant speed descent until reaching $AL6_f$ in the sixth segment. Then, the aircraft decelerates to EAS8 keeping a levelled flight in the seventh segment and descents again in the last segment at constant EAS8 until reaching the minimum altitude. The selection of EAS3=164.6m/s provides a Mach number M=0.8 at the assigned flight level $AL4_i$=7,724m. Hence, the optimization of the $AL4_i$ and EAS3 is constrained in such a way to fly the aircraft at a constant fixed Mach number in the fourth and fifth segment. Outbound values are penalized with a high value of the fitness function as specified in the former case analysis. Additionally, the EAS6 will be equal to the EAS3 as the flight level at the beginning of the descent (sixth segment) and the end of the climb (third segment) is the same. The EAS1 and EAS8 selection is intentionally sub-optimal and the two speeds can vary in the range 89-128.6m/s, where the lower limit represents the stall speed while the upper limit the EAS constraint of 250 knots below 10,000 feet. The EAS3 can vary in the range 117.1-184.6m/s to allow a Mach number M=0.8 at $AL4_i$ with an altitude range 6,096-12192m (20,000-40,000ft). Finally, the AL3 and AL8 can vary between 3,048m and 12,192m (10,000-40,000ft). Two additional features are introduced in the framework for the complete profile analysis: the limitation of the maximum TET to 1,900K and the computation of the thrust at „idle' for the descent phase. The climb phase optimization is expected to provide results in agreement with the constrained climb analysis.

COMPLETE FLIGHT PROFILE RESULTS

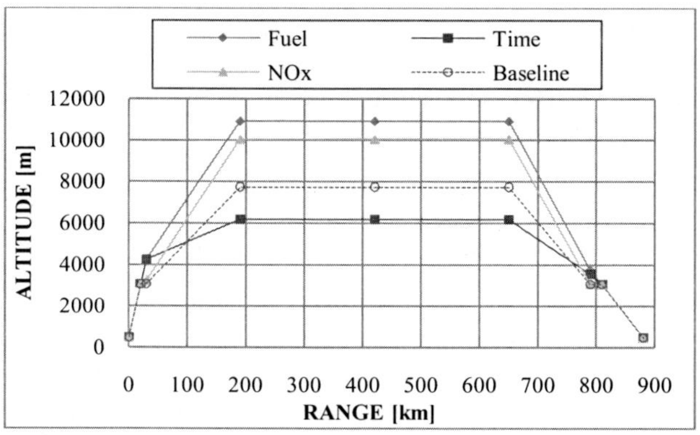

Figure 6: Complete flight profile - Optimum trajectories

The optimized trajectories in the complete flight profile analysis are presented in Figure 6 together with the baseline trajectory defined previously. The objective values relative to each trajectory are given in Table 8 with the percentage deviation from the baseline flight profile plus the relative optimized variables. The results obtained for the complete flight profile optimisation are in agreement with the trend observed in the constrained climb case. Consistent reductions in the fuel burned (-17.5%) and NOx emitted (-31.5%) were made possible with the selection of specifically optimized trajectories. The minimum time trajectory achieves a reduction in the flight time of about 8% at the expense of a 100%+ NOx production and 14% fuel consumption increase. The aircraft tries to climb at the maximum admissible EAS to perform the two cruise segments at M=0.8 close to the minimum cruise flight level of 6,096m. The altitude at the beginning of the third segment, $AL3_i$, is limited by the maximum TET allowed of 1900K (Figure 10), as the thrust available in the second segment is also used to accelerate from 128m/s to 184m/s. The cruise altitude is chosen in order to maximize the TAS by flying at the maximum EAS3 at the minimum altitude required (Figure 6). In the descent phase, the EAS6 equals the EAS3 while the EAS8 is once again close to the maximum of 128.6m/s. The selection of $AL7_i$ is optimized to provide a feasible descent profile as variations from this value result in a thrust required lower than the idle thrust in the sixth or seventh segment. It should be noticed that the selection of a specific climb phase affects the descent phase as the weight of the aircraft and the thrust requirements will vary as a consequence of it.

In the minimum fuel trajectory, the climb phase features an EAS1 and AL3 similar to the minimum time case while the EAS3 is optimized to provide Mach 0.8 at a cruise altitude $AL4_i$ of 10,915m. It is interesting to notice that the maximum altitude of 12,192m is not reached, although it corresponds to a lower EAS3=117.1m/s.

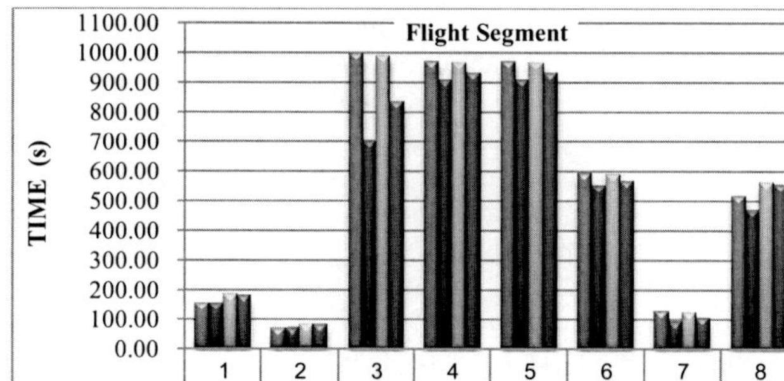

	1	2	3	4	5	6	7	8
Fuel OPT	154.29	67.68	995.13	969.59	969.59	590.94	127.62	516.70
Time OPT	153.66	67.40	700.35	906.19	906.19	551.69	90.75	468.36
NOx OPT	183.02	79.74	988.62	965.41	965.41	588.35	124.50	561.58
Baseline	180.98	78.84	835.60	929.73	929.73	566.26	104.41	552.25

Figure 7: Complete flight profile – Flight time per segment

As the aircraft flies the first segment at an EAS1 of 128m/s, it should decelerate in the second segment in order to reach the maximum altitude. Hence, this strategy would result in increasing the overall fuel consumption even if the cruise contribution would be lower. As shown in Figure 8, the aircraft consumes an amount of fuel in the third segment higher than in the minimum time case, but this extra amount is compensated from the lower fuel consumption of the fourth and fifth cruise segments. The descent phase features an $AL7_i$=3,740m that is optimised to minimise the thrust-fuel required in the sixth segment (the longest in the descent phase). The selection of EAS8=116m/s provides the minimum fuel in the eight segment. It is worth noticing that the minimum time case, with a higher EAS8=128m/s, and the minimum NOx case, with a lower EAS8=107m/s, provide a higher fuel consumption in this last segment.

Table 8: Complete flight profile – Objective functions and decision variables summary

	Flight Time		*Fuel Burned*		*NO_x Emitted*		DECISION VARIABLES					
	S	%	KG	%	KG	%	EAS1 (m/s)	EAS3 (m/s)	EAS8(m/s)	$AL3_i$(m)	$AL4_i$(m)	$AL7_i$(m)
Baseline	4177.80	-	4830.20	-	21.40	-	109.00	164.60	109.00	3050.00	7724.00	3050.00
MIN TIME	3844.60	-7.98	5491.85	13.70	49.19	129.83	128.39	184.59	128.52	4227.70	6161.50	3548.80
MIN FUEL	4391.53	5.12	3986.68	-17.46	16.23	-24.18	127.86	130.01	116.52	4230.79	10915.15	3739.71
MIN NO_x	4456.61	6.67	4193.90	-13.17	14.66	-31.51	107.79	137.96	107.19	3221.40	10037.70	3060.10

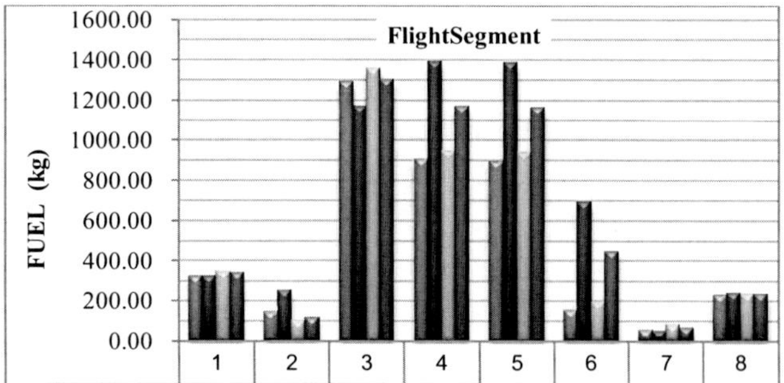

	1	2	3	4	5	6	7	8
Fuel OPT	322.65	145.59	1288.4	902.32	894.55	148.51	54.25	230.32
Time OPT	321.91	248.86	1166.7	1390.7	1386.8	692.22	45.90	238.58
NOx OPT	344.41	100.58	1357.3	945.81	937.68	192.73	82.15	233.15
Baseline	341.95	113.37	1302.5	1166.2	1162.7	445.15	66.57	231.57

Figure 8: Complete flight profile – Fuel burned per segment

In the NOx minimization case, the aircraft performs a slower climb in the first segment with a constant EAS1= 108m/s and a modest acceleration in the second segment up to EAS3=138m/s at an altitude $AL3_i$=3,221m. As shown in Figure 9, the second segment is crucial for the NOx production as the aircraft must climb and accelerate at the same time over a small range (10km). The climb variables selection for the second segment enables to produce less NOx than the minimum fuel case as well as the minimum time case. In this last case, the high acceleration coupled with the high rate of climb causes the TET to reach almost 1900K. This peak in the TET results in an exponential increase of the EINOx that, multiplied for the fuel burned, gives 17.2kg of NOx emitted only in the second segment.

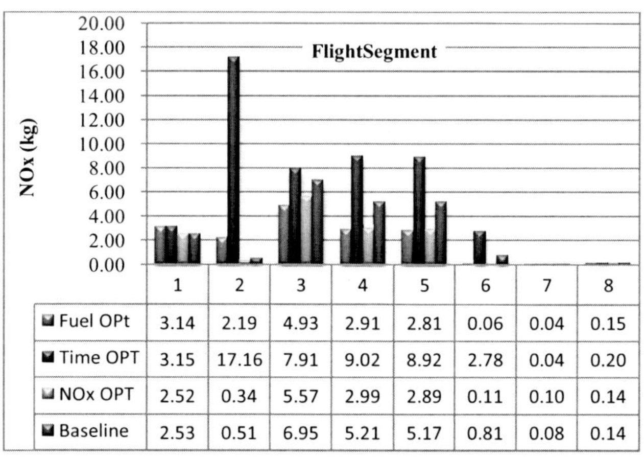

	1	2	3	4	5	6	7	8
Fuel OPt	3.14	2.19	4.93	2.91	2.81	0.06	0.04	0.15
Time OPT	3.15	17.16	7.91	9.02	8.92	2.78	0.04	0.20
NOx OPT	2.52	0.34	5.57	2.99	2.89	0.11	0.10	0.14
Baseline	2.53	0.51	6.95	5.21	5.17	0.81	0.08	0.14

Figure 9: Complete flight profile – NOx emitted per segment

Nevertheless, the NOx saved in the second segment (due to the small acceleration and rate of climb) results in a consistent amount of NOx spent in the third segment to climb up to $AL4_i$=10,037m. The cruise emissions are close to the minimum fuel case even if the EAS3 of 138m/s is higher than in the latter. The selection of the cruise flight level is a trade-off between the NOx emitted in the climb to reach the cruise and the NOx emitted along the cruise. Increasing the cruise altitude further would result in the cruise emissions to reduce even more but not enough to compensate the increase in the climb emissions. The descent phase features an $AL7_i$=3,060m close to the minimum allowed and an EAS8=107.2m/s. Reducing the speed further would provide a benefit in terms of NOx emitted but it is not feasible due to the TET limitations imposed (TET ≥ 1010K).

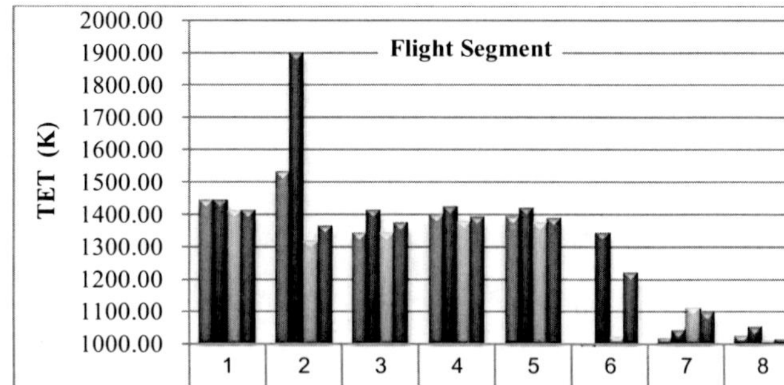

	1	2	3	4	5	6	7	8
Fuel OPT	1444.0	1529.9	1342.2	1396.9	1392.3	974.88	1018.0	1024.6
Time OPT	1444.7	1899.3	1411.6	1420.8	1419.6	1342.1	1041.3	1051.6
NOx OPT	1410.4	1318.5	1341.1	1377.0	1372.6	1020.2	1111.3	1010.8
Baseline	1411.5	1362.8	1372.3	1390.0	1388.6	1220.5	1095.5	1012.8

Figure 10: Constrained flight profile – TET per segment

The results obtained in this optimisation study highlight two different trajectories eligible to minimize the impact of aircraft operations on the environment. Figure 11 presents the percentage reductions of pollutants emitted as to baseline trajectory. The minimum fuel trajectory provides reductions of CO_2 and H_2O in excess of 18%, while the NOx reduction is close to 24%.

The NOx optimized trajectory performed well as the amount of CO_2 and H_2O emitted is reduced by 13%, which is 5% less than the former. In this case, the oxides of nitrogen show a reduction 8% higher than the minimum fuel case close to 32%. Eventually, it would be possible to perform a double-objective optimization to identify trajectories that present acceptable combinations of the three pollutants via a 2D Pareto Front. According to the requirements of each specific mission it would then be possible to select the most appropriate trajectory. Nevertheless, it should be noticed that the percentage values do not represent a reliable measure of the effects produced by the pollutants. The amounts of carbon dioxide and water vapour produced are four orders of magnitude higher than the oxides of nitrogen. Even assuming that the oxides of nitrogen have the same effect on the atmosphere as the carbon dioxide, the huge difference between the amounts of the two pollutants would give priority to the latter. Hence, at first glance it could be possible to select the minimum fuel trajectory as a possible „greener' trajectory, i.e. the one to fly in order to minimise the impact of aircraft operations on the environment.

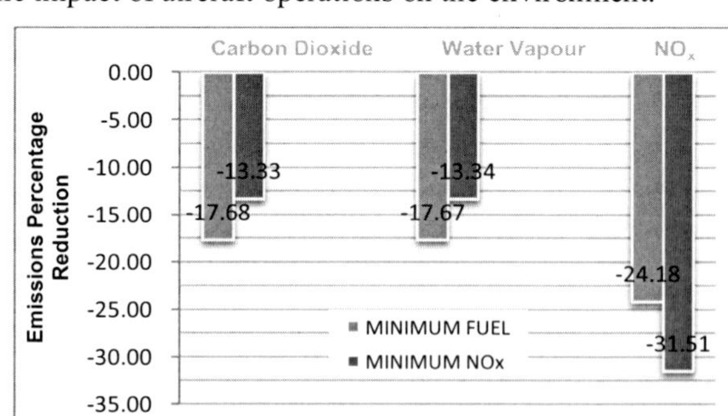

Figure 11: Complete flight profile - Emissions relative to the baseline trajectory (expressed in %)

Nevertheless, further work will be done in the future to increase the flexibility of the methodology proposed thereby implementing additional models in the framework to take into account new factors. For instance, a noise model would be really useful to better understand the noise produced in the case of „greener' trajectories. Likewise, a model to simulate weather conditions and contrail formation could increase the accuracy of the simulation. The simulation of the take-off and approach phase will be added as well to provide a better estimate of the total amount of pollutants emitted along the flight path. Finally, the flight path will be divided into a far greater number of elements in order to determine solutions which represent more accurately actual aircraft trajectories.

CONCLUSIONS

In this work, a multi-disciplinary simulation framework which allows the modelling of the performance of aircraft-engine systems plus the prediction of the pollutants emitted along a given flight profile has been developed and implemented. Three CU in-house programs have been deployed for this purpose: APM for the aircraft performance, TurboMatch for the engine performance, and Hephaestus for the gaseous emissions prediction. The simulation framework was combined with a commercial available optimizer to create an „optimization framework'. By doing so, it was possible to investigate the potential of „greener' trajectories as a future mitigation policy for the minimisation of the environmental impact of commercial aircraft operations. The optimisation study was focused on the minimization of three different objectives: flight time, fuel burned and NOx emitted. The CO_2 and H_2O emissions are directly proportional to the fuel burned as they are final products of the combustion process. To understand the strategy adopted by aircraft for the minimisation of the three objectives, the simulation framework was deployed on a climb schedule with operational constraints. A detailed comparison of the time and fuel minimisation trajectories highlighted that the optimization with respect to one criterion lead to solutions that exhibit a poor performance with respect to the other. However, the minimum fuel trajectories showed a trend of the design variables close to the minimum NOx case. Finally, to evaluate the pollutants emitted along a hypothetic mission the complete flight profile was considered. Operational constraints were introduced as before, and the flight path was divided in eight elements - three for the climb/descent and two for the cruise. Significant reductions in the pollutants emitted were possible for specifically optimised trajectories. Depending on the relevance of the two criteria – minimum fuel and minimum NOx, the selection of a unique greener trajectory will be up to the final user. In conclusion, the developed optimisation framework proved to be very useful for identifying the characteristic features of aircraft trajectories with a minimum environmental impact.

ACKNOWLEDGMENTS

The research leading to these results has received funding from the European Union's Seventh Framework Programme (FP7/2007-2013) for the Clean Sky Joint Technology Initiative under grant agreement n° CSJU-GAM-SGO-2008-001. Additionally, the authors would like to extend their gratitude to Mr Richard Long who developed the aircraft performance model utilised in this simulation framework.

REFERENCES

1. Sir N. Stern, 2006, Stern Review: *The Economics of climate change.*
2. PARTNER, http://web.mit.edu/aeroastro/partner/index.html, (accessed 8th March, 2010).
3. ACARE, 2002, *Strategic Research Agenda*, Volume 1.
4. ACARE, 2004, *Strategic Research Agenda*, Volume 1.
5. Clean Sky JTI (Joint Technology Initiative), http://www.cleansky.eu, (accessed on the 9th March, 2010).
6. Report of Group of Personalities, January 2001, *European Aeronautics: a vision for 2020.*
7. Antoine, N.E., Kroo, I.M., 2005: Framework for Aircraft Conceptual Design and Environment Performance Studies, AIAA Journal, Vol.43, No.10.
8. P. Brooker, 2009, Aviation and climate change: II - Air traffic management and aviation non CO2 issues, Air Traffic International 2009, 26-30.
9. Report of the Science & Technology Sub-Group, 2005: Air Travel – Greener by Design, Mitigating the environmental Impact of Aviation: Opportunities and Priorities. Published by the Royal Aeronautical Society, UK.
10. Rogers, H.L., Lee, D.S., Raper, D.W., Foster, P.M. deF., Wilson, C.W. and Newton, P.J. *The impacts of aviation on the atmosphere*, Aeronautical Journal, October 2002, 104, (1064), pp521-546.
11. Celis, C., Long, R.F., Vishal, S., Zammit-Mangion, D., Zolata, H., Singh, R., and Pilidis, P., 2009: *On trajectory Optimisation for Reducing The Impact of Commercial Aircraft Operations on the Environment*, ISABE-2009-1118, School of Engineering, Cranfield University, UK.
12. MATLAB®, October 2009, *Optimization Toolbox TM User's Guide 4,* The MathWorks, Inc., www.mathworks.com.
13. MATLAB®, March 2009, *Genetic Algorithm and Direct Search Toolbox TM 2 User's Guide,* The MathWorks, Inc., www.mathworks.com.
14. Gas Turbine Engineering Group, 2008: *The TURBOMATCH Scheme*, TURBOMATCH, User's Manual, School of Engineering, Cranfield University.
15. Long, R.F., 2009: *Tal Linja Aircraft Performance Model Version 0.1*, Department of Aerospace Engineering, Cranfield University, UK.
16. Celis, C., Moss, B., Pilidis, P., 2009: Emission Modelling for the Optimisation of Greener Aircraft Operations. In: *Proceedings of ASME Turbo Expo 2009: Power for Land, Sea and Air,* GT2009, Orlando, Florida, USA, June 8-12 2009, GT2009-59211.
17. Schwefel, H.P., 1981, *Numerical Optimization of Computer Models*, John Wiley, Chichester, UK.

Proceedings of the ASME 2010 International Mechanical Engineering Congress & Exposition
IMECE2010
November 12-18, 2010, Vancouver, British Columbia, Canada

IMECE2010-38530

PRELIMINARY DESIGN OF A HYBRID PROPULSION SYSTEM FOR HIGH-ENDURANCE UAV.

Roberto Capata
Dept. of Mechanical and Aerospatial Engineering
University of Roma "Sapienza", Italy

Luca Marino
Dept. of Mechanical and Aerospatial Engineering
University of Roma "Sapienza", Italy

Enrico Sciubba
Dept. of Mechanical and Aerospatial Engineering
University of Roma "Sapienza", Italy

ABSTRACT

In recent years, a renewed interest in the development of unmanned air vehicles (UAV's) led to a wide range of interesting applications in the fields of reconnaissance and surveillance.

In these types of mission, the noise produced by propeller driven UAVs is a major drawback, which can be partially solved by installing an electric motor to drive the propeller. The evolution of high performance brushless motors makes electric propulsion particularly appealing, at least for small and medium size UAVs. All electric propulsion systems developed to date are though characterized by the limited range/endurance that can be obtained with a reasonably sized battery pack. In this paper we propose a hybrid propulsion system based on recently developed, high efficiency micro-turbines which can be used to power an electric generator. The UMGT is under evaluation in our department, to achieve the optimal configuration and performances. For this scope a new compact regenerative combustion chamber has been developed and several tests has been carried out, with the aim to reduce weight and dimension and increase vehicle payload. In a high range/endurance mission the ultra-micro-turbine can provide the energy required for the cruise phase (the so-called "transfer to target"), while in the final approach, in which a quiet flight attitude is a demanding item, the battery pack drives the motor.

The mission requirements adopted in the preliminary aircraft design presented here consist mainly of a long endurance (> 12 hours) step, with a cruise speed of 33.3 m/s and a dash speed of 45 m/s at an altitude of 5000 meters. The maximum take-off weight is 500 N, with a payload of 80 N.

Under the above assumptions, a flying wing configuration for the UAV was defined, with a length of 1.6 meters and a span of 2.5 meters. A system of elevons assures the pitch and roll motion while a double vertical tail, in which a pusher propeller is lodged, guarantees the yaw stability and control.

INTRODUCTION

In recent years, a renewed interest in the development of unmanned air vehicles (UAV's) led to a wide range of interesting applications in the fields of reconnaissance and surveillance.

In particular, the attention is being focused on the propeller as propulsion device, and in these types of mission, the noise produced is one of the most disturbing problems. The limits that many airports in the world impose on the noise level, and the detection of military UAV's as a consequence of their acoustic signatures, are examples of one of the possible limitations related to the adoption of a propeller. In addition to the design of a quiet and efficient propeller, the noise problem can be partially solved by installing an electric motor to drive the propeller. The evolution of high performance brushless motors makes electric propulsion particularly appealing, at least for small and medium size UAVs. All electric propulsion systems developed to date are though characterized by the limited range/endurance that can be obtained with a reasonably sized battery pack.

The choice of a hybrid propulsion system can overcame this last deficiency and here we propose a solution based on recently developed, high efficiency micro-turbines which can be used to power an electric generator. The possible strategy to extend the capabilities of a high range/endurance mission of the

UAV, foreseen the ultra-micro-turbine as the primary source of energy for the cruise phase (the so-called "transfer to target"), while in the final approach, in which a quiet flight attitude is a demanding item, the battery pack can drive the motor.

NOMENCLATURE

A	Aspect Ratio
CC	Combustion Chamber
C_D	Drag Coefficient
C_L	Lift Coefficient
DMA	Department of Mechanics and Aerospatial Engineering
GT	Gas Turbine
S	Wing Surface
T	Temperature
UMGT	Ultra Micro Turbo Group
UAV	Unmanned Aerial Vehicle
Λ	Leading Edge Sweep Angle
b	span
b_t	tail span
c_r	root chord
c_t	tip chord
λ	taper ratio
ρ	density

1. AERODYNAMIC AND PERFORMANCE ANALYSIS [1,2]

The main target of this section is the aerodynamic analysis of the preliminary designed UAV, to size the power plant. The aircraft conceptual design here presented, relies on the mission requirements, which mainly consist in a long endurance (>12 hours) cruise phase at an average speed of 33.3 m/s and an operative altitude of 5,000 meters. In addition a dash speed is 45 m/s is mandatory. The maximum take-off weight is 500 N, with a payload of 80 N.

Under the above assumptions, a flying wing configuration for the UAV was defined, with a length of 1.6 meters and a span *b* of 2.5 meters. A system of elevons assures the pitch and roll motion while a double vertical tail, in which a pusher propeller is lodged, guarantees the yaw stability and control. The choice of a flying wing, reduced the drag of the fuselage, but this configuration could present the drawbacks of a more difficult trim and, in some cases, a reduced useful volume for the payload. Later we will show that, for the characteristics of the payload and the propulsion system, the cited limitations are easily avoided.

The wing has a surface $S = 2.37$ m^2, a taper ratio $\lambda = c_t/c_r = 0.2$ and an aspect ratio $A = b^2/S = 2.64$.

The leading edge sweep angle Λ is 45.9 degree.

To these parameters correspond a root chord c_r=1.6 m, and a tip chord c_t=0.3 m.

A 3D sketch of the geometry of the wing is reported in Fig.1 The twin tail is made by two identical trapezoidal surfaces with area 0,2 m^2, a span b_t=0.4, a root and tip chords equal to 0.8 meter and 0.2 meter, respectively. The aspect ratio of the tail is 0.8, its taper ratio 0.25 and the leading edge sweep angle is 56 degree.

The low mean value of the Reynolds number suggest the choice of low speed airfoil for the wing, in particular the NASA/LANGLEY LS(1)-0417 (GA(W)-1) and NASA/LANGLEY LS(1)-0413 (GA(W)-1), for the root and the tip section, respectively, show a good behaviour in term of aerodynamic efficiency and pitching moments.

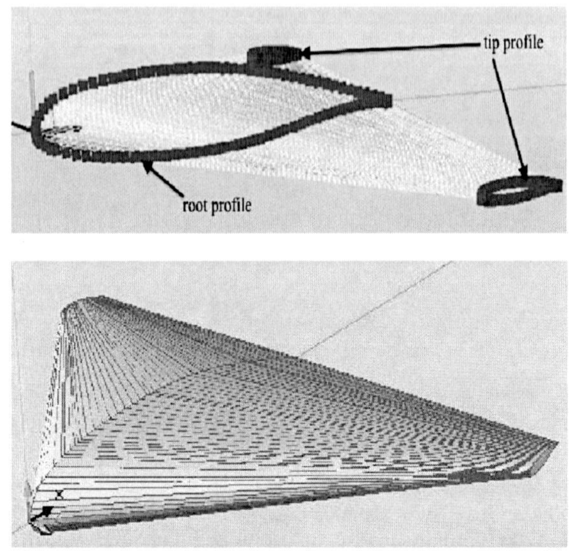

Figure 1: Geometry of the wing

From the geometrical and aerodynamic features of the aircraft the lift and drag characteristics can be summarized with the polar equation which gives the drag coefficient distribution as a function of the lift coefficient.

For the UAV here presented, and at the operative condition described, we have

$$C_D = 0.012+0.18\, C_L^2 \qquad (1)$$

We remind that, in level flight condition, $C_L=W/(1/2\, \rho\ V^2 S)$ and $C_D= D/(1/2\, \rho\ V^2 S)$.

The following tables summarize the main parameters which drive the design the propulsion system.

Table 1: Cruise speed aerodynamic and propulsion characteristics

C_L	C_D	Drag (N)	Power (W)
0.54	0.064	59	2476

Table 2: Dash speed aerodynamic and propulsion characteristics

C_L	C_D	Drag (N)	Power (W)
0.29	0.027	48.8	2636

The mission requirements (Italian Army HQ) fix, for the volume necessary to the payload, a box of sizes

(0.5 m x 0.5 m x 0.2 m). The maximum thickness of the wing, at the longitudinal symmetry plane, is about 0.272 m. A check on the spanwise decrease of the thickness of the wing, guaranties enough space to accommodate the payload box. The remaining volume is then available to fit the propulsion plant, the fuel, the flight control and guidance system.

2. THE ULTRA-MICRO TURBO GROUP AND ITS COMPONENTS

2.1 THE ULTRA-MICRO TURBO GROUP COMPONENTS [5,6,9]

This section describes the general engineering characteristics of the ultra-micro turbo group (UMTG) object of the research. The compressor and turbine that will be analyzed and simulated in detail in several previous IMECE papers and , here, are simply presented from the perspective of a mechanical engineer. The 2.5 kW UMTG is composed of a compressor, an air pre-heater, a combustion chamber and a turbine, that together compose the thermal section; a double effect (reversible) electrical engine for the electric part; a shaft and the bearings for the mechanical, and obviously all the auxiliaries components required, as valves, controllers, and ducts, including the management unit. For safety reasons the fuel tank, a gas cylinder, is, in this preliminary phase, external to the metallic UMTG enclosure.

2.1.1 AIR PRE-HEATER [9]

The air pre-heater and the combustion chamber are components entirely designed, developed, patented and manufactured by the DMA of University La Sapienza-Roma. They have been conceived in a unique body: the air pre-heater spirals along the outer face of the cylindrical combustion chamber. Doing so it is possible to achieve many advantages: the pre-heater is pretty compact, it gives an optimal space solution, and it also performs a thermal insulation, providing a bonus to its primary function and helping to prevent a dangerous over heating inside the metal box.

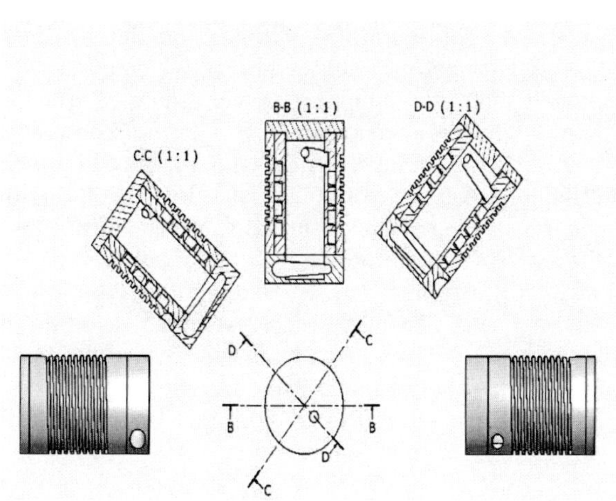

Figure 2: Cad pictures with a detailed view of the pre-heater.

The previous cad figures (See figure 2) show the concept of pre-heater spiralling around the cylindrical combustion chamber, as can be appreciated in the following photos (See figure 3).

As shown by the data obtained by the process simulator confirmed by several experimental tests, the exhausted gases have still a large enthalpic content (1278 K). The flow exiting from the compressor increases its temperature by a Δ T= 393 K, but the stack gases are too hot to be delivered in the environment for two reasons: the efficiency loss and the dangerous hot spot nearby the exhaust pipe. This is a problem has been solved inserting a duct spiralling around the exhausted pipe. This solution allows a large increase in the efficiency and keep under control the over heating danger.

Figure 3: Photos of the spiralling pre-heater.

2.1.2 COMPRESSOR

The compressor working in this turbo group has been modified (rebuilt under permission by the Garrett® [11]) and represented in a previous paper [6, 9] The overall blades number is 12, but just 5 of them are completely developed, the other 6 are splitter blades. The blades are 0° overlapped and generate a radial flow outlet, with an axial flow inlet, engaged by the inducer presents in the complete blades. The outlet velocity is a little backward, because of the backward bladed small angle, a solution that should lower the maximum velocity magnitude at exit of the rotating wheel.

2.1.3 COMBUSTION CHAMBER [5,6,8,9]

As told in the air pre-heater sub-section, the whole body of the combustion chamber has been designed, patented and performed by the Mechanical and Aeronautics Department of University La Sapienza-Roma. A first order thermo-fluid-dynamics simulation has been carried out, and a more detailed study is in progress, to assess the correct thermal field and the completeness of combustion. The test has allowed to elaborate the temperature map of the CC and the correct analysis of all thermal flows

The hole on the top surface of the combustion chamber is the bay for the spark plug to activate the ignition of the air-fuel mixture. On the cylindrical external surface, the second hole from the bottom is the entrance of the air to be pre-heated, that flows in the spiralling duct and enters the combustion chamber from its upper internal side of the combustion chamber. The combustion air is injected with an intense swirl motion, in order to enhance mixing with the fuel that is in turn injected from a second port in the upper half of the cylinder. The combustion progresses and the exhausted gases exit from a port located on the opposite (bottom) side of the chamber.

Further developments of the combustion chamber are being explored, and include several possibilities: from an increased number of spiralling turns of a smaller cross area, to grooved internal surfaces. The scope of such improvements is clearly to enhance the heat exchange and provide a lower fuel consumption and a higher overall efficiency. The option of conceiving and developing a secondary air pre-heater (thus implementing in effect a two-zone combustion) is also being considered.

The overall length of the assembled and modified combustion chamber is 2.2 cm, the outer diameter 4.2 cm and the overall height is 13.0 cm and can be seen on the fig. 4

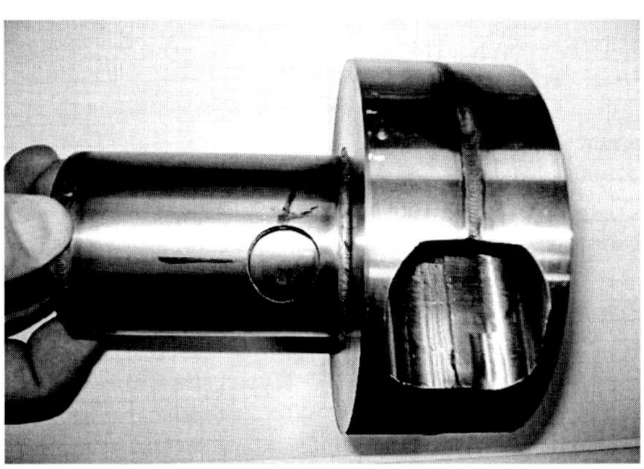

Figure 4: the realized and tested combustion chamber

2.1.4 TURBINE [11]

The turbine used in this ultra-micro turbo group is a machine manufactured by Garrett/Honeywell®. This is a centripetal radial machine, its body is cast Ni-based alloy with 9 highly twisted blades. In line with usual construction practice, the gas has a purely radial inlet and an almost axial outlet velocity, known to achieve the maximum specific work from radial turboexpanders.

2.1.5 DUAL EFFECT ELECTRIC ENGINE

Volume constraints have suggested in this phase to adopt a configuration with a single shaft for compressor and turbine: the alternative solution with a double shaft, that splits the power requested from that one delivered, demands a more complicated arrangements of the electrical devices, but has a higher efficiency due the possibility of selecting the optimal ranges for the two machines. One of the advantages of the single shaft configuration is that it permits an easier way to arrange the volume distribution and automatically ensures the matching between the machineries. There is of course a drawback: the switching of the electrical driving engine to an electrical generator when the turbine generates enough power to drive the compressor is controlled by an ad hoc electronic device on the basis of a real-time evaluation of the amperometric absorption of the engine. Furthermore, the electrical efficiency of this dual effect device will be obviously lower than this one of a single shaft machine, deteriorating the overall efficiency of the turbo group.

The currently installed electrical engine-generator is manufactured by the C.I.S. s.r.l.- located in Busto Arsizio (Varese - Italy)- and is capable of reaching the angular velocity of 18,850 rad/s (180,000 rpm). A sufficiently lower angular velocity is sufficient to allow the turbine to drive the compressor during the start up phase and so deliver useful electrical output. The overheating danger for the electrical device is averted by external cooling fins.

2.1.6 BEARINGS [10]

The choice of correct type of bearings is of vital importance for the correct operation of this machines. The angular velocity range is very high, and there are one or two classes of bearings, actually designed for ultra-micro turbo group up to a range of 500,000 rpm. The better choice is the air bearing, capable of withstanding the severe conditions of high speed and high temperature. However air bearings and aerodynamic bearings are prone to dynamic instabilities, therefore unconventional bearings such as foil bearing could present an interesting solution. In this study the bearing is of the sleeve type, highly stable and with an acceptable efficiency. The oil lubricated bushing is coupled with a thrust bearing. The journal bearing (See figure 5) consists of a steel housing with a coated, tilting and pivoting pad

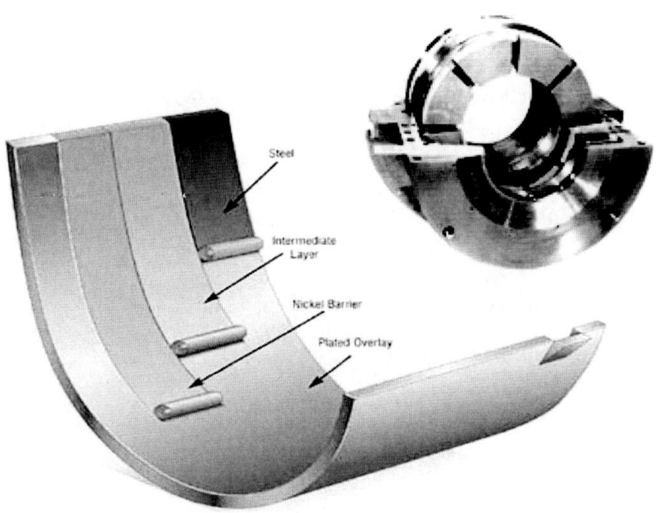

Figure 5: A radial and thrust pad bearing and the layer system in an axial journal bearing.

2.1.7 MANAGEMENT UNIT

The management unit has many tasks: i.e. to control and grant the right behaviour of every components during the start up and shut down phases. This is the actual brain of the turbo-group, ensuring the halting of the fuel injection in case of overheating or any other malfunction. It involves the control over the inlet gas valve but also over the thermocouples monitoring the temperature of various streams. This management unit has been adapted from commercially available pre-wired electronic boards, by properly reprogramming.

2.1.8. OVERALL DIMENSIONS AND WEIGHT

The adopted configuration previews parallelism between the axis of the GT group and the axis of the combustion chamber. In fact, this disposition guarantees a minimization of overall dimensions and allow an easy component installation. The dimensions of the configuration are visible in figure 6.

Then the weights of the several components of the proposed configuration will be estimated. Also in this case, the obtained values serve to giving an indication of maximum total weight of the GT-group, vital for the calculation of the payload of the UAV.

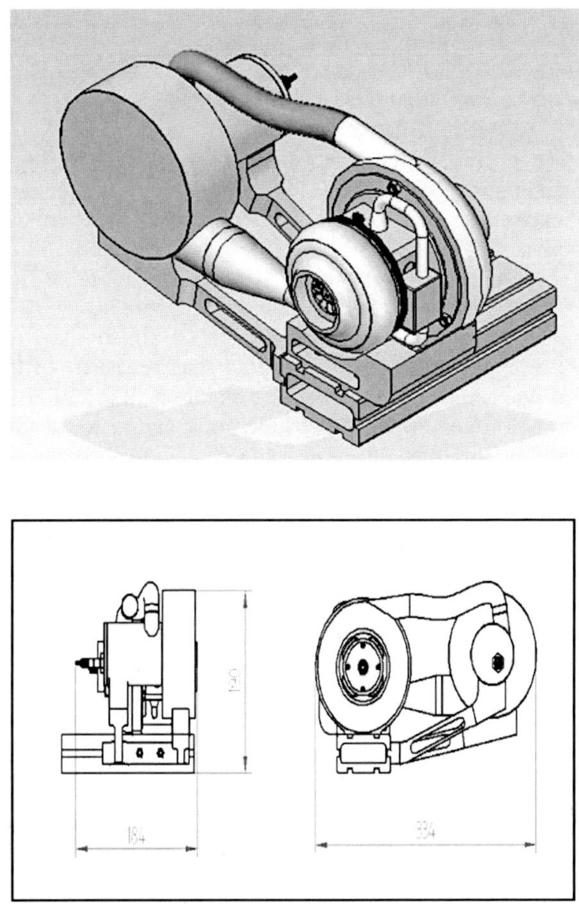

Figure 6: assembled configuration dimensions and layout (dimensions in mm).

Table 3. Configuration dimensions

Proposed assembling configuration		
L (mm)	b(mm)	h (mm)
184	334	190

Table 4: weight evaluatuion

Component	*Material*	*Weight (Kg)*
Shaft	Steel	0,03
Journal bearings	Bronze	0,02
Combustion	Stainless steel	4,23

Chamber		
Plug	Stainless steel	0,05
Compressor	Steel	0,05
GT main body (includine inlet turbine volute)	Various materials (avarage density)	3,3
Bearings (x 3)	Steel	0,006
Guides	Aluminium	0,90
Oil pump	Steel	0,5
Gear	Steel	negelecting
Tubine support	Aluminium	0,24
CC supports	Aluminium	0,49
Turbine	Stainless steel	0,06
Total: ≈ 10 kg		

3. CONCLUSIONS

In this paper a conceptual design of an high-endurance UAV and the preliminary design and experimental tests on a hybrid propulsion system are presented and discussed.

The drone has a flying wing configuration with a twin tail and a propeller located in the middle.

The aerodynamic analysis was aimed at the investigation of the lift and drag characteristics of the drone and its thrust and power requirement to fulfill the mission conditions. The choice of a flying wing gave the possibilities to reduce the wetted surface and, as consequence, its drag, improving the performance of the UAV. The volume available to fit the payload box and the propulsion plant was checked and a layout of the components is given. The UMGT design satisfis all previous requirements (payload and dimensions). The feasibility of the project has been validated and the further design step will be to realize a mock up of the drone and then a flying prototype for the performance tests.

ACKNOWLEDGEMENT

The work presented in this paper has been sponsored by the Technical Division of the Italian Army Head Quarter, within the frame of contract n. 714.

REFERENCE

[1]. O. Gur, A. Rosen, “*Design of a Quiet Propeller for and Electric Mini Unmanned Aerial Vehicles*”, J. Prop. Power, Vol. 25, pp. 717, 2009.

[2]. O. Gur, A. Rosen, ”*Optimizing Electric Propulsion Systems for Unmanned Aerial Vehicles*”, J. Aircraft, Vol. 46, pp. 1340, 2009.

[3]. R. Capata, E. Sciubba, “*Preliminary consideration on the thermodynamics feasibility and possible design of Ultra-, Micro- and Nano-gas turbines*” IJOT, vol. 9 (n. 2) , pp 81-91, June 2006

[4]. R. Capata, E. Sciubba, “*Design and Performance Prediction of a Ultra-Micro Gas Turbine for Portable Power Generation*” di R. Capata e E. Sciubba. IMECE2007-42118, Seattle – Washington, 12-15 Nov. 2007

[5]. R. Capata, E. Sciubba, “*Preliminary Design Of A Nano Gas Turbine For Portable Power Generation*” ISROMAC12-2008-20086, The 12th International Symposium on Transport Phenomena and Dynamics of Rotating Machinery – Honolulu, Hawaii, February 17-22, 2008

[6]. R. Capata, E. Sciubba, “*Design and Performance Prediction of a Ultra-Micro Gas Turbine for Portable Power Generation*” IMECE2008-67155, Boston – Washington, 30 Oct-6 Nov. 2008

[7]. R. Capata, E. Sciubba, “*Progress in the development of a prototype “Ultra-Micro” Gas Turbine Set for Portable Power Generation*” ECOS 2008, Gliwice, Polonia, Giugno 24 – 27, 2008.

[8]. R. Capata, E. Sciubba, “*The Ultra-Micro Gas Turbine Generator project at UDR1: experimental assessment of the compressor map and of the regenerative combustion chamber efficiency*” ECOS 2009, Foz DoIguaçù, Brasile, 31 Agosto 3 settembre 2009.

[9]. R. Capata, E. Sciubba, “*The α -prototype of an ultra-micro-gas turbine at the University of Roma 1. Final assembly and tests*”, IMECE2009-12281 November 13-19, 2009 , Lake Buena Vista, Florida USA

[10]Vleugels at Al. , “*High-speed bearings for micro Gas turbines: stability analysis of foil bearings*”, J. of Micromechanics & Microengineering, Vol 16 (2006) s282-s289

[11]Garret ®, www.turbobygarret.com

Proceedings of the ASME 2010 International Mechanical Engineering Congress & Exposition
IMECE2010
November 12-18, 2010, Vancouver, British Columbia, Canada

IMECE2010-38594

STUDY OF A THERMAL BARRIER COATING IN A PULSED DETONATION ENGINE

ERIC K. ANDERSON
National Research Council
Dayton, OH
eric.anderson@wpafb.af.mil

JOHN L. HOKE
Innovative Scientific Solutions
Dayton, OH
john.hoke@wpafb.af.mil

FRED R. SCHAUER
Air Force Research Laboratory
WPAFB, OH
frederick.schauer@wpafb.af.mil

ABSTRACT

A pulse detonation engine (PDE) works by establishing regular purge-detonation cycles. During a cycle, a point inside the detonation tube downstream of the ignition source is exposed to extreme high temperature briefly as the detonation wave passes, followed by near room temperature purge gas. As is the case for many combustion systems, minimizing unwanted heat loss in a PDE is important to maximizing efficiency. For this reason, the effect of coating the inner wall of a detonation tube with a thin, low thermal conductivity coating was studied experimentally and computationally. A three inch diameter 304 stainless steel detonation tube with 1.65 mm wall thickness was coated on the inside with a 0.635 mm thick layer of 93% Zirconium Oxide and 7% Yttrium Oxide.

Experimental results were obtained operating the PDE at a frequency of 10 Hz. Temperature was measured at four points on the exterior of the detonation tube between 33 and 45 inches downstream of the ignition source. At all locations measured, the exterior tube temperature would rise from an initial ambient temperature faster with the uncoated tube until a crossover point approximately 1000-1500 cycles after the start of the experiment. After this cross over point, outer wall temperature was higher for the coated tubes, which at this point was increasing faster for the tube with the thermal barrier.

To better understand the physical reason why the insulated tube always reached a higher temperature after roughly 1000 cycles, a simplified computational model was created using a time-dependent, one-dimensional, implicit finite difference method to calculate the temperature at 20 evenly spaced nodes. The temperature profile of the tube was assumed to vary only radially, so the interior points were solved using the transient one-dimensional heat conduction equation. One dimensional convection boundaries were applied at the inner and outer walls of the tube. Boundary conditions of increasing complexity were applied, beginning with a constant temperature, natural convection condition at the outer wall and a forced convection boundary condition at the inner wall with a step change in temperature to simulate the detonation wave.

Experimentally measured wall temperatures are presented and compared to computationally obtained temperature profiles. The focus is on understanding the physical reason for the behavior of the insulated tubes.

INTRODUCTION

The PDE has recently gained interest both as a possible stand-alone aircraft propulsion system with potentially high efficiency due to constant volume combustion [1], and as an efficient combustor in a gas turbine engine [2]. One potential problem is the large amount of heat generated by the detonation process, which is both transferred to the materials in the engine, and carried out with the exhaust stream. Compared to conventional combustion systems which burn the air-fuel mixture at subsonic speeds, the cyclic nature of PDE tends to generate higher peak temperatures, but high temperatures are experienced over a small fraction of the operating cycle [3].

Most previous computational models of PDE engines have neglected heat losses, assuming the PDE cycle is isentropic due to the relatively short residence time of hot gases. However, as early as the late 1960's, measurements of pressure, temperature, and heat flux in detonation waves showed significant departure from expected isentropic profiles [4]. These heat loses can result in a loss of specific impulse from a PDE engine [5]. A 2005 paper by Radulescu et al. [6] describes a computational model to compute the PDE flow field for a single cycle while accounting for heat loss, but assumes constant tube wall temperature (298 K), citing large heat capacity and thermal conductivities of steel. However, for a large number of cycles, the tube wall has been shown to reach temperatures exceeding 800 K [3], and with a tube coated with a low thermal conductivity material, the temperature of the tube wall varies with radius.

As an attempt to alleviate the heat load on the detonation tube of a PDE, a Yttria stabilized ZrO_2 coating was applied to the inside of the tube. This thermal barrier coating has significantly lower thermal conductivity and thermal diffusivity than the stainless steel of the tube. However, the results showed that for a hydrogen fueled PDE operating at 10 Hz, the tube with the thermal barrier coating would reach a higher temperature after approximately 1000 cycles. The computational model described here offers insight into why this occurs as well as determining the operating conditions where a thermal barrier is helpful and where it is detrimental. The purpose of the model is to offer insight into the effect of engine operating parameters rather than achieving a close match with experimental data.

EXPERIMENTAL SETUP

Experimental detonation tube temperature measurements were performed in the Air Force Research Laboratory (AFRL) Pulse Detonation Research Facility at Wright Patterson AFB, Ohio. The PDE consisted of a GM quad four cylinder head with one detonation tube attached to the #1 combustion chamber. For the baseline experiments the detonation tube was a 7.62 cm diameter, 1.2 m long stainless steel tube with a structurally modified 61 cm Schelkin spiral [7] to promote detonation. The wall thickness of the tube was 1.65 mm. Temperature measurements were provided by thermocouples clamped to the outside of the tube 83.8 cm, 94.0 cm, 104.1 cm, and 114.3 cm downstream of the cylinder head.

To test the effect of the thermal barrier coating, an identical detonation tube was prepared with a 0.635 mm coating of Yttrium Oxide stabilized Zirconium Oxide (93 % ZrO_2) on the inside of the tube.

The PDE cycle can be divided into three segments. The first is the filling of the tube with fresh air-fuel mixture. The second is the detonation initiation and blow-down of the mixture. The third is purging the tube of exhaust gases with approximately 294 K air. With the engine operating at 10 Hz, all three processes occur in 0.1 s, and each takes approximately 1/3 of the cycle.

For all tests the fill gas was a well premixed hydrogen and air mixture with an equivalence ratio of 1.0 and a temperature of approximately 310 K. Further details of the engine and facility are provided in [8].

COMPUTATIONAL METHOD

Model Geometry

A one dimensional model for the detonation tube was created as shown in Fig. 1. On the left, the figure depicts a cross-section of the tube perpendicular to its length. On the right, the 1-d model is depicted which assumes that tube temperature varies only with the distance from the center of the tube, and with time after start of the engine. Experimental results presented below show that tube wall temperature only varies about 175 K moving downstream axially. As shown in the figure, the temperatures inside the tube and outside the tube are denoted T_{in} and T_{out}, respectively. Heat transfer coefficients are denoted h_{in} and h_{out} for the inner and outer tube walls.

The tube wall is modeled by dividing the thermal barrier coating into 5 regions bounded by 6 nodes (1 at the inner surface and 1 at the interface between the coating and the stainless steel tube). The distance between nodes is 0.127 mm. The 304 stainless steel tube wall is divided into 13 equally spaced regions with 14 nodes (1 at the interface and 1 at the outer tube wall).

Material Properties

Constants describing relevant material properties values are shown in Table 1 for both the thermal barrier coating and the 304 stainless steel tube. Thermal diffusivity, α, is calculated from thermal conductivity, k, as

$$\alpha = \frac{k}{\rho c_p}, \tag{1}$$

where ρ is the material density and c_p is the constant pressure specific heat. Thermal diffusivity, α, is used to calculate the mesh Fourier number, σ, using the equation

$$\sigma = \frac{\alpha(\Delta t)}{(\Delta x)^2}, \tag{2}$$

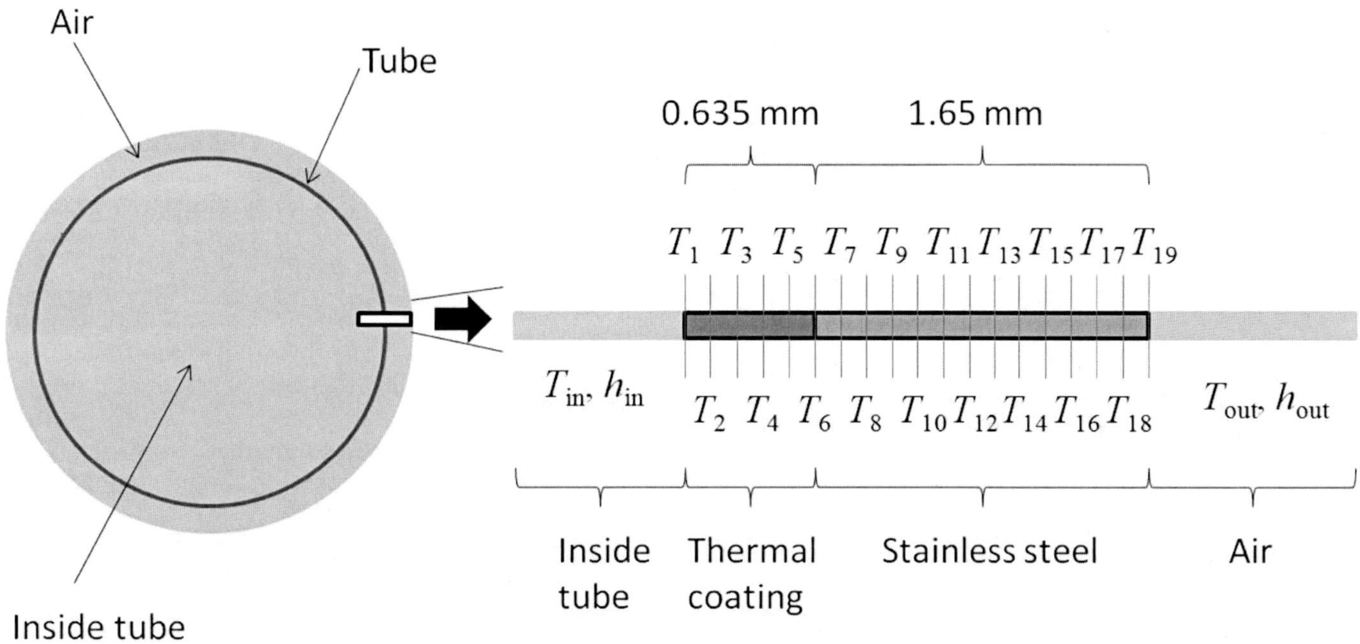

Fig. 1: Diagram of 1-d tube wall model.

	304 Stainless Steel	ZrO_2 Coating
k [W/(mK)]	Equation (4)	0.5-3
α [m^2/s]	Equation (5)	1.0 x 10^{-6}
σ [unitless]	~0.25	0.062
Bi_{in} [unitless]	0.0127	0.0127
Bi_{out} [unitless]	4.2x10^{-4}	N/A
$h_{in,hot}$ [W/(m^2K)]	250	300
$h_{in,cold}$ [W/(m^2K)]	50	60
h_{out} [W/(m^2K)]	25	N/A

Tab. 1: Relevant material properties. Properties of 304 stainless were obtained from White [9] and properties of the thermal barrier coating were obtained from [10].

where Δt is the time step for the model and Δx is the spacing between nodes (0.127 mm in our case). The mesh Biot number, needed to model the convection boundaries inside and outside the tube, is computed as

$$\text{Bi} = \frac{h\Delta x}{k}, \tag{3}$$

where h is the heat transfer coefficient. For the outside of the tube, the values for heat transfer coefficient shown in Tab. 1 are estimates for a flat plate exposed to natural convection, and for the inside of the tube, values are estimated assuming forced convection [9]. The dependence of h on gas velocity is neglected. Note that the Zirconia coated tube has 20% higher h values due to greater surface roughness, and the cold h values are much lower than the hot values due to much lower gas velocity during purge and fill.

Results from [9] and [10] show that all of the above values vary less than 10% across the range of tube temperatures encountered, with the exception of the thermal conductivity and diffusivity of stainless steel. For 304 stainless steel, the thermal conductivity roughly doubles from 273 K to 1073 K, so a linear fit was applied to the temperature – conductivity data in [9], resulting in the relationship

$$k = .015\,T + 10.505, \tag{4}$$

which was used to compute the stainless steel thermal conductivity in the model. The computational model also incorporated the following linear relationship between the thermal diffusivity of steel and temperature:

$$\alpha = (1.37 \times 10^{-9})T + 3.75 \times 10^{-6}. \tag{5}$$

The above relationship was generated from data provided in [11].

Governing Equations

In order to satisfy the requirement for stability of the explicit model for transient conduction, a time step on the order of 0.001 s or smaller is required, with accuracy of the solution improving with smaller time steps. To achieve sufficient accuracy with minimal computation time an implicit model was used instead.

For the thermal barrier coated tube, the model was constructed using the implicit nodal equations for unsteady heat conduction described by White [9]. Referring to Fig. 1, the temperature T_1 for time step i+1 is described by the equation:

$$T_1^{i+1} = \frac{2\sigma\left(T_2^{i+1} + \mathrm{Bi}_{in} T_{in}^{i+1}\right) + T_1^i}{1 + 2\sigma + 2\sigma \mathrm{Bi}_{\mathrm{in}}}, \tag{6}$$

where the constants for the coating are used. The interior points for the coating, T2-T4, are described by the equation

$$T_j^{i+1} = \frac{\sigma\left(T_{j-1}^{i+1} + T_{j+1}^{i+1}\right) + T_j^i}{1 + 2\sigma}, \tag{7}$$

where the constants for the barrier are used and the lower subscript is used to denote the physical location (j=2,3,4,5) of the node.

At the boundary between the coating and the steel tube, the temperature was computed using the following equation, derived from the first law:

$$T_6^{\mathrm{i}+1} = \frac{\sigma_S\left[T_7^{i+1} + \frac{k_b}{k_S} T_5^{i+1} + \frac{1}{2} T_6^i\left(1 + \frac{k_b}{k_S \sigma_b}\right)\right]}{\sigma_S + \frac{k_b}{k_S}\sigma_S + \frac{\sigma_S}{2} + \frac{k_b \sigma_S}{2 k_S \sigma_b}}. \tag{8}$$

For the interior points of the stainless steel tube, equation (7) was used with the mesh Fourier number, σ, for steel and j=7-18. Finally, at the outer surface of the tube, a convection boundary was applied with the equation

$$T_{19}^{\mathrm{i}+1} = \frac{2\sigma\left(T_{18}^{i+1} + \mathrm{Bi}_{out} T_{out}^{i+1}\right) + T_{19}^i}{1 + 2\sigma + 2\sigma \mathrm{Bi}_{\mathrm{out}}}, \tag{9}$$

using the Biot and Fourier numbers for steel.

The non-coated tube was modeled by simply removing nodes j=1-5 and applying the following equation for node 6:

$$T_6^{i+1} = \frac{2\sigma\left(T_2^{i+1} + \mathrm{Bi}_{in} T_{in}^{i+1}\right) + T_1^i}{1 + 2\sigma + 2\sigma \mathrm{Bi}_{\mathrm{in}}}, \tag{10}$$

with Biot and Fourier numbers for steel.

Computational Method

A Matlab® script was written to solve the above equations by iterating at each time step 10-1000 times. The computation would start at time t=0 and march forward by Δt after completing the specified number of iterations. For each iteration, the temperature of all 19 nodes was recalculated before proceeding to the next iteration. The initial temperature distribution at each successive time step was assumed to be equal to that of the previous time step before starting iteration. Since the PDE was operated at 10 Hz, one engine cycle would occur in 0.1 s, and during that time T_{in} was held at 310 K during fill and purge, and at 2500 K during detonation. T_{out} was held at a constant 294 K. At the start of computation the T_1-T_{19} were set to 294 K.

Model Assumptions and Limitations

It is important to note the limitations of the model employed here, which include:

- Gas temperature vs. time at the inside surface of the tube is assumed to follow a step change. However, the actual temperature will rise and fall over finite periods of time. Results indicate that at time scales greater than the cycle time, the actual temperature profile is less important than the overall heat flux during the cycle.
- Gas temperature outside the pipe is assumed constant. Actual outside gas temperature is likely to rise during operation of the PDE.
- Temperature dependence of thermal properties is neglected for the ZrO_2 coating, but data from [4] shows that this is a reasonable assumption.
- The model assumes a constant gas velocity and therefore heat transfer coefficient inside the tube. In the PDE tube the velocity will be O(1000) during detonation and blowdown and O(-1) during purge. This would have the effect of increasing heat transfer during the hot portion of the cycle [3].

Results below show the effect of altering some of these assumptions.

RESULTS AND DISCUSSION

Experimental Results

Experimental results obtained with the pulse detonation engine running on hydrogen fuel, an equivalence ratio of 1, and a frequency of 10 Hz are shown in Fig. 2. The temperatures reported are outer wall temperature measurements made 83.8 cm downstream of the cylinder head for both the coated and uncoated tubes. Note that the tube temperature initially rises faster for the uncoated tube, but between 50 and 100 s after engine start, the rate of temperature rise of the coated tube becomes greater than that of the uncoated tube. This results in a crossover point around 100 s and a higher peak temperature when the engine is stopped (seen as a sudden drop in temperature). This behavior was unexpected, so the computational model in subsequent

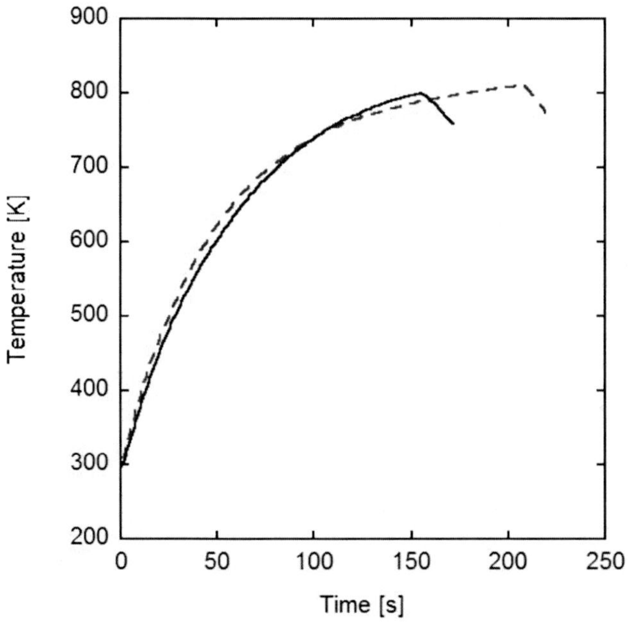

Fig. 2: Experimental tube outer wall temperature measurements. Tube coated with Zirconia is shown with a solid line. Non-coated tube is shown with a dashed line.

sections was designed to explain the physical reason for this temperature crossover and lower higher final temperature for the Zirconia coated tube. Temperature profiles were similar for the other equivalence ratios and downstream temperature measurement locations.

Computational Results

The gas temperature (T_{in}) profile applied at the inside convection boundary during an engine cycle is shown in Fig. 3. Note that this profile is much simpler than the actual profile, with a step change from near-ambient temperature to a temperature close to the adiabatic flame temperature of hydrogen. This simplified temperature profile is sufficient to capture the temperature crossover seen in the experimental results.

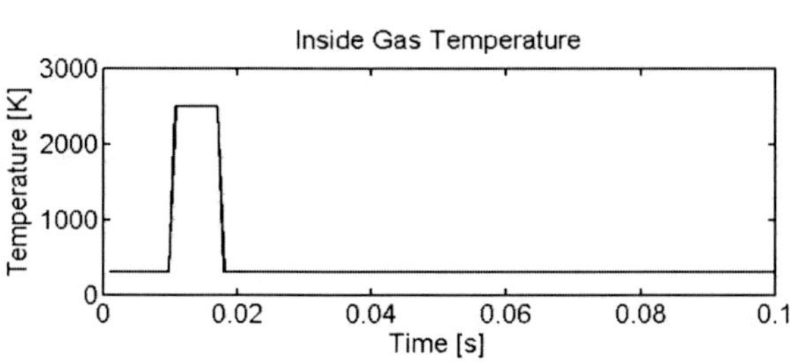

Fig. 3: Gas temperature profile used in computational model for the inside of a detonation tube during a PDE engine cycle.

Effect of the Inside Heat Transfer Coefficient

Using the image in figure 4 to estimate the surface roughness of the coated tube, a value of 280 Ra (μm) is estimated by comparing feature size to pipe inner diameter. This roughness value is greater than 15 Ra (μm) commonly reported stainless steel tube. For this reason, the inside hot heat transfer coefficient ($h_{in,hot}$) and inside cold heat transfer coefficient ($h_{in,cold}$) were both increased by 20% for the Zirconia coated tube. The values used for $h_{in,cold}$ were 50 and 60 for the non-coated and coated tubes, respectively, and 250 and 300 for $h_{in,hot}$. These values are estimates, but are within the generally accepted range for forced convection. The values are much higher for detonation and blow-down due to higher gas velocities.

Outer wall temperatures shown in Fig.5 were obtained computationally using the same inner and outer heat transfer coefficients for the coated and uncoated tubes. The model does not capture the temperature crossover with these heat transfer coefficients. The outer wall temperatures of Fig. 6 were obtained using 20% higher inner heat transfer coefficients for the barrier coated tube (values shown in Tab. 1). With these heat transfer coefficients, the temperature crossover is captured, albeit at a later time after engine start when compared to experimental data. This difference is not surprising given the simplified T_{in} profile and estimated heat transfer coefficients.

These plots suggest that the difference in conductivity of the thermal barrier and the added thickness of the tube is not enough to capture the physics of this tube heating problem. When both tubes have the same heat transfer coefficient the coated steel tube outer wall temperature rises faster due to the lack of a low conductivity barrier, but once the temperature has stabilized (at times > 500 s), the temperatures are the same. By simply increasing the heat transfer coefficient of the coating by 20%, the model captures this temperature crossover.

Heat Flux at Inner and Outer Surfaces

With the assumed convective boundary conditions at the inner and outer surfaces, heat flux across the boundary can be calculated using

$$q = h(T_w - T_\infty), \tag{11}$$

where the appropriate h values from Tab. 1 are used and q is the heat flux per unit area and T_∞ is the free stream temperature [9]. As the equation shows, a greater difference between the wall temperature and the free stream temperature results in greater heat flux. Once a stable tube temperature is reached, the net heat flux into the tube should be zero when averaged over an engine cycle. Figure 7 shows the sum of the inner surface heat flux and the outer surface heat flux. This

Fig. 4: Image showing the surface roughness of the Yttria stabilized Zirconia coated tube. The non-coated tube is smooth by comparison.

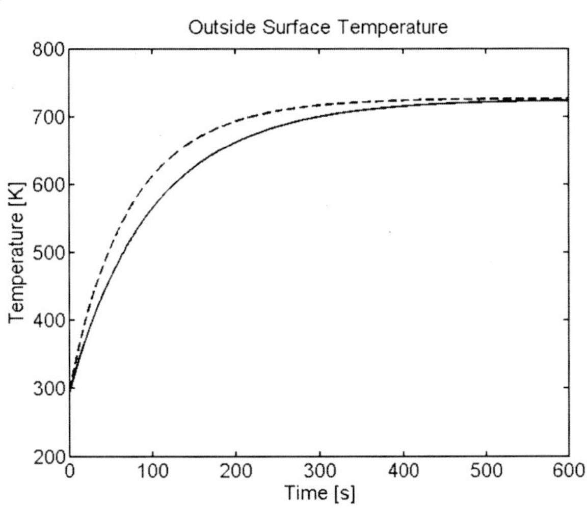

Fig. 5: Calculated outside surface temperature using $h_{in} = 50$ during purge and fill and 250 during detonation and blowdown for both coated and uncoated tubes. Non-coated tube is shown with a dashed line.

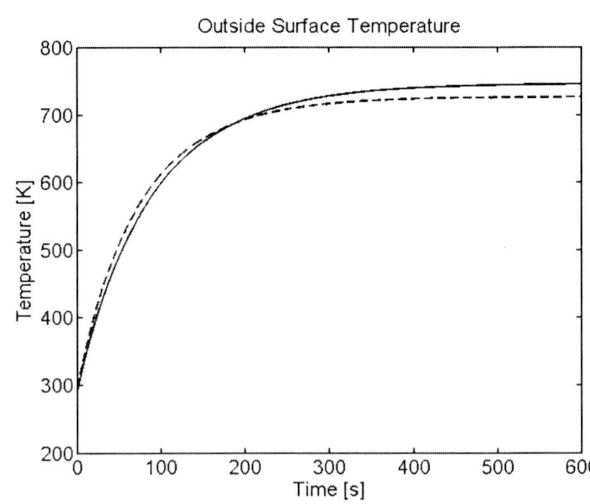

Fig. 6: Calculated outside surface temperature using the *h* values presented in Tab. 1. Non-coated tube is shown with a dashed line.

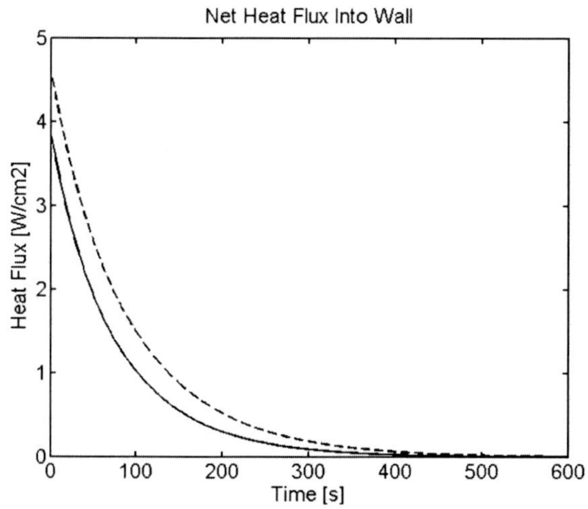

Fig 7: Net heat flux for the model with 20% higher heat transfer coefficient for the coated tube. The non-coated tube is shown with a dashed line.

net heat flux approaches zero once the tube temperature is stable.

The results of Fig. 7 are averaged over an entire cycle of the PDE so that variations within a cycle are not shown. Not surprisingly, the non-coated tube has roughly 10% greater net heat flux into the tube material than the coated tube, which explains its more rapid initial temperature rise. However, once the temperature of the tube has stabilized, both coated and non-coated tubes approach zero net heat flux.

The equilibrium temperature reached depends on the ratio of cycle averaged heat flux at the inner surface to heat flux at the outer surface. Figure 5 shows that with equal heat transfer coefficients for the coated and non-coated tubes, the same equilibrium temperatures are reached. However, increasing the heat transfer coefficient for the inside of the coated tube on the inside by 20%, the tube temperature at which cycle averaged heat flux at the inside the surface is balanced with heat flux at the outside surface shifts upward slightly. This is reflected in the higher final temperature for the coated tube seen in Fig. 6.

Temperature Variation within the Tube Wall

Further understanding of the phenomenon responsible for the temperature crossover can be gained by examining the temperature profiles within the tube at several points in time after the start of engine operation. Note that the direction of heat flux at the inside surface changes during each engine cycle, with heat flowing into the tube when the free stream temperature is at 2500 K and heat flowing out of the tube when the free stream temperature is at 310 K. For this reason, the temperature profile within the tube depends greatly on timing relative to the start of an engine cycle. Figure 8 shows temperature profiles near the end of a 2500 K exposure, and the results show higher temperature gradients near the inside surface of the coated tube than for the non-coated tube.

This provides a second explanation for the slower initial outside wall temperature rise of the coated tube. For less than 1000 cycles (t<100 s), more heat is stored near the inside wall of the tube. Figure 9 shows temperature profiles near the end of exposure to ambient temperature. As one would expect the temperature gradient near the inside surface of the tube has reversed.

CONCLUSIONS

Experimentally measured PDE tube wall temperatures of thermal barrier coated and non coated tubes were unexpectedly higher for the Yttria stabilized Zirconia coated tube after roughly 1000 engine cycles. The tube was modeled computationally and results suggest that a slightly higher heat transfer coefficient for the thermal barrier is responsible for a higher final temperature for the coated tube. For less than 1000 engine cycles, experimental results showed higher outside wall temperatures for the non-coated tube than for the coated tube. Examination of computationally generated temperature profiles as a function distance from the inside surface of the tube showed that this is likely due to the low thermal conductivity of the Zirconia coating. These results suggest that the use of a thermal barrier coating may not always result in lower temperatures of metal engine components if the heat transfer coefficient for the coating is greater than that of the uncoated surface. A plausible explanation for higher heat transfer coefficient of the barrier in this study is greater surface roughness compared to the non-coated tube.

ACKNOWLEDGEMENTS

The authors would like to extend their thanks to Curtis Rice and Christopher Stevens for technical support. Funding was provided by the National Research Council, the Air Force propulsion directorate, and the Air Force Office of Scientific Research.

REFERENCES

[1] Kuo, K., Principles of combustion, Wiley, New York.

[2] Naples, A.G., Hoke, J.L., 2010, "Study of heat loads from steady deflagration and pulsed detonation combustion", AIAA Paper No. 2010-956, 48th Aerospace Sciences Meeting and Exhibit, 4-7 January 2010, Orlando, FL.

[3] Hoke, J., Bradley, R., and Schauer, F., "Heat transfer and thermal management in a pulsed detonation engine", AIAA Paper No. 2003-6486, 41st Aerospace Sciences Meeting and Exhibit, 6-9 January 2003, Reno NV.

[4] Edwards, D.H., Brown, D.R., Hooper G., Jones, A.T., "The influence of wall heat transfer on the expansion following a C-J detonation wave", Journal of Physics D: Applied Physics, Vol. 3, 1970, pp. 365-376.

[5] Laviolette, J.P., Kiyanda, C.B., Higgins, A.J., 2002, "The effect of friction and heat transfer on impulse in a detonation tube", Canadian Section of the Combustion Institute, Spring Technical Meeting, Windsor, Ontario, Canada, May 2002.

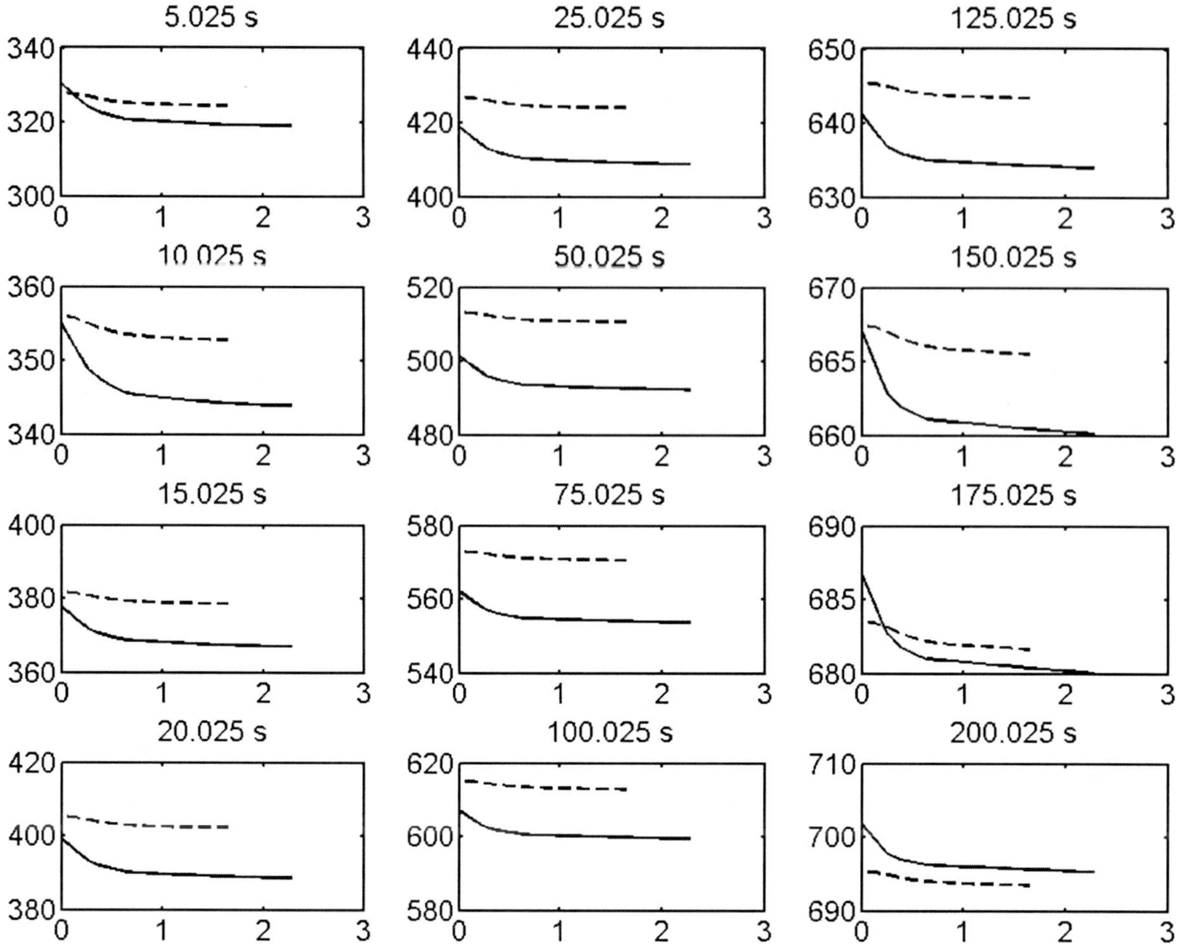

Fig 8: Temperature profiles [K] as a function of distance from the inside surface (x-axis, [mm]) of the detonation tube. Non-coated tube is shown with a dashed line. Titles indicate time after start of the engine (end of exposure to 2500 K gas).

[6] Radulescu, M.I., Morris, C.I., Hanson, R.K., 2004, "The effect of heat loss on the flow fields in a pulse detonation wave engine", AIAA Paper No. 2004-1124, 42nd Aerospace Sciences Meeting and Exhibit, 5-8 January 2004, Reno, NV.

[7] Tucker, K.C., 2005, "A flash vaporization system for detonation of hydrocarbon fuels in a pulse detonation engine", *PhD Dissertation*, Air Force Institute of Technology, AFIT/DS/ENY/04-07.

[8] Schauer, F., Stutrud, J., Bradley, R, 2001, "Detonation initiation studies and performance results for pulsed detonation engine applications", AIAA Paper No. 2001-129, 39th AIAA Aerospace Sciences Meeting & Exhibit, 8-11 January 2001, Reno, NV.

[9] White, F.M., 1988, Heat and mass transfer, Addison-Wesley, New York.

[10] Khor, K.A., Gu, Y.W., 2000, "Thermal properties of plasma-sprayed functionally graded thermal barrier coatings", Thin Solid Films, **372**, pp. 104-113.

[11] BNL 10200-R, Vol. 1, 1980, "Brookhaven National Laboratory Selected Cryogenic Data Notebook".

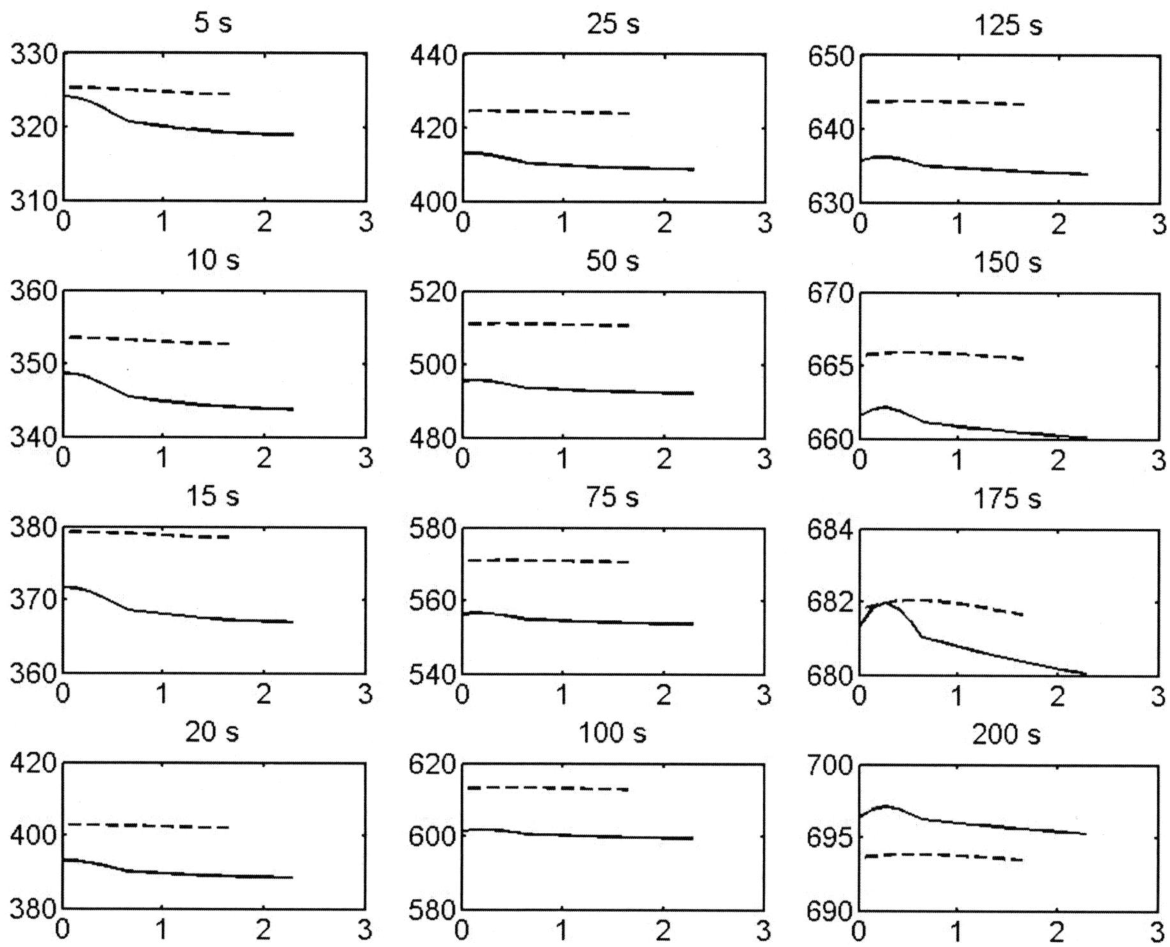

Fig 9: Temperature profiles [K] as a function of distance from the inside surface (x-axis, [mm]) of the detonation tube. Non-coated tube is shown with a dashed line. Titles indicate time after start of the engine (end of exposure to ambient temperature gas).

Proceedings of the ASME 2010 International Mechanical Engineering Congress & Exposition
IMECE2010
November 12-18, 2010, Vancouver, British Columbia, Canada

IMECE2010-38600

CIRCULATION CONTROL SPAN-WISE BLOWING LOCATION OPTIMIZATION FOR A HELICOPTER ROTOR BLADE

Alan M. Didion, Jonathan Kweder, Mary Ann Clarke, James E. Smith
Center for Industrial Research Applications
West Virginia University
Morgantown, West Virginia, USA

Abstract

Circulation control technology has proven itself useful in the area of short take-off and landing (STOL) fixed wing aircraft by decreasing landing and takeoff distances, increasing maneuverability and lift at lower speeds. The application of circulation control technology to vertical take-off and landing (VTOL) rotorcraft could also prove quite beneficial. Successful adaptation to helicopter rotor blades is currently believed to yield benefits such as increased lift, increased payload capacity, increased maneuverability, reduction in rotor diameter and a reduction in noise. Above all, the addition of circulation control to rotorcraft as controlled by an on-board computer could provide the helicopter with pitch control as well as compensate for asymmetrical lift profiles from forward flight without need for a swashplate.

There are an infinite number of blowing slot configurations, each with separate benefits and drawbacks. This study has identified three specific types of these configurations. The high lift configuration would be beneficial in instances where such power is needed for crew and cargo, little stress reduction is offered over the base line configuration. The stress reduction configuration on the other hand, however, offers little extra lift but much in the way of increased rotor lifespan and shorter rotor length. Finally, the middle balanced configuration offers a middle ground between the two extremes. With this configuration, the helicopter benefits in all categories of lift, stress reduction and blade length reduction.

Nomenclature

ρ_{air}	Density of air	kg/m^3
ω	Angular velocity	RPM, rad/s
R	Rotor length	m
c	Chord length	m
E	Young's modulus	GPa
b	Span	m
A_b	Blade section area	m^2
I	Moment of inertia	kg·m^2
V	Linear velocity	m/s
L	Lift force	N
S	Stress	GPa
C_L	Lift coefficient	-

Introduction

Previous research into the use of active circulation control applied to helicopter rotors found that no one perfectly optimal configuration is apparent, therefore, this research focuses on finding multiple configurations, each with its own benefits. Because of this wide array of possibilities and the fact that one craft may require use of more than one circulation control configuration profile over the course of a diverse flight, it is proposed here that a standard circulation control blade be created that can alternate through profiles on computer command.

Circulation control refers to a technology that utilizes high velocity jets of air blown or sucked through the leading or trailing edge of an airfoil to augment the circulation around the airfoil and thus the lift coefficient. In this application, air is blown over the rounded trailing edge of the blade, adhering to the surface inducing the Coandă effect and entraining the flow in the existing boundary layer. The jet pulls more air down with it, increasing circulation, and therefore increasing the lifting capability.

The application of circulation control technology to rotorcraft proves more difficult than to fixed-wing aircraft, because of the nature of the blades themselves. The fundamental rules of rotational bodies demonstrate that the outer edge of a rotating body is traveling at a faster speed than the internal sections. This means that the outer sections of the rotor blade are generating more lift than are internal sections due to the higher free speed velocity. This induces a "coning" effect on the rotor blades that can produce excess stress, as shown in Figure 1. Because of this characteristic, circulation control jets will have varying degrees of effect on the overall lift depending on their location and use along the span of the blade.

To compensate for this effect, the radially spaced exit jets will need to vary in strength and concentration depending on their distance from the rotor hub. This research explores the various sets or "profiles" of lift coefficients that may be beneficial. It will be the subject of subsequent research to design the proper slots to create such lift coefficients through blowing.

Figure 1: The MH-53J Pave-Low III Heavy Lift Helicopter in Flight Demonstrating a Rotor "Coning" Effect (aerospaceweb.org)

Procedure

For a baseline model for this analysis, the American Sikorski UH-60A *(Blackhawk)* utility helicopter was chosen because of its widespread use. The Sikorsky UH-60A (*Blackhawk*) is a helicopter used mostly by the U.S. Military and allies for tactical troop transport. It has been in service since 1979 and is a widely recognized symbol of American aerospace technology prowess and military power. The relevant Blackhawk helicopter specifications can be found in Table 1.

Table 1: UH-60A Blackhawk Specifications (air-attack.com)

Rotor diameter	16.36 m
Normal takeoff weight	7375 kg
Maximum takeoff weight	9185 kg
Maximum speed	295 km/h
Rate of climb	3.6 m/s
Main rotor airfoil	SC1095R8

The UH-60A Blackhawk's rotor blades make use of the SC1095R8 airfoil, (see Figure 2). This airfoil's cross-sectional shape has a blade thickness of 5% of the chord length. A composite material, Zoltek Panex® 33 48K Continuous PAN Carbon Fiber, was chosen as the rotor blade material. This material has a Young's modulus of 228 GPa.

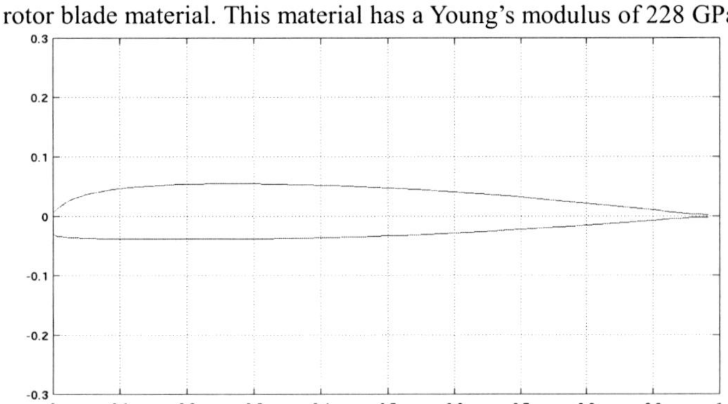

Figure 2: Profile of the SC1095R8 Airfoil (Abbott, 1959)

Table 2 displays the constants used in these calculations, as derived from the given specifications, airfoil diagram, and chosen rotor material.

Table 2: Constant Values Used in Calculation

Density of air	$\rho_{air} = 1.225$ kg/m^3
Angular velocity	$\omega = 400$ RPM $= 41.89$ rad/s
Rotor length	R = 8.18 m
Chord length	c = 0.183 m
Rotor thickness	d = .00915 m
Young's modulus	E = 228 GPa

The idea behind this study was to calculate the increase in overall stress in the helicopter blade with and without circulation control. The first task in completing this study was to find the area of the rotor blade sections at set intervals along the blade. A_b is the same for each segment studied, see Eq. 1.

$$A_b = c \cdot b \qquad \textbf{Eq. 1}$$

The next calculation yields the linear velocity of each segment along the span of the blade (b) by multiplying its angular velocity (ω) with its distance from the rotor hub, as shown in Eq. 2.

$$V = \omega \cdot b \qquad \textbf{Eq. 2}$$

The equation for lifting force (L) calculated the load experienced by the blade segment and is dependent on the value of air density (ρ), lift coefficient (C_L), the previously calculated blade section area (A_b) and linear velocity (V) (Bertin, 2002, Thwaites, 1987). When calculated at set intervals along the span and plotted, a parabolic lift curve shows an increasing trend toward the outer edge of the rotor blade, which suggests a greater stress level as the radius increases (Eq. 3).

$$L = \frac{1}{2} \cdot \rho_{air} \cdot V^2 \cdot A_b \cdot C_L \qquad \textbf{Eq. 3}$$

The statics equation for cantilever beam stress (S) makes use of calculated values of lift (L) as the force experienced, span (b), chord length (c), moment of inertia (I) and the given Young's modulus (E) of the particular material (Hibbler, 2004). This yields a parabolic stress curve similar to the lift curve, increasing toward the outer edge of the blade (Eq. 4). For this particular blade, the moment of inertia was calculated using an elliptical cross-section, seen in 6 (Shigley, 2004).

$$S = \frac{Lc^2}{6EI} \cdot (3b - c) \qquad \textbf{Eq. 4}$$

$$I = \frac{\pi \cdot \left(\frac{c}{2}\right) \cdot \left(\frac{d}{2}\right)^3}{4} \qquad \textbf{Eq. 6}$$

With the baseline data calculated, it was then explored how manipulating the lift coefficient over the span affected total lift and stress. It was discerned that a stronger coefficient was needed toward the rotor hub, decreasing toward the blade tip. A "stair step decay" pattern was set that changed the lift coefficient with each span interval to be a certain percentage of the previous value. These percentages decreased after a set interval to produce "stair step" increases in the rate of decay. This was done to simplify the process of manual trial and error experimentation with different decay percentage sets. It was found that steady decay could also create a reliable curve that was balanced in all of the criteria examined here (lift, stress, blade length).

Results

After gathering the aircraft's specifications, calculations were done concerning the magnitude and location of stresses experienced by the rotor blade of constant chord length spinning through air. Assuming the blade has a constant chord length is a simplification to make the calculations easier. A graphic representation of the amount of stress in the blade was created with respect to the distance from the center of the rotor disk (span). The lift force applied to the blade was found to be parabolic, increasing toward the outer edge of the blade due to higher linear speed than in the center, resulting in increased stresses toward the outer edge as well, as shown in Figure 3.

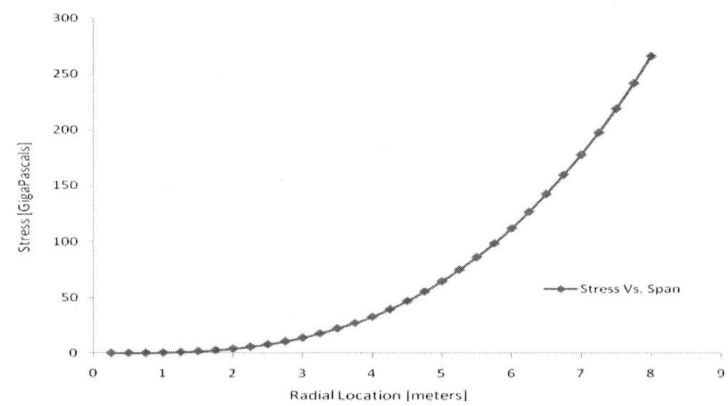

Figure 3: Calculated Stress Distribution in the Un-Augmented Rotor Blade

As helicopter blades experience a vortex action at the tip and loss of lift, approximately the last 10% of the blade span experiences greatly reduced lift and stress forces (almost to zero), meaning that at the point approximately 90% from the rotor hub is the site of greatest lift and stress. So for the current purpose, the last few points of data are ignored and the greatest stress is found at 7.25 m of 1.98 Gpa.

From this trend, the final lift and stress curves can be modified for optimal balance by adjusting the lift coefficient, $C_{L,}$ by the use of circulation control jets. For the sake of current technology and physical limitations, the variable lift coefficient was bounded by a minimum of 1, meaning that the jet is off completely, and a maximum of 3. It was found that there is no one universally optimal coefficient augmentation trend. Instead, three different trends were selected from an infinite number of possibilities, each having its own unique advantages and disadvantages. Three lift coefficient sets were designed and new stress curves were graphed for each (Figure 4, Figure 5, Figure 6). Each figure shows the stress with respect to span.

The first configuration, (Figure 4) was created using a stair step decay in lift coefficient with various step lengths, the longest of which being toward the outward edge. This proved beneficial in stress reduction while keeping even with the total lift generated by the un-augmented blade data. Blade stress can be reduced through circulation control by over 80 GPa while retaining the same total lift. This configuration also allows the blade to be shortened by an entire 1 m. Such a profile would be beneficial in the private sector, providing a longer blade lifespan and shorter blade length while only minimally affecting performance.

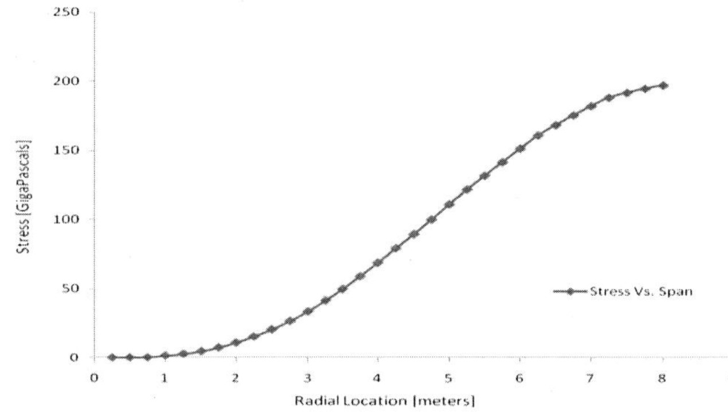

Figure 4: Calculated Stress Distribution for Stress Reduction Optimization

The second configuration, (Figure 5) was created also using a stair step decay in lift coefficient with various step lengths, the longest of which being toward the inward edge. This proved beneficial in creating increased lift while breaking even with the stress produced in the un-augmented blade. At the same rotor length, it was found that total lift could be increased by 20 kN while still reducing stress by 2 GPa. Such a profile would be optimal for commercial heavy-lift applications and military cargo and troop transports as lift would be greatly enhanced, increasing load capacity and speed while not affecting blade stress levels.

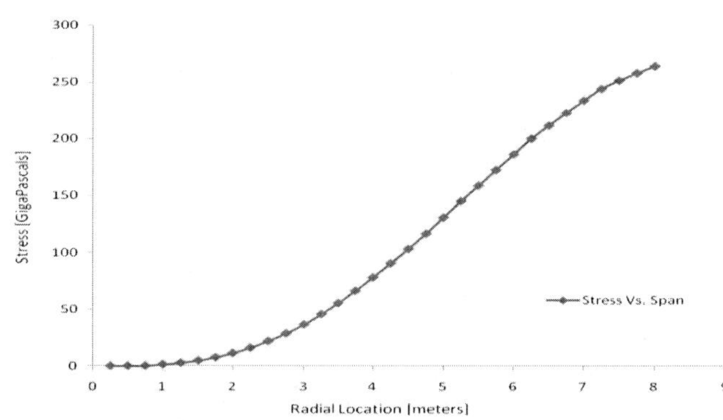

Figure 5: Calculated Stress Distribution for High Lift Optimization

The third configuration, (Figure 6) was created using a constant rate of coefficient decay and was found to provide a balance between the previous two sets. It was found that if lift was equaled with the un-augmented data, stress was reduced by 80 GPa and the blade could be shortened by 1.5 m. If the blade ended where the lift coefficient reaches 1, lift could be increased by 11 kN, stress was reduced by 50 GPa and the blade could be shortened by .5 m. This profile represents a medium between the stress reduction and high lift profiles and demonstrates the capability of the technology to be tailor made for any application.

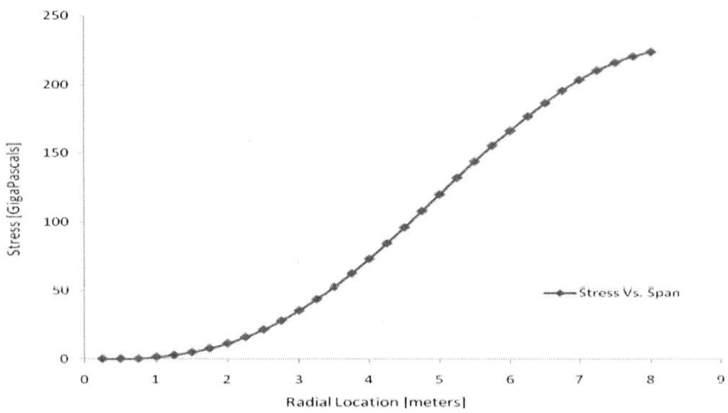

Figure 6: Calculated Stress Distribution for Balanced Optimization

Conclusions and Recommendations

There are seemingly an infinite number of blowing slot configurations, each with separate benefits and drawbacks. As the high lift configuration would be beneficial in instances where such power is needed for crew and cargo, little stress reduction is offered over the control curve. The stress reduction configuration however offers little extra lift but much in the way of increased rotor lifespan and shorter rotor length. Finally, the middle balanced configuration offers a middle ground between the two extremes. With this configuration, the helicopter benefits in all categories of lift, stress reduction and blade length reduction.

When superimposed onto one another, the differences in each profile are shown in Figure 7. Seeing that the stress curve is derived from the lift curve, the total lift generated by each profile is shown as the area under each curve. It is noticed that all profiles increase total lift over the control curve and none exceed the control's maximum stress.

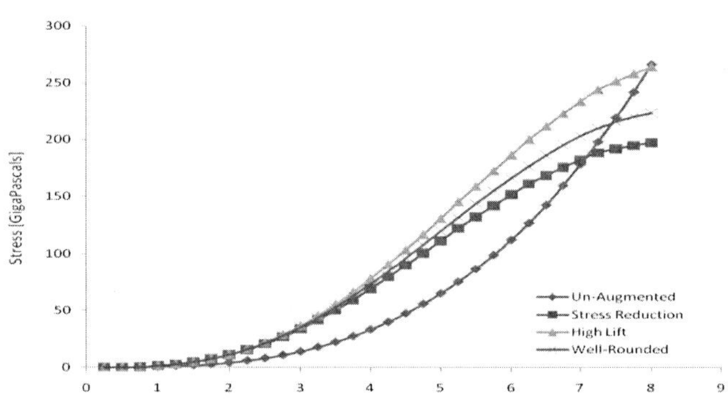

Figure 7: Comparison of Optimized Distribution Curves

An optimal circulation control system in this situation will be a "custom" type of technology, meaning that the exit slots will have to be planned out specifically for the individual craft's specifications and duties. Specifically, since the Black Hawk already makes use of an extra-long chord rotor blade for stress reduction and since agility is important, the balanced circulation profile would prove most beneficial. This way, more lift is gained while retaining the agility and speed it requires. More generally however, since one helicopter may require the use of several lift configurations over the course of a flight, such as a medical helicopter needing increased lift at takeoff and optimal speed during forward flight, it would be advantageous to be able to cycle circulation control profiles mid-flight. For this reason, it is proposed that a standard circulation control blade be made that has similar exit jets at regular intervals along the entire span of the blade. These jets can be controlled individually by computer to provide lift to produce the effect desired. Subsequent research regarding the implementation of such a system will continue.

References

Aerospaceweb.org, *UH-60 Black Hawk.* 2009. http://www.aerospaceweb.org/aircraft/helicopter-m/uh60/.

Abbott, Ira, von Doenhoff, A. *Theory of Wing Sections*, Dover Publications, Inc. New York, NY, 1959.

Air-attack.com, *MH-54 Pave Low Factsheet.* 2010. http://www.air-attack.com/page/60/MH-53-Pave-Low.html.

Anderson, John D. *Introduction to Flight (5th edition.).* McGraw-Hill, Inc. New York, NY, 2004.

Bertin, John J, *Aerodynamics for Engineers*, Prentice Hall, Inc. Upper Saddle River, NJ, 2002.

Didion, Alan M. "Analysis of the Structural Loads Developed Through the Use of the Circulation Controlled Helicopter Blade." NASA Undergradute Research Fellowship Final Report, Morgantown, WV. 2010.

Hibbler, R.C. *Engineering Mechanics, Statics*, Prentice Hall, Upper Saddle River, NJ, 2004.

Shigley, Joseph, E, *Mechanical Engineering Design*, McGraw Hill, Inc. New York, NY, 2004.

Thwaites, Bryan, *Incompressible Aerodynamics*, Dover Publications, Inc. New York, NY. 1987.

Proceedings of the ASME 2010 International Mechanical Engineering Congress & Exposition
IMECE2010
November 12-18, 2010, Vancouver, British Columbia, Canada

IMECE2010-38603

Investigation into the Feasibility of an Augmented Propeller Design with the Use of a Passive Circulation Control System

Jonathan Kweder, Mary Ann Clarke, James E. Smith
Center for Industrial Research Applications, West Virginia University
Morgantown, WV USA

ABSTRACT

Circulation control is a high-lift methodology that can be used in a variety of fluid dynamic systems, such as, on the wing of an aircraft. Circulation control increases the near surface velocity of the airflow over a rounded surface of an object, typically a slightly modified airfoil. This is primarily achieved though the addition of a jet of air to a specially designed aircraft wing using a series of blowing slots that eject pressurized high velocity (above the free-stream velocity) jets of air over the trailing edge and/or leading edge. Studies have also been conducted into the addition of circulation control to bodies such ad propellers and rotors. In the early years of circulation control there were three main critical design issues with the addition of circulation control to a rotating body. The first being the exposure of the rotors angles of attack between 0-35° caused by the inflow of air through the propeller plane. Through the study of high angles of attack in wind tunnel testing, it is possible to predict the behavior of the rotor blade at these higher angles of attack. The second obstacle in the prior applications of circulation control to a propeller was the inability to achieve the response times necessary to effectively use circulation control during the rotation of the propeller.

A further requirement of circulation control applications to propeller powered aircraft is the power required to supply the airflow. An active circulation control system uses an internal pumping system which can use power from the aircraft or from an additional power source, such as a generator, to pressurize the air plenums in order to use circulation control on the aerodynamic body. With the development of unmanned aerial vehicles (UAVs), propeller performance enhancement is desirable in order to increase the thrust, and/or the overall range of the aircraft. The application of the active circulation control to the propeller, though potentially beneficial, is currently envisioned as creating technical difficulties in the supply of air to the circulation control blowing slot.

A passive system in which air can be supplied to a strategically placed circulation control blowing slot can also enhance the performance of a propeller. The proposed passive system will take advantage of the pressure differential upstream and downstream of the propeller plane, forward air velocity, stagnation pressure, and centripetal acceleration to pressurize the internal plenum of the circulation control system and thus not require an additional power source to augment the propeller of the aircraft. Also, because the system will not need to be pressurized from an outside source, no additional weight or requirements will be necessary for the aircraft other than the implementation of an updated propeller. It has been shown that through the addition of a pressure capture device on the front of a propeller, a six to fourteen percent increase in lift coefficient can be achieved simply by allowing the stagnation air ahead of the propeller to pressurize the internal plenums. Although not significant for use in larger propeller driven aircraft, for UAV applications, this can lead to a two to five percent increase in range.

INTRODUCTION

Applications of this CC technology have been in the research and development phase since the 1960's with primary emphasis on fixed wing aircraft. The initial implementations of this technology were referred to as "blown flaps." Circulation control increases the near surface velocity of the airflow over a rounded surface of an object, typically a slightly modified airfoil (Maskel, 1959, Kind, 1968, Williams, 1969). This is primarily achieved though the addition of a jet of air to a specially designed aircraft wing using a series of blowing slots that eject pressurized high velocity (faster than the free-stream velocity) jets of air over the trailing edge and/or leading edge. The most prevalent design for a circulation controlled wing, and the primary application of this study has a rounded trailing edge, and ejects the air tangential to the surface, inducing the Coandă effect (Rae, 1972). This phenomenon keeps the boundary layer jet attached to the wing surface longer than a conventional wing and thus increases the virtual length, camber, and the angle of attack of the airfoil by entraining the free-stream air near the jet. The jet blowing over the rounded trailing edge increases the circulation of the free-stream around the airfoil, or similar object, increasing the aerodynamic forces experienced. As the circulation control airflow entrains free stream air, the interaction between the object and the surrounding fluid alters the pressure distribution along the surface of the airfoil, another indication of the enhanced forces experienced by the object.

The principle motivation in the use of circulation control has been to increase the lifting force when large lifting forces and/or slow speeds are beneficial, such as at take-off and landing (Englar, 1972, Ambrosaiani, 1971). On current fixed wing aircraft, wing flaps and leading edge slats are used during landing and on take-off. The benefit of the circulation control wing is that no additional skin friction drag is created by the increased surface area created by the movement of conventional surfaces into the airflow around the wing, and the lift coefficient is greatly increased (Myer, 1972, Loth, 1976, 1984). However, as with any lifting surface, the use of circulation control increases the induced drag of the airfoil in proportion to the square of the lift coefficient (Harness, 1970, Holt, 1972, Gibbs, 1975, Angle, 2008). Additionally, these active circulation control systems require energy to pump the fluid, along with the additional hardware, weight and system complexity which can reduce the advantages achieved by the additional lift.

Circulation control has also been adapted for use on rotary-wing aircraft. In the early years of circulation control there were two main critical design issues with the addition of circulation control to a rotating body such as a helicopter rotor. The first being the exposure of the rotors on a helicopter to a moderate range of angles of attack, 0 – 50°· (Stepniewski, 1981) caused by the inflow of air through the rotor plane. Through the study of high angles of attack in wind tunnel testing, it is possible to predict the behavior of the rotor blade at these higher angles of attack (around 20-35°). The second obstacle in the prior applications of circulation control to a helicopter main rotor was the inability to achieve the response times necessary to effectively use circulation control on a rotary wing aircraft to accommodate the asymmetry of lift experienced during maneuvering flight. With the recent development, in the last 10 years, of smart materials, a specially designed near surface piezoelectric valve (Smith, et al, 2002) has been developed, enabling the concept of using circulation control on the blades of a rotary wing aircraft. Through the use of these and similar types of smart materials, the response time of actuation of circulation control can be reduced thus lending the phenomena to a wider array of applications including rotorcraft and propellers.

A further requirement of circulation control applications to rotorcraft is the power required to supply the airflow. An active circulation control system uses an internal pumping system which can use power from the original aircraft power-plant or from another additional power source, such as a generator, to pressurize the air plenums in order to use circulation control on the aerodynamic body. In the past, for some of these applications, the additional weight cost of components far outweighed the benefits derived from the use of circulation control. While new material and control strategies have made available a great variety of applications, its use for some simpler systems can still be problematic.

With the development of numerous unmanned aerial vehicles (*UAVs*), wing and propeller performance enhancement is desirable. The application of the active circulation control to either the wing or the propeller, though potentially beneficial is currently envisioned as creating technical difficulties in the supply of air to the circulation control blowing slot. Thus, a passive system in which air can be supplied to a strategically placed circulation control blowing slot could be used to enhance the performance of these aircraft. Although the use of circulation control slots on the wing will require the addition of an auxiliary pumping system, it is envisioned that the use of circulation control on the main propeller would not require this extra system and might be able to create a usable air supply on its own.

This proposed passive system will take advantage of the pressure differential upstream and downstream of the propeller plane, forward air velocity, and centripetal acceleration to pressurize the internal plenum of the circulation control system and thus not require an additional power source to augment the propeller of the aircraft.

The goal of this study was to apply a passive circulation control system inlet to an already existing propeller design, and optimize the system dimensions (i.e. slot exit, exit length, exit placement, operating rotational speed, and internal plenum dimensions) and calculate a trend study for various types of propellers and propeller geometries. Through the study of these trends, the overall change in thrust and range can then be calculated in order to show the overall effectiveness of adding a passive circulation control system to a propeller.

NOMENCLATURE

Symbol	Description	Units
A	Area	m^2
c	Chord	m
C_L	Lift Coefficient	-
C_μ	Blowing Coefficient	-
h	Height	m
L	Lift	N
P	Pressure	atm or N/m^2
r	Radial Location	m
R	Ideal Gas Constant	N-m/kg-°K
T	Temperature	°K
V	Velocity	m/s
V_j	Jet Velocity	m/s
V_o	Downstream Velocity	m/s
V_s	Upstream Velocity	m/s
z	Altitude	m
μ	Dynamic Viscosity	$kg\text{-}s/m^3$
Δ	Change in	-
ρ	Density	kg/m^3
γ	Specific Heat Ratio	-
Γ	Circulation	
∞	Freestream	-

ANALYSIS

Due to the need for improvements in propeller performance, independent of the application of a passive circulation control system, an aircraft propeller is the currently envisioned preferred embodiment of the technology. The use of a passive system for circulation control, and the corresponding system to provide air to the blowing slot, is the distinguishing feature of this proposed research. The use of the term passive system indicates that mechanical and/or electrical power is not used to supply the airflow directly to the circulation control sub-system. The use of this system can be applied to any rotating object which generates a fluid dynamic force in any fluid medium, however, for simplicity it will be described based on the first envisioned application, a propeller operating in air for an UAV.

In traditional flow control applications blowing/suction slots are located on the surface, on the leading and/or trailing edge of an airfoil, which are/is coupled to a device to create a pressure differential with the surrounding air. In order for an aircraft to gain the amount of lift needed at take-off, and potentially landing, an increase in the thrust potential of the propeller can decrease the runway length required with a relatively small amount of circulation control enhancement. It is currently believed that the rotational nature of a propeller can provide limited pumping capability to supply the circulation control sub-system, mitigating the need for the additional weight of an external pumping system and the associated ducting.

In conventional propeller theory, the free stream air that is "pulled" into the propeller imparts a pressure on the hub of the propeller (Theodorsen, 1948). This pressure on the front side of the propeller is greater than the local atmospheric pressure. In fluid dynamics, pressure flows from high pressure to low pressure, thus for air, or similar fluid, a local velocity will be generated. By strategically placing a channel within the hub and through the blades of the propeller the higher pressure at the center of the hub can be utilized to drive a flow to the circulation control blowing slot. In addition, the fluid will be accelerated due to the centripetal forces applied by the rotational nature of the propeller.

The higher pressure at the hub is loosely dependent on the propeller performance, only from the standpoint that a higher static thrust propeller will impart more velocity to the surrounding fluid (Glauert, 1943). Thus, any airfoil shape and any propeller diameter can be utilized with the passive circulation control propeller, however the mass flow rate required by the circulation control sub-system will vary. The variation in mass flow may require slightly different sized blowing slots and ducting configurations, however the passive circulation control air, or fluid, supply concept is unchanged. By implementing an inlet to a fluid flow channel near the hub of the propeller, the elevated pressure supplies the circulation control sub-system with air.

Preliminary genesis for a passive circulation control system installed on a propeller was to address a need for increased range of unmanned aerial vehicles, as well as the possibility for applications to autonomous underwater vehicles.

An acceptable way of studying the design factors as components in the design as well as how they affect one another is to conduct a sensitivity analysis including all of the variables which are predicted to have an influence in order to find a predicted trending from previously known historical data. Although this type of analysis highlights the variables which have the greatest degree of impact on the system and eliminates the variables which have little impact, this form of study cannot take the place of conventional testing of a new system. However, it can be powerful in using theoretical data and predictions to obtain an idea of what may happen when a prototype is built and examined. This method is also helpful in predicting the necessary hardware needed in order to complete testing and analysis of a prototype, including data acquisition components, testing schedule, and experimental setup.

SENSITIVITY ANALYSIS

In order to begin the process of designing and conducting an experiment, it is necessary to look at all of the variables that are considered to have a potential impact on the final result of the propeller performance. In the case of a passive circulation controlled propeller, as in this research, there are several factors which need to be studied. Among these important variables, are parameters such as free stream velocity, rotational speed of the propeller, propeller diameter, sweep, area, and volume of the circulation control ducts and plenums, slow exit area, exit pressure/velocity, mass flow rate, and fluid density. Also needing to be researched are certain blade characteristics, such as, twist, chord length, and cross-section. The final area of concern in this research is the design of a capture device which will need to efficiently capture free stream air and to pressurize the internal circulation control plenum of the propeller. Capture device areas of interest include inlet and outlet areas, mechanical design, and rate of fluid rejection. This section explains some of the preliminary work conducted in order to determine the impact these variables will have on the entire system.

EFFECT OF FORWARD VELOCITY ON INLET AND OUTLET PRESSURES

The inward velocity through the rotor plane can be converted to a pressure force on the propeller hub, and through an analysis of the Bernoulli Equation the relationship between pressure and velocity is defined by Equation 1 (Dommasch, 1953, Anderson, 2001, Bertin, 2002). With an inlet on the propeller hub feeding to a channel in the propeller blade, pressurized air induced from forward thrust (T) can move through the inlet at the high pressure stagnation location, on the front of the propeller, to a low pressure region (atmospheric, or less) at the channel exit. The Bernoulli Equation can be used to estimate the inlet pressure based on the assumption that there is a stagnation point at the hub and the initial flow conditions at an elevation (z) are seen as atmospheric pressure (P), density (r), and temperature and the forward speed (V) of the inflow of air is zero. Conversely, when the propeller operates in a compressible fluid, similar to air above a speed where the Reynolds number reaches 0.3 or more, it is necessary to calculate the effect inside the plenum using Equation 3, where the gravitational constant (g) can be taken as the rotational component of the mass spinning with the propeller and the radial distance (r) is the distance away from the center hub along the length of the propeller blade. Using this equation, the pressure can be predicted along the length of the blade at several different radial locations incorporating the effect of centripetal forces on the mass of air (Thwaites, 1931, Kueth, 1967, Katz, 2001).

$$L = \rho_\infty V_\infty \Gamma \qquad \text{Equation 1}$$

$$P + 0.5\rho V^2 + \gamma z = \text{constant} \qquad \text{Equation 2}$$

$$\int \rho g r dr \qquad \text{Equation 3}$$

As an estimate of the impact of the forward flight speed on the pressure changes on the hub of the propeller, the total pressure as determined from the Bernoulli Equation was used to generate Figure 1 which indicates that at a stationary forward flight speed, atmospheric pressure is achievable throughout the propeller plenum. As the forward flight speed increases, so does the available plenum pressure predicted due to stagnation forces directly upstream of the propeller plane. This trend is shown for a blade diameter of 0.4 meters (1.3 feet) (Delano, 1945). This increase in pressure due to the flight speed allows for a greater flow rate of the captured air for circulation control uses.

In Figure 1, several propeller rotational speeds are shown having similar trends. Rotational speeds of one, two, four, eight and sixteen thousand rpm's, also shows a similar predicted trend of pressure addition to the atmospheric value due to the forward flight speed. At a forward velocity of 120 m/s, and 8,000 rpm's, there is a predicted increase from 1.1 atm to 1.2 atm. Similarly, the study conducted at 16,000 rpm also predicts an increase from 1.4 atm to 1.51 atm. Differences in the predicted pressure increase is due to suction forces on the propeller at each rotational speed as well as a change in the stagnation force on the propeller traveling at a forward speed of 120 m/s. A higher rotational speed produces a greater suction force on the upstream face of the propeller. Each of these three studies was performed using a blade diameter of 0.4 meters. This estimation does not, however, take into account the effect that the centripetal acceleration has on the mass of air in the circulation control slot.

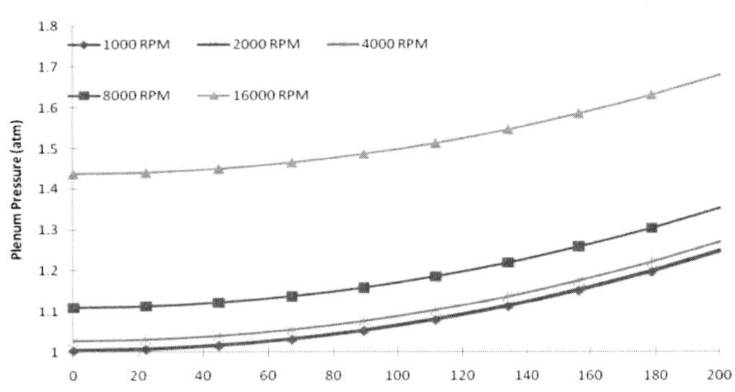

Figure 1. Effect of Forward Speed on Plenum Pressure

In order to verify the results of the trend study found above, it was necessary to use previous experimental data taken in 2004, (Englar, 2004). The results of the study used to verify the predicted results are given in Figure 2, below. In the historical study, the results, obtained in 2004, show the slot exit velocity over a range of airspeeds. Assuming similar ducting areas and rotational speeds, Figure 2, shows an approximate fit to the data studied by Englar in 2004. Englar's work, as in this study, shows that as the forward speed increases, the exit velocity of air used in circulation control applications also increases, therefore having a greater effect on the streamlines of the airfoil. Using the results of this 2005 study, it is possible to predict that at higher forward velocities, higher exit velocities are available for circulation control.

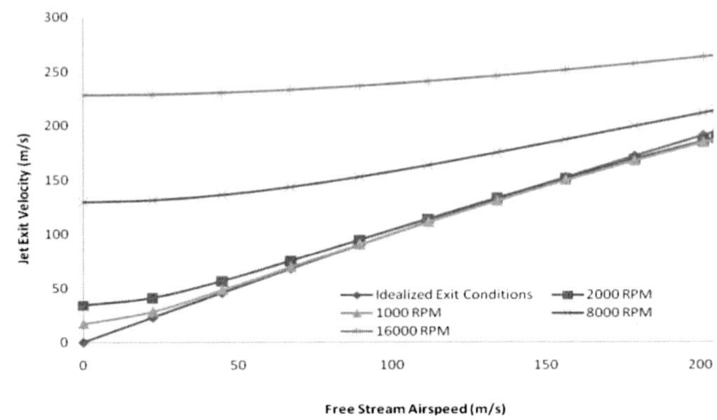

Figure 2. Effect of forward speed on exit velocity effect of propeller diameter on change in pressure

According to propeller theory, the thrust of a specified propeller can be related to a change in pressure between the upstream and downstream pressures of the propeller disc. Idealizing the propeller plane as a thin disk in a slipstream field of flow, the thrust (T) produced by the propeller is directly proportional to the frontal area on the propeller (A) and the squares of the upstream (V_s) and downstream (V_o) velocities as shown in Equation 4 (Nelson, 1944).

$$T = 0.5A\rho(V_s^2 - V_o^2) \qquad \text{Equation 4}$$

In order to relate the thrust to the pressure differential, the Bernoulli equation, Equation 4, is solved for the pressure difference (ΔP) in Equation 5, which, when substituted into Equation 4, results in Equation 6 for finding the thrust output (T) of the propeller (Nelson, 1944).

$$\Delta P = 0.5(V_s^2 - V_o^2) \qquad \text{Equation 5}$$

$$T = A\Delta P \qquad \text{Equation 6}$$

With a value for thrust related to the pressures upstream and downstream of the propeller disk, Figure 3 was created to show the effect that the propeller diameter had on the pressure change across the disc. As the propeller diameter increases, the difference in pressure upstream and downstream of the blade decreases significantly. This suggests the pressure differences that are most useful in passive circulation control applications are found toward the lower end of the propeller diameters. This is due to the fact that this particular circulation control application takes advantage of the pressure difference between the upstream and downstream sides of the propeller in order to pressurize the plenums inside the blades. In keeping with UAV applications, the diameter of the propellers used are generally between 0.2 and 0.5 meters in diameter. According to this trend, the propellers in this area see around 0.75-0.50 atm (7.35 – 11.02 psi) pressure change. This change is predicted to be an adequate air pressure to run boundary layer control and/or circulation control applications assuming the exit slot can be designed to provide the required exit slot ejection velocity.

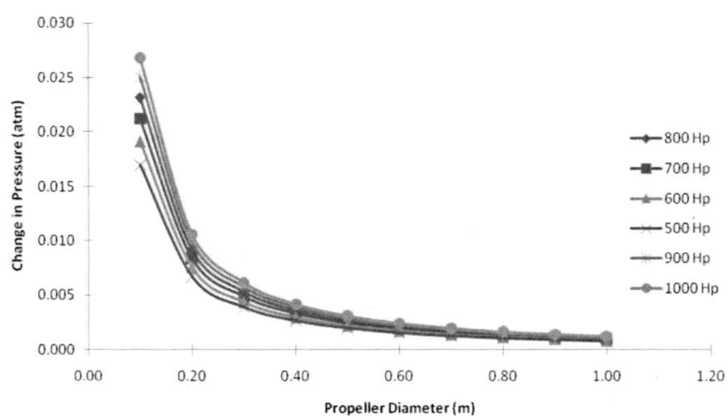

Figure 3. Change in Pressure as it Relates to Varying Propeller Diameter

EFFECT OF BLADE DIAMETER AND CENTRIPETAL ACCELERATION ON PLENUM PRESSURE

It is currently envisioned that a constant cross-section capillary will be used initially throughout the blades of the propeller, minimizing the regions in which pressure can build up and restrict the flow. When a full design is undertaken, it will be necessary to attempt to make the exit slot the limiting dimensions of the channel in order to prevent "choked" flow through the capillaries of the propeller blades. This can become a challenging task as there is only so much room available for use on the interior of a propeller blade.

It is important to note that the use of a constant cross section is not a requirement of the design, it was chosen for simplicity and ease of fabrication. The static pressure outside the blowing slot was considered to be standard atmospheric conditions, thus the stagnation pressure is elevated and the blowing slot can be analyzed as a nozzle with a specified back pressure (inside the flow system) and surroundings. Another form of the Bernoulli equation, Equation 15, was used to determine the exit velocity of the passive circulation control augmented propeller. Equation 2, from (Englar, 2004), which is a derived expression based on Bernoulli, Ideal Gas, and Isentropic Equations, was also used to verify the magnitude of the jet exit. Equation 7 is based on the specific heat ratio (g), and the gas constant (R) of the fluid as well as the temperature (T) and internal pressure (P) of the plenum. As noted above, it is necessary to take into account the pressurization of the plenum due to centripetal forces using Equation 3 when the propeller operates above sonic speeds and a normally incompressible fluid, such as air, becomes a compressible fluid.

$$V=\{[2\gamma RT/(1-\gamma)][1-(P_\infty P)^{(\gamma-1)/\gamma}]\}^{0.5} \qquad \text{Equation 7}$$

Below in Figure 4, a trend study of plenum pressurization capabilities due to centripetal acceleration as well as forward velocity was completed on variety of blades ranging from 0.1 meters to 1.0 meter. Also, for simplicity, the channel area is assumed to be square at 0.05 by 0.05 meters. The trend shows a linear-like increase in the amount of pressure in the plenum as the rotational speed increases as well as when the propeller diameter increases. It should be noted that as the tip speed reaches transonic or supersonic speeds, that is the tip speed surpasses the local speed of sound, the performance no longer increases.

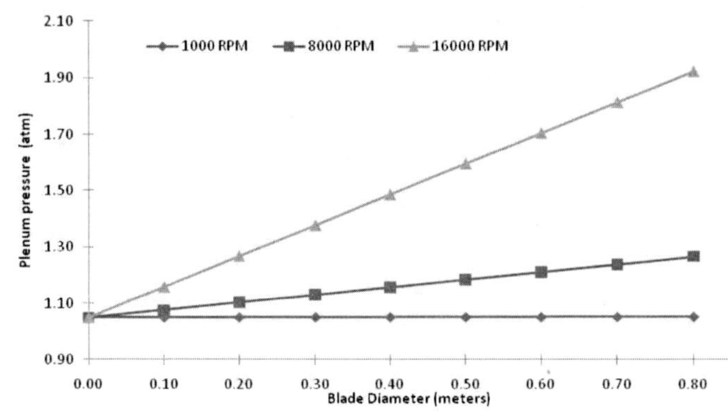

Figure 4. Plenum Pressure versus Blade Diameter Function of Propeller Rotational Speed

This preliminary study shows that the centripetal force experienced by the fluid while rotating has a large impact on the internal pressurization of the circulation control plenum of the proposed augmentation. This trend study was completed using predicted data from a 16 inch diameter propeller spinning at three different rotational speeds which predicts that the majority of the internal pressure useful as a circulation control air supply will be provided by the centrifugal accelerations acting on the volume of air while the blades are turning and not due to the forward velocity of the aircraft, as expected.

INLET DESIGN

Another potentially important component of this proposed design is the centrically mounted flow capture device, as shown in Figure 5, directly upstream of the propeller. The sides of the capture mechanism effectively form a cone around the propeller centerline. This image represents the edge of the flow capture device as a straight line, this is merely for simplicity and the device could be a continuously variable curve to promote the smooth flow of air into the high pressure, stagnation zone of the propeller, which is located at the base of the conical structure. To minimize the influence of the drag force created by the flow capture device, it is currently predicted that the maximum diameter of the „cone' will be much smaller than the diameter of the original propeller hub. The width, distance the device extends in front of the propeller is envisioned to be between ¼ and 1 hub diameters, depending on the design flight speed of the craft.

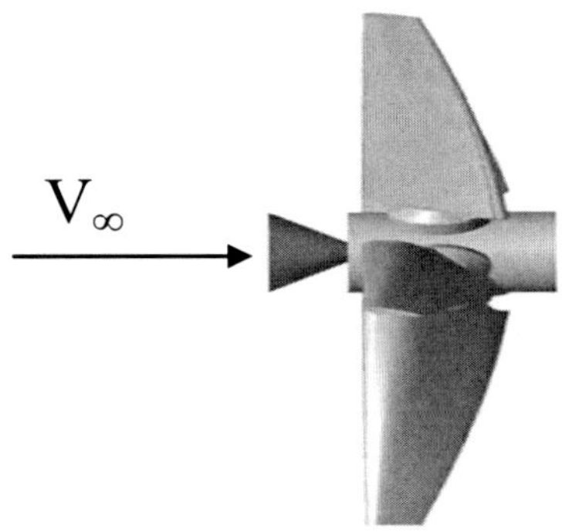

Figure 5. Pro-ENGINEER Representation on a Circulation Control Augmented Propeller (Not to Scale)

There are several important aspects which will need to be considered when designing the pressure inlet for the passive circulation control system. Among these variables are the inlet diameter, inlet length, rotational speed, inlet-to-propeller interface, and the effect of forward velocity on the pressure or volume of fluid rejection due to the rotation of the propeller apparatus. Figure 6 shows a rendering of a fundamental capture nozzle for the proposed passive circulation control propeller apparatus. The preliminary design considerations are expected to be found using a flow visualization/computational software application (such as Fluent) in order to be able to easily change the inlet and outlet dimensions and to predict the trends in effectiveness and performance of the design before it is applied to an actual model. The largest concern of this type of apparatus is the rejection of the fluid due to the centripetal forces experienced by the fluid as the body rotates. Figure 7 shows an initial conceptual design to try to reduce the amount of fluid rejection experienced during rotation. A conical shaped set of veins are located on the inner face of the capture device in order to direct the airflow into the pressurized plenums inside the propeller airfoils. Again, the testing of this concept will be completed in a computational fluid dynamics solver software package before it is experimentally employed in order to predict the optimum design dimensions and reduce the investments expended on building many different physical models.

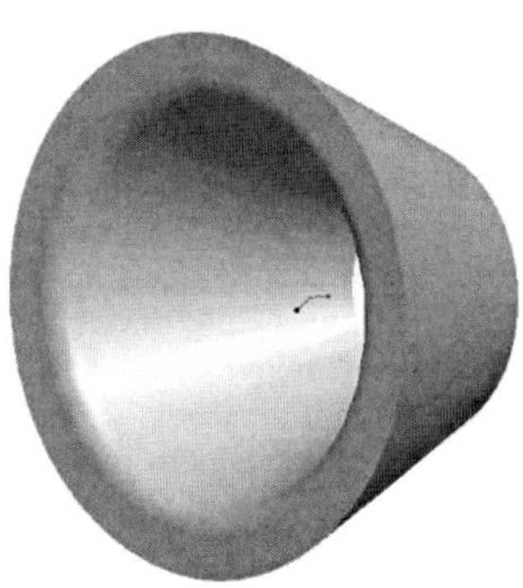

Figure 6. Fundamental Design of Capture Apparatus (not to scale)

Figure 7. Initial Capture Apparatus Design for Reduction of Air Rejection (not to scale)

The feasibility of this device is closely related to several areas of interest including the pressurization availability, the increase of drag forces due to the addition of a rounded trailing edge, and finally the addition of a mass capture device on the front plane of the propeller.

The rounded trailing edge of circulation control devices has been shown to increase the drag in comparison to that of non-circulation control bodies of the same basic shape, with a sharp trailing edge. In general circulation control cases, the increase in lift greatly overcomes the increase in drag on the aerodynamic body and the increased thrust needed to drive the circulation control system is not significant in comparison to the advantages of the greater capacity for lift. It is also a well studied fact that the drag increase on the rounded trailing edge or the circulation control body is greatest when the circulation control system is first turned on or cycled between on and off positions. The proposed system, again, is a passive system and is never turned off. With this idea in mind, it is predicted that without needing to turn the system on and off it will perform in a better capacity to increase lifting forces of the propeller, even with the reduced jet exit velocity, and with a minimal increase in the drag forces.

The addition of a mass capture device on the frontal stagnation area of the propeller hub at this point of the research is merely a concept and needs to be studied further. In the preferred design, the capture device would be situated inside the existing hub space and not add an additional frontal area to the propeller hub. The addition of drag forces due to this apparatus is predicted to be less than even the current cone design because of the suction forces predicted to siphon mass and pressure into the internal plenum of the circulation control system which is predicted to reduce the overall stagnation force on the hub. It is the pressure force from this stagnation area added to the centrifugal forces from the rotating propeller that provides the motive force for implementation of this passive system

According to previous studies and Equation 9, a jet momentum coefficient (C_μ) of 0.1 can be used to provide a lift coefficient increase of 3.0 at a plenum pressure to freestream pressure ratio of 1.25 (P/P_∞), similar to the maximum pressure calculated due to stagnation forces (Englar, 2004).

$$C_\mu=2h_jV_j^2/(cV_\infty^2) \quad \text{Equation 9}$$

It is clear that this system requires a systems approach for utilization and will force a redesign of the entire propeller blade and the way they are used. Numerous studies have been conducted in the past on propellers including the use of active blowing slots to create large increases in thrust delivery. As with most active circulation control systems used, the hardware, weight, and pumping penalties far outweigh the benefits for all but a few unique applications. In contrast, and particularly regarding fuel consciousness, any improvements, even single digit improvements will provide value to the industry. It is believed that this technology will provide such a benefit.

Preliminary research, conducted in this sensitivity analysis, into this passive circulation control technology shows that a maximum increase in lift coefficient of 0.2 is achievable at 75% of the radius of the propeller blade without drastically changing any of the existing hardware on the aircraft. A retro-fitted propeller is currently envisioned as the only hardware change to the entire aircraft. The use of this passive circulation control system avoids the need for excess weight additions, such as pumping systems, to control the augmented jet velocity. Although the research is not complete, the selection of the position along the radius and the jet exit dimensions will allow for optimization of the entire system, as well as making the exit area the limiting dimension of the system to provide consistent pressurization for the jet velocity. Currently, it is believed that with the increased lift coefficient and the siphoning of the stagnation pressure, the overall drag force on the system will be reduced and a total thrust increase can be seen of 4 to 5 percent.

This technology, currently envisioned for unmanned aerial vehicle applications, suggests a single digit percent increase in the overall range of the aircraft. Although small in conventional standards, this increase could greatly increase the overall versatility of the aircraft and is particularly attractive because there is no need for additional hardware additions to the currently used aircraft.

PRELIMINARY COMPUTATIONAL RESULTS

In order to see the effect of adding circulation control to a rotating body, such as a propeller, the aerodynamic forces were studied. Examining a propeller blade is similar to an airfoil in geometry, however, a propeller blade sees an increased angle of attack as the blade location approaches to the central hub. This imparts a "twist" on the blade.

A computational model has been built using Fluent software, shown in Figure 8. This model, of a two-dimensional airfoil was studied to show the increase in lift across a full spectrum of angles of attack. For this study, a NACA 0018 cross section airfoil was chosen, after finding that CC is most effective on a non-elliptic airfoil. The model was built to have a span of 8 inches matching the maximum span the actual propeller has. Both the unmodified version and modified version (addition of circular trailing edge) were analyzed. Below, Figure 9, shows one contour study analyzed on the unmodified airfoil.

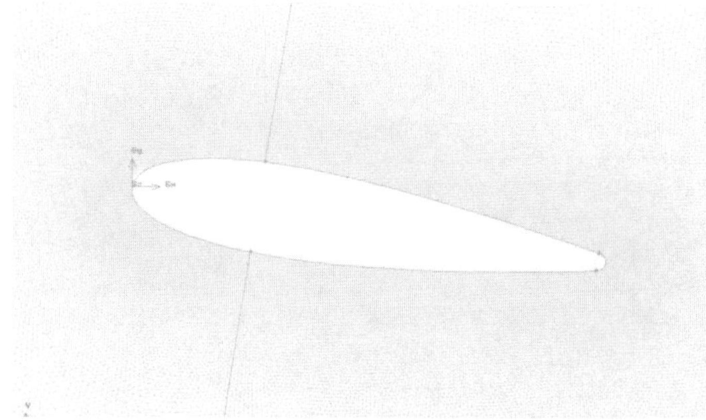

Figure 8. Fluent Airfoil Model

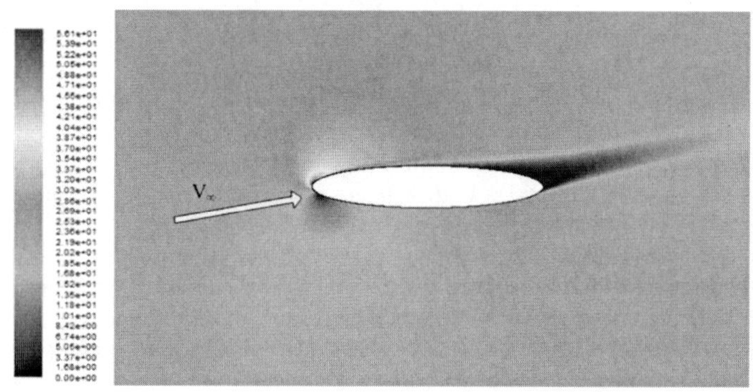

Figure 9. Velocity Contour Plot at 15° AoA

The NACA 4-(5)(08)-03 two bladed propeller has a maximum twist angle of 66 degrees at a radial location of 0.3 (*r/R*) (Delano, 1945). The computational study was done throughout a series of angles of attack ranging from 0-100 degrees, in order to predict the lift increase throughout the entire radial location of the blade. As it has yet to be determined where the CC ejection slot will be located to best augment the propeller, each available location needed to be analyzed throughout the radius. In Figure 10, the lift force profile for both the un-augmented and augmented models are shown as it corresponds to each angle-of-attack. This study was conducted at a Re of 40k which corresponds to a freestream velocity of 50 mph, at sea level. The C_μ values of 1% and 10% (corresponding to 17 and 87 mph, respectively) are shown on the plot below and can be seen to increase the lifting force available to the model by a maximum of 20% in the region of angle of attack the propeller blade will experience. Although some of the dimensions of this model are ideally modeled (i.e. exit slot and chord) it illustrates an effective trend that the CC technology is predicted to have on the propeller.

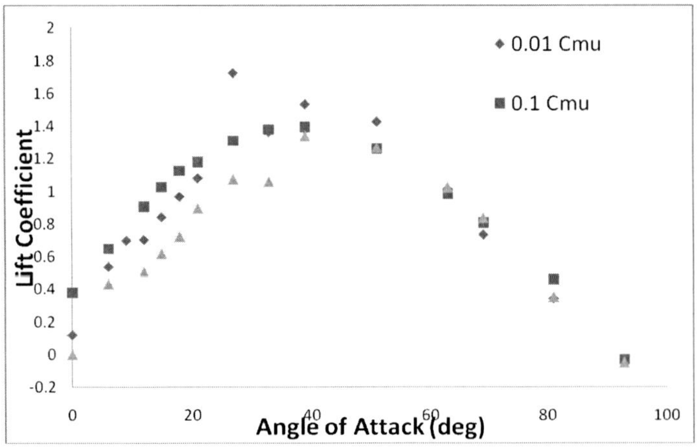

Figure 10. Lift Coefficient Augmentation

CONCLUSIONS AND RECOMMENDATIONS

It is clear that this design requires a systems approach for utilization and will force a redesign of the entire propeller blade and the way they are used. Numerous studies have been conducted in the past on propellers including the use of active blowing slots to create large increases in thrust delivery. As with most active circulation control systems, use of the hardware, weight, and pumping penalties far outweigh the benefits for all but few unique applications. In contrast, and particularly regarding fuel consciousness, any improvements, even single digit improvements will provide value to the industry. It is believed that this technology will provide such a benefit.

Preliminary research into this passive circulation control technology shows that a maximum increase in lift coefficient of 0.2 is achievable at 75% of the radius of the propeller blade without drastically changing any of the existing hardware on the aircraft. A retro-fitted propeller is currently envisioned as the only hardware change to the entire aircraft. The use of this passive circulation control system avoids the need for excess weight additions, such as pumping systems, to control the augmented jet velocity. Although the research is not complete, the selection of the position along the radius and the jet exit dimensions will allow for optimization of the entire system, as well as making the exit area the limiting dimension of the system to provide consistent pressurization for the jet velocity. Currently, it is believed that with the increased lift coefficient and the siphoning of the stagnation pressure, the overall drag force on the system will be reduced and a total thrust increase can be seen of 4 to 5 percent. This technology, currently envisioned for unmanned aerial vehicle applications, suggests a single digit percent increase in the overall range of the aircraft. Although small by conventional circulation control application standards, this increase could greatly increase the overall versatility of the aircraft and is particularly attractive because there is no need for additional hardware additions to the currently used aircraft.

REFERENCES

Abbott, I.H., Von Doenhoff, A. E., Theory of Wing Sections, Dover Publications, Inc, New York, NY, 1959.

Ambrosiani, J.P., "Analysis of a Circulation Controlled Elliptical Airfoil," Ph.D. Dissertation, Department of Mechanical and Aerospace Engineering, West Virginia University, Morgantown, WV, 1971.

Anderson Jr, J.D., Fundamentals of Aerodynamics, McGraw Hill, Inc., New York, NY, 2001.

Angle II, G.M., "Aerodynamic Benefits of Near-Surface-Actuated Circulation Control Blowing Slots for Rotorcraft Use," Ph.D. Dissertation, Department of Aerospace Engineering, West Virginia University, Morgantown, WV, 2008.

Bertin, J.J., Aerodynamics for Engineers, Prentice –Hall, Inc., Upper Saddle River, NJ, 2002.

Delano, J.B., "Investigation of the NACA 4-(5)(08)-03 and NACA 4-(5)(10)-03 Two-Bladed Propellers at Forward Mach Numbers to 0.725 to Determine the Effects of the Camber and Compressibility on Performance," National Advisory Committee for Aeronautics Report No. 1012. Langley Field, VA. 1945.

Dommasch, D.O., Elements of Propeller and Helicopter Aerodynamics, Pitman Publishing Corporation, New York, New York, 1953.

Englar, R.J., Williams, R.M. "Test Techniques for High Lift, Two-Dimensional Airfoils with Boundary Layer and Circulation Control for Application to Rotary Wing Aircraft." Canadian Aeronautics and Space Journal, Vol. 19, No. 3, 1973.

Englar, R.J., "Overview of Circulation Control Pneumatic Aerodynamics: Blown force and Moment Augmentation and Modification as Applied Primarily to Fixed Wing Aircraft," Proceedings of the 2004 NASA/ONR Circulation Control Workshop, NASA CP-2005-213509, Pages 23-66, 2004.

Gibbs, E.H., "Analysis of Circulation Controlled Airfoils," Ph.D. Dissertation, Department of Mechanical and Aerospace Engineering, West Virginia University, Morgantown, WV, 1975.

Glauert, H. The Elements of the Aerofoil and Airscrew Theory, Cambridge University Press, New York, NY, 1943.

Harness, G.S., "An Experimental Investigation of a Circulation Controlled Cambered Elliptical Airfoil," Masters Thesis, Department of Aerospace Engineering, West Virginia University, Morgantown, WV, 1970.

Holt, D.J., "Circulation Controlled Airfoil with Pulsed Flow." Masters Thesis Department of Mechanical and Aerospace Engineering, West Virginia University, Morgantown, WV, 1972.

Katz, J., Plotkin, A. Low Speed Aerodynamics, Cambridge University Press, Cambridge, NY, 2001.

Kind, R.J., Maul, D.J., "An Experimental Investigation of a Low-Speed Circulation-Controlled Aerofoil," The Aeronautical Quarterly, Vol. XIX, May, 1968.

Kueth, A.M., Schetzer, J.D., Foundations of Aerodynamics, John Wiley and Sons Inc., New York, NY, 1967.

Loth, J.L., Fanucci, J.B., Roberts, S.C., "Flight Performance of a Circulation Controlled STOL Aircraft." Journal of Aircraft, Vol. 13, No. 3, 1976.

Loth, J.L., Boasson, M. "Circulation Control STOL Wing Optimization," *Journal of Aircraft*, Vol. 21 No. 2, February, 1984.

Maskell, E.C., Spence D.A., "A Theory of the Jet Flap in Three Dimensions," *Proceedings of the Royal Society of London, A Mathematical and Physical Sciences Series*, Vol. 251, No. 1266, 1959.

Myer, D.P., "An Investigation of a Circulation Controlled Cambered Elliptical Airfoil with a Rounded Trailing Edge," Masters Thesis, Aerospace Engineering Department, West Virginia University, Morgantown, WV, 1972.

Nelson, W.C. *Airplane Propeller Principles*, Johan Wiley and Sons, Inc. New York, NY, 1944.

Rae Jr., W.H., Shindo, S., "Comments on V/STOL Wind Tunnel Data at Low Forward Speeds." Washington university, Seattle, WA, 1969.

Smith, J.E., Loth, J.L., Craven, R.P.M., Bond, R., "Piezoelectric Actuators for Circulation Controlled Rotorcraft," US Patent 6,425,553 B1. 2002.

Stepniewski, W.Z., Kweys, C.N., Rotary Wing Aerodynamics, Dover Publications, Inc., New York, NY, 1981.

Theodorsen, T, Theory of Propellers, McGraw-Hill Book Company, Inc. New York, NY, 1948.

Thwaites, B., Incompressible Aerodynamics – An Account of the Theory and Observation of the Steady Flow of Incompressible Fluid Past Aerofoils, Wings, and Other Bodies, Oxford University Press, New York, NY, 1960.

Williams, R.M., "Some Aspects on Rotor Circulation Control." CAL/AVLABS 3rd Symposium of Rotary Wing and V/STOL Aircraft, Buffalo, NY, 1969.

Proceedings of the ASME 2010 International Mechanical Engineering Congress & Exposition
IMECE2010
November 12-18, 2010, Vancouver, British Columbia, Canada

IMECE2010-38604

EFFECT OF LEADING EDGE BLOWING SLOTS ON STALL ANGLES OF A 10:1 ELLIPTICAL AIRFOIL

Jonathan Kweder, Mary Ann Clarke, James E. Smith
Center for Industrial Research Applications, West Virginia University
Morgantown, WV USA

ABSTRACT

Traditional uses of circulation control have been studied since the early 1960's and have been developed primarily using trailing edge slots over a rounded trailing edge in order to take advantage of the Coandă effect. The leading edge activated slots allow jets of air to enter the freestream flowing around the airfoil thus enhancing the energy of the lift force. The main purpose of circulation control for fixed wing aircraft is to increase the lifting force when large lifting forces and/or slow speeds are required, such as at take-off and landing. While there is a significant increase in the lifting forces achievable through the use of circulation control, there is also an inherent increase in the drag force on the airfoil (Abramson, 2004, Loth, 1976, 1984).

Current effects of circulation control on stall angles of airfoils are not well documented and thus needs to be studied. Stall occurs when a sudden reduction in lift occurs caused by a flow separation between the incoming air flow and the lifting surface. The angle at which this happens is commonly called the critical angle of attack, and is typically between eight and twenty degrees depending on the wing profile, aspect ratio, camber, and planform area.

For this study, a 10:1 aspect ratio elliptical airfoil with a chord length of 11.8 inches and a span of 31.5 inches was inserted into the West Virginia University Closed Loop Wind Tunnel and was tested at varying wind speeds (80, 100, and 120 feet per second), angle of attack (zero to sixteen degrees), and blowing coefficients, ranging from 0.0006 to 0.0127 depending on internal plenum pressure. By comparing the non-circulation controlled wing with the active leading edge slot circulation control data, a trend was found as to the influence of the circulation control exit jet on the stall characteristics of the wing. For this specific case, when the circulation control is in use on the 10:1 elliptical airfoil, the stall angle decreases, from eight degrees to six degrees, while providing up to a 46% increase in lift coefficient.

INTRODUCTION

In recent years there has been a renewed interest in high lift airfoil sections, as well as lift augmentation devices. This interest has been generated mainly for use on both fixed and rotary wing vertical/short take-off and landing (V/STOL) aircraft. In addition to the small areas that these aircraft need to land, they are incredibly flexible in their fight envelope. A problem with these aircraft is the amount of lift that is required to get the aircraft to take-off vertically or on small runways.

A way of increasing the lift on an airfoil is to use circulation control (CC) technology. Circulation control is a form of lift enhancement that relies on the blowing of high velocity air around the leading and/or trailing edge of the lifting body. Circulation control works by increasing the near surface velocity of the airflow over a specially designed aircraft wing. For use with rear blowing slots, the wing normally has a rounded trailing edge, which ejects the air tangentially, through these slots, inducing the Coandă effect. This phenomenon keeps the boundary layer jet attached to the wing surface longer than a conventional wing and thus increases the lift generated on the wing surface due to the relaxation of the Kutta condition for the rounded trailing edge (Myer, 1972, Ambrosiani, 1971). The leading edge slots are positioned tangential to the wing surface, adding energy to the boundary layer. The circulation control airflow adds to the lift force through conventional airfoil lift production, by altering the circulation of stream lines around the airfoil.

Circulation can be described as the line integral around a closed path of the velocity component tangential to the path (Kuethe and Schetzer, 1967). Also, the Kutta-Joukowski Theorem states that the lift per unit span of an airfoil in moving air is equal to the product of the circulation, free stream velocity and the density of the fluid that the lifting body is moving through (Anderson, 2001). This shows that the total lift force on the wing can be significantly controlled by restricting the circulation around it. As with changing the angle-of-attack (AOA) of the wing to increase lift, it is necessary to analyze how CC affects the stall angles of the wing. It is the purpose of this study to identify and define the angles at which stall occurs without circulation control and compare them to the experimental stall angles using circulation control for leading edge blowing slot applications.

NOMENCLATURE

Symbol	Description	Units
A	Area	ft^2
b	span	ft
c	Chord	ft
C_L	Lift Coefficient	-
C_μ	Blowing Coefficient	-
h	Height	ft
L	Lift	lb
P	Pressure	atm or psi
r	Radial Location	ft
R	Ideal Gas Constant	ft-lb/slug-°R
S	Wing Area	ft^2
t	Slot thickness	ft
T	Temperature	°R
V	Velocity	ft/s
V_j	Jet Velocity	ft/s
V_o	Downstream Velocity	ft/s
V_s	Upstream Velocity	ft/s
w	Error Term	-
z	Altitude	ft
μ	Dynamic Viscosity	$slug\text{-}s/ft^3$
Δ	Change in	-
ρ	Density	lb/ft^3
γ	Specific Heat Ratio	-
Γ	Circulation	-
∞	Freestream	-

CIRCULATION CONTROL HISTORY

Circulation control (CC) as a high-lift device is normally used on the main wing of an aircraft. This technology has been in the research and development phase for over sixty years, primarily for fixed wing aircraft when the early models were referred to as "blown flaps." The first reported use of blowing slots to control lift is attributed to H. Hagedorn and P. Ruden, in 1938, who noticed an unaccountable increase in lift at high blowing rates during investigations into boundary layer control on a flap (Maskell and Spence, 1959). Interest in active blowing systems increased with the arrival of the turbojet engine, initially in Great Britain and France on a jet flap configuration.

The flow of a fluid over curved surfaces has since been studied for a variety of applications. Circulation control works by increasing the near surface velocity of the airflow over the leading edge and/or trailing edge of a specially designed aircraft wing using a series of blowing slots that eject high velocity jets of air (Kind and Maull, 1968). This keeps the boundary layer jet attached to the wing surface longer than a conventional wing and thus increases the lift generated on the wing surface. The circulation control airflow adds to the lift force through conventional airfoil lift production by altering the circulation of stream lines around the airfoil.

The main purpose of the circulation control for fixed wing aircraft is to increase the lifting force when large lifting forces and/or slow speeds are required, such as at take-off and landing. Wing flaps and slats are currently used during landing on almost all fixed wing aircraft and on take-off by larger jets. While flaps and slats are effective in increasing lift, they do so with a penalty of increased drag. The benefit of the circulation control wing is that no extra drag is created from the movement of surfaces into the airflow around the wing and the lift coefficient is greatly increased. However, with the use of circulation control to increase lift coefficient, there is an inherent increase in the induced drag over the airfoil.

In past trials, a circular airfoil shape was used to analyze the potential fluid flow, (Theodorsen, 1931, Katz, 2001). The original methods are for frictionless, incompressible fluids, thus not truly valid for circulation control airfoils, especially those using active circulation control, due to the compression of air at the blowing slot. In 1975, more experimentation was done using the Theodorsen method in a potential flow analysis of circulation controlled airfoils (Gibbs, 1975). A Coandă simulation was then created which under-predicted the decay of the maximum jet velocity.

From previous experimentation, four main benefits were achieved by using an active circulation control method on fixed wing aircraft to control moment augmentation (Ambrosiani, 1971 and Englar, 2005). These benefits are:

1. Very small movement, or even non-moving, control surfaces are required,
2. Lift augmentation can be achieved, independent of the airfoil angle of attack,
3. Jet turning angle is no longer limited by physical jet exit angle or flap deflection angle,
4. Very high force augmentation can be achieved per unit blowing momentum input.

The pitching moment (C_m) and center-of-pressure (C_p) of the model are very important parameter when testing the stall angles in a wind tunnel environment. In 1970, two models were examined for the change in the pitching moment and center of pressure, the results concluded that at three different angles of attack, neither of these two parameters are significantly affected by a change in lift coefficient (Harness, 1970).

Very few stall experiments were conducted in past work with a circulation control airfoils, however, an experiment was conducted on a 20% elliptical airfoil with 5% camber where the blowing velocity (C_μ) was varied between 0.006 to 0.119 with a constant free stream velocity of 100 m/sec (Harness, 1970). Stall angles of this airfoil were increased from nine to fourteen degrees through this range of C_μ. This stall study did not, however, take into account leading or trailing edge separation effects, only the effects of the entire suction surface of the airfoil. Unlike the model used in this current experimentation, Harness's airfoil had an unaltered trailing edge, so the Kutta Condition still held true.

WIND TUNNEL MODEL

For this experiment, an airfoil was designed and constructed to have both a leading edge and trailing edge blowing slot, completely independent from one another. Figure 1 shows a computer aided drawing of the inside of the wing model.

Figure 1. Inside Structure of Circulation Control Wing (Computer Aided Drawing)

The center spar which runs along the span of the wing is sealed to both the upper and lower skins so that the two air plenums are separate from one another. There are also five porous walls situated before both the leading and trailing edges, used in an attempt to create a uniform exit jet pressure. The orientation of the wing model in the wind tunnel with respect to the free stream velocity is shown in Figure 2. It can also be seen that the leading blowing slot has exits which are tangential to the wing surface.

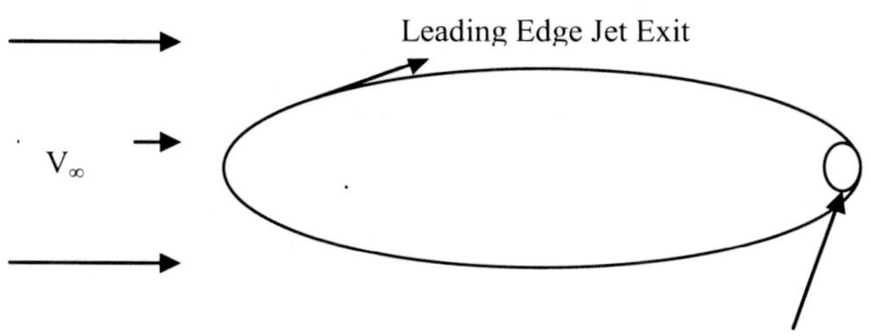

Figure 2. Airfoil and Slot Location

The circulation control model was suspended in West Virginia University's Closed Loop Wind Tunnel via a support structure which reached around the top and bottom of the test section, but did not allow the model to touch the wind tunnel. This reduced the amount of errors potentially caused by vibration that might be transferred to the model through the tunnel. The loading on the model was recorded using four linear load cells and because the model was also supported by these load cells, it was necessary to record the tare values at each angle of attack.

BLOWING SLOT CONTROLS

A cylindrical cam was used to control the jet outflow through the circulation control exit slot. This cam had the ability to be in three different positions, simply by turning them and was designed to seal one side of the wing while opening the other side. The "top open position" allows the circulation control jet to exit on the top side of the airfoil. Similarly, the "bottom open position" allows the airflow to exit over the bottom surface of the wing. The third position is the "closed position" which restricts any airflow from exiting both the top and bottom blowing slots. Figure 3 shows the cross-sectional area, as well as the dimensions, of the leading edge control cam.

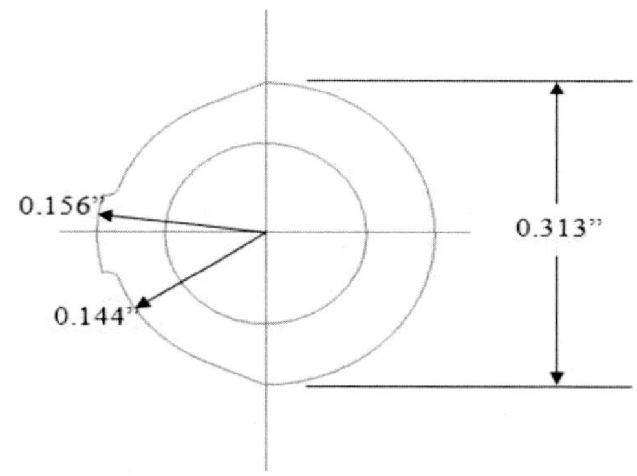

Figure 3. Leading Edge Jet Control Cam Cross Sectional Area

LOAD CELLS

To ascertain where the stall of the airfoil occurred in a wind tunnel environment, the lifting force needed to be measured. For this experiment, two linear regression type load cells were attached to the tunnel stand holding the model. The load cells were positioned on-axis in relation to the lift force direction to prevent having to find the force components based on the angle of attack for each data series of the experiment, as well as on both the top and bottom of the model to avoid saturating the load cells at high angles of attack. The rated loading for these gauges were 25 pounds. With one set at the top and one set at the bottom of the model, a total of 50 pounds could be measured without over stressing the load cells.

These load cells used a set of multi-directional strain gauges to read a voltage into the data acquisition software. Using a calibration curve, the voltage could then be converted to a force. A top lift and drag measurement was taken in conjunction with a bottom lift and drag readings. The resultant forces were added together to come up with the total loading applied to the wing in the lift and drag directions.

A linear calibration was required to mathematically change the read voltage from the strain gauges to a physical loading. This calibration needed to be done for each of the four load cells used for this experimentation. To get a suitable set of data points, known weights, were hung in all 4 directions of loading, and the voltages were read by the data acquisition software. The loadings were applied and recorded in both the negative and positive directions of lift and drag.

The samples were taken at 200 Hertz for 3 seconds. After the samples were reduced, and averaged, it was shown that there was a drag force cross-loading when weight was applied to the lift direction only, as well as lifting force cross-loading when load was applied in the drag direction only. Consequently, several multi-loading data fields were recorded to add to the calibration.

TESTING SCHEDULE

Several different parameters were varied to obtain a set of usable data for the CC wing model. One parameter that was varied was tunnel speed, from 80 fps to 100 fps, and finally to 120 fps. No slower speeds were tested because of wind tunnel limitations. At each of these three velocities, three different circulation control jet velocities were also examined. These jets velocities were dependent upon the regulated pressure of the storage tank. The three regulated plenum pressures which fed to the leading edge slots were ten, twenty, and thirty psi. Each of these tests was run at angles of attack from zero degrees to sixteen degrees in two degree increments. Table 1 shows the testing schedule used for data acquisition in this experiment.

Table 1. Testing Schedule

Tunnel Velocity (fps)	80			100			120		
Plenum Pressure (psi)	10	20	30	10	20	30	10	20	30
Angle of Attack	0-16 Degrees								
# of Runs	3	3	3	3	3	3	3	3	3

STALL ANGLES WITHOUT USING CIRCULATION CONTROL

The first task in testing the model was to obtain the baseline data without the circulation control activated. This produced the expected values of lift and drag to be used as a reference when circulation control was turned on. Three freestream air velocities were tested, 80, 100, and 120 feet per second. The formula for the section lift coefficient (C_L) is given by Equation 1 (Anderson, 2001 and Bertin, 2002).

$$C_L = \frac{2L}{\rho V_\infty^2 S} \qquad \text{Equation 1}$$

These lift equation, in addition to depending on lift (L) also relies on the density of the fluid (ρ), freestream velocity (V_∞), and the wing area (S). The values for these model parameters were measured over a series of testing runs. The data acquisition software was set to record the voltage readings of the load cells in a text file at a rate of 10,000 Hz for a three second duration. This sample rate gave 30,000 recorded points for each of the four load cells which were then averaged to get a voltage reading for the forces in each load cell direction. The calibration data completed for the load cells was then used to transfer the voltages read to a force value. Figure 4 shows the lift coefficient as a function of angle of attack of the model compared for each of the three velocities tested. The stall characteristics of this particular airfoil, Figure 4 shows that stall would be predicted to occur between a four and six degree angle of attack.

In the section drag coefficient plot, there is an increase in the value found for the negative angles of attack. This is attributed to the placement of a bead of sealant added under the skin to prevent airflow between the two circulation control plenums. This sealant was unavoidable, and needed to be installed to prevent circulation control flow from the leading edge slot from interfering with flow to the trailing edge slot. While there is a small lack of symmetry in the results, the plots were consistent and predictable over the range of tunnel velocities chosen.

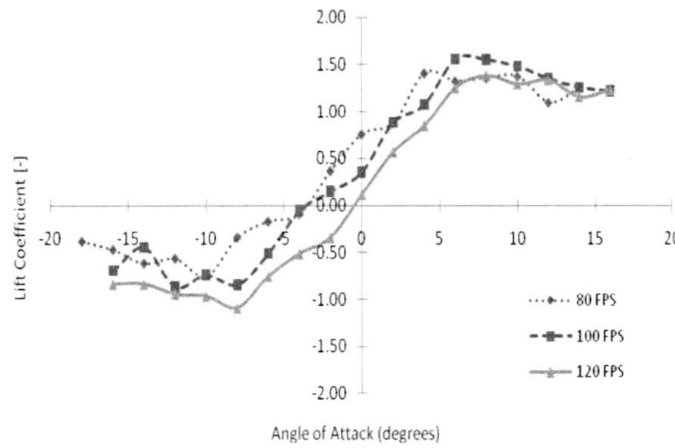

Figure 4. Section Lift Coefficient with No Blowing Conditions

The resulting value for section lift was then compared with historical data in previous research. Below, in Equations 2- Equation 5 are the formulae for the computed uncertainty values which were used to calculate the instrumentation used in the experiment. Each of these equations is based on the partial derivatives of each of the measured quantities, as well as the instrumentation uncertainty. As an example, the error in lift coefficient is based on the lift force uncertainty (w_L), the density uncertainty (w_ρ), freestream velocity uncertainty ($w_{V\infty}$), and the wing area uncertainty (w_A).

$$W_{C_L} = \left[\left(\frac{2w_L}{\rho V_\infty^2 S}\right)^2 + \left(\frac{2Lw_\rho}{\rho^2 V_\infty^2 S}\right)^2 + \left(\frac{4Lw_{V_\infty}}{\rho V_\infty^3 S}\right)^2 + \left(\frac{2Lw_A}{\rho V_\infty^2 S^2}\right)^2 \right]^{0.5} \qquad \text{Equation 2}$$

$$W_\rho = \left[\left(\frac{w_{P_\infty}}{RT}\right)^2 + \left(\frac{P_\infty w_T}{RT^2}\right)^2 \right]^{0.5} \qquad \text{Equation 3}$$

$$W_{V_\infty} = \left[\left(\frac{w_{\Delta P}}{2} \sqrt{\frac{2}{\Delta P * \rho}}\right)^2 + \left(\frac{w_\rho}{2} \sqrt{\frac{2\Delta P}{\rho^3}}\right)^2 \right]^{0.5} \qquad \text{Equation 4}$$

$$w_A = [(bw_c)^2 + (bw_b)^2]^{0.5} \qquad \text{Equation 5}$$

With instrumentation uncertainties calculated and plotted along with the tested and historical data comparisons, it is shown in Figure 5 that the values of the lift coefficient for the tested model fall within the error values. The previous trial only experimented at angles of attack ranging from zero to six degrees. The results of this study were tested out to sixteen degrees angle of attack. Also compared in Figure 5 are the results from wind tunnel work done on a NACA 0010-34 airfoil (Abbott and von Doenhoff, 1949). Because the airfoil tested in this experimentation does not have any historical work completed, the closest fit was chosen for comparison. This data does not fall within the error calculations for this experiment mainly because of the small radius Coandă surface on the trailing edge of the model

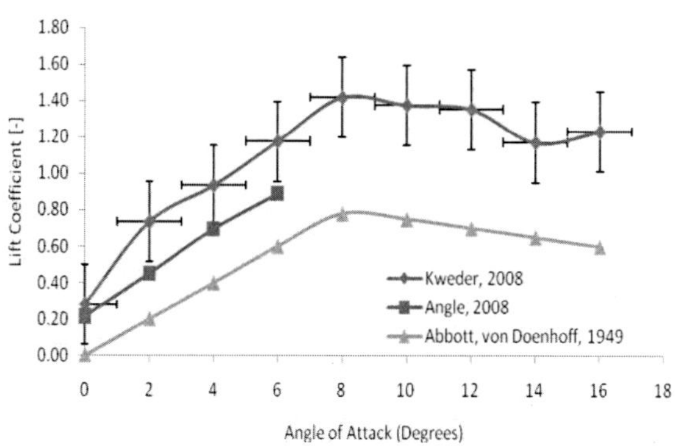

Figure 5. Lift Coefficient Comparison to Previous Experimentations (Angle, 2008, Abbott and von Doenhoff, 1949)

STALL ANGLES USING CIRCULATION CONTROL

When investigating the effect of circulation control on an aerodynamic body, it is important to note that the results are based on a blowing coefficient (C_μ). In past experiments conducted in circulation control theory, the blowing coefficient has been defined in several different ways. Most experimenters define C_μ in terms of velocity (V), slot thickness (t), chord length (c) and density (ρ) of the free stream and jet exit. The equation for this relationship is shown below in Equation 7 (Harness, 1970). For this experiment, the blowing coefficient ranged from 0.0006 to 0.0127 by varying the freestream velocity in the tunnel, as well as varying the air pressure in the blowing slot plenum.

$$C_\mu = \frac{\rho_j v_j^2 t}{\frac{1}{2}\rho_\infty v_\infty^2 c} \qquad \text{Equation 7}$$

Flow characterization of the velocity of the jet exiting the blowing slots of the airfoil was necessary. This data was taken using a water manometer and a pitot static pressure probe. A sample was taken one third of the distance between each of the five exit jets on the leading edge. The three indicated jet pressures that were examined were 10, 20 and 30 psi.

Jet velocity was then calculated through incompressible relations using Equation 8. In this equation, jet velocity (V_j) is dependent on, the gas constant (R), the ration of freestream pressure (P_∞) to plenum exit pressure (P_d), specific heat ratio (γ = 1.4 for air) and plenum exit temperature (T_d).

$$V_j = \sqrt{\frac{2\gamma R T_d}{\gamma - 1}\left[1 - \frac{P_\infty}{P_d}^{\frac{\gamma-1}{\gamma}}\right]} \qquad \text{Equation 8}$$

The leading and trailing edge jet velocities were calculated and normalized according to the maximum across the entire exit slot. The results are found in Figure 6.

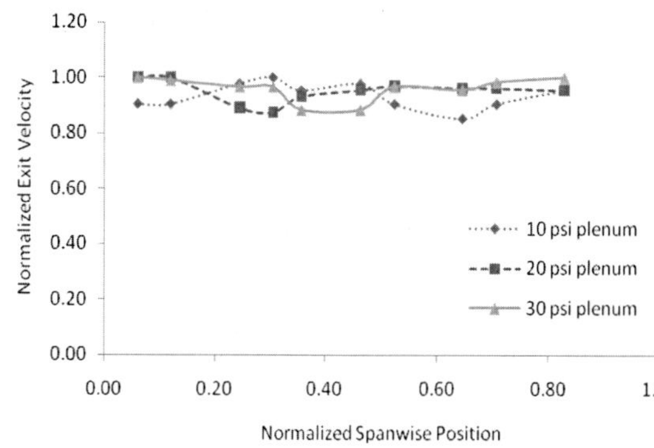

Figure 6. Jet Exit Velocity Profile

Following the completion of the test cases at angles of attack from zero to sixteen degrees, the section lift coefficient for the various tunnel speeds could be compared to the results of the airfoil without active circulation control to show the effect they had on the stall angles of the airfoil. As an example, Figure 7 shows the effect of leading edge circulation control on the airfoil. The stall angle is decreased from eight degrees without CC to six degrees with the active circulation control. This trend continues as the blowing coefficient (C_μ) varies from 0.0006 to 0.0011 and finally to 0.0019. The plot also shows as the blowing coefficient decreases, that is as the tunnel velocity increases or the jet velocity increases, the circulation control becomes less effective. Similarly, in Figure 7, as the blowing coefficient increases, the drag coefficient goes down. This is predicted since the more lift added to a particular airfoil, the more drag force it experiences.

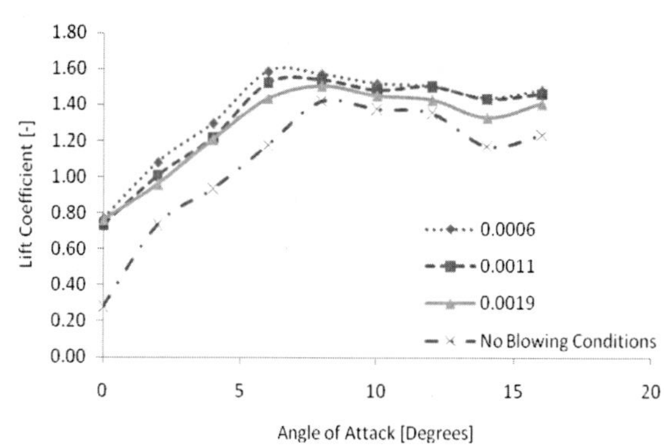

Figure 7. Section Lift Coefficient versus Angle of Attack (Tunnel Speed 120 FPS)

At other tunnel speeds tested (80 and 100 FPS), the results are similar, with the stall angle occurring between six and seven degrees. The same trends also occurs in the lift coefficient, the maximum increase is when the blowing coefficient is lowest, suggesting that the higher the tunnel speed or exit jet velocity the less of an effect that CC has on the airfoil.

A second data comparison was conducted to examine the relationship between lift coefficient and blowing coefficient at certain selected angles of attack, also known as the lift augmentation ratio. These values were compared to previous experimental data, which indicated for that experiment that a maximum lift augmentation ratio achievable was 80 (Englar, 2005). A linear fit was applied to the plot to show the effective lift augmentation ratio. According to the results, the lift augmentation ratio is over 100 at the model's stall angle. This is a considerably higher than the results predicted to give a lift augmentation ratio of 80.

The previous experiments were able to study blowing coefficients ranging from 0.02 to 0.16 and did not focus on low speed results. This current study focused on slower tunnel velocities and smaller plenum pressures, and as a result, the maximum blowing coefficient experienced was 0.0042 on the trailing edge.

CONCLUSIONS

This research set out to study the effect that leading edge, circulation control has on the stall angles of a 10:1 elliptical airfoil with leading edge blowing slots. A variety of different test scenarios were completed which examined blowing coefficients varying between 0.0006 and 0.0127 by fluctuating wind tunnel speed between 80, 100 and 120 feet per second, as well as changing the jet exit plenum pressures between 10, 20, and 30 pounds per square inch. The result indicates that the stall angle of this particular airfoil was found to occur at six degrees for leading edge, active blowing control.

The section lift coefficient of the airfoil was greatly increased from a maximum of 1.4 at stall without circulation control to a maximum value of 2.4 using CC, a 70% increase. With this increase in lift coefficient, the section drag coefficient was also raised from a maximum of 0.5 without CC to a value of 1.4 with active controls to achieve a 70% increase in drag. This increase in lift and consequently, in drag as well, comes at a reduction of stall angle of two degrees.

RECOMMENDATIONS

To continue with future research in the area of circulation control, several aspects need to be studied. The first of these topics is low speed results. It has been shown that circulation control is less effective as the free-stream velocity of the air moving around the wing increases. Because of tunnel limitations, velocities lower than 80 feet per second were excluded from this experiment. Slower speeds must be examined to be able to use the circulation control wing in other applications, such as low speed rotors and for windmills.

The second facet needing future consideration is to look more closely at the leading edge effects of circulation control on aerodynamic forces, including the drag. Because load cells were the chosen method of measuring in this particular study, the drag forces that were experienced could not be accurately measured. This is due to the fact that the use of load cells do not account for wall interference with the exit jet. Future experiments are suggested to employ a wake rake collection survey to analyze the drag forces experienced by the model while the leading edge slots are activated.

Another suggestion for future research is to experiment with different exit jet angles to see what effect this jet has on the virtual camber of the airfoil shape. It has been shown that because the jet exits at an angle different from the chord length, this changes the camber characteristics of the airfoil, and changes an elliptical airfoil into an effective cambered airfoil.

REFERENCES

Abbott, I.H., Von Doenhoff, A. E., Theory of Wing Sections, Dover Publications, Inc, New York, NY, 1959.

Abramson, J. "Two-Dimensional Subsonic Wind Tunnel Evaluation of a 20-Percent-Thick Circulation Controlled Airfoil." David W. Taylor Naval Ship Research Center, Bathesda, MD, 1975.

Abramson, J. "Characteristics of a Cambered Circulation Control Airfoil Having Both Upper and Lower Surface Trailing Edge Slots." Naval Surface Warfare Center – Carderock Division, Bathesda, MD, 2004.

Ambrosiani, J.P., "Analysis of a Circulation Controlled Elliptical Airfoil," Ph.D. Dissertation, Department of Aerospace Engineering, West Virginia University, Morgantown, WV, 1971.

Anderson Jr, J.D., Fundamentals of Aerodynamics, McGraw Hill, Inc., New York, NY, 2001.

Angle II, G.M., "Aerodynamic Benefits of Near-Surface-Actuated Circulation Control Blowing Slots for Rotorcraft Use," Ph.D. Dissertation, Department of Aerospace Engineering, West Virginia University, Morgantown, WV, 2008.

Bertin, J.J., Aerodynamics for Engineers, Prentice –Hall, Inc., Upper Saddle River, NJ, 2002.

Englar, R.J., "Overview of Circulation Control Pneumatic Aerodynamics: Blown force and Moment Augmentation and Modification as Applied Primarily to Fixed Wing Aircraft," Proceedings of the 2004 NASA/ONR Circulation Control Workshop, NASA CP-2005-213509, Pages 23-66, 2005.

Gibbs, E.H., "Analysis of Circulation Controlled Airfoils," Ph.D. Dissertation, Department of Aerospace Engineering, West Virginia University, Morgantown, WV, 1975.

Harness, G.S., "An Experimental Investigation of a Circulation Controlled Cambered Elliptical Airfoil," Master's Thesis, Department of Aerospace Engineering, West Virginia University, Morgantown, WV, 1970.

Katz, J., Plotkin, A. Low Speed Aerodynamics, Cambridge University Press, Cambridge, NY, 2001.

Kind, R.J., Maul, D.J., "An Experimental Investigation of a Low-Speed Circulation-Controlled Aerofoil," The Aeronautical Quarterly, Vol. XIX, May, 1968.

Kueth, A.M., Schetzer, J.D., Foundations of Aerodynamics, John Wiley and Sons Inc., New York, NY, 1967.

Loth, J.L., Fanucci, J.B., Roberts, S.C., "Flight Performance of a Circulation Controlled STOL Aircraft." Journal of Aircraft, Vol. 13, No. 3, 1976.

Loth, J.L., Boasson, M. "Circulation Control STOL Wing Optimization," Journal of Aircraft, Vol. 21 No. 2, February, 1984.

Maskell, E.C., Spence D.A., "A Theory of the Jet Flap in Three Dimensions," *Proceedings of the Royal Society of London, A Mathematical and Physical Sciences Series*, Vol. 251, No. 1266, 1959.

Myer, D.P., "An Investigation of a Circulation Controlled Cambered Elliptical Airfoil with a Rounded Trailing Edge," Master's Thesis, Aerospace Engineering Department, West Virginia University, Morgantown, WV, 1972.

Theodorsen, T., "Theory of Wing Sections of Arbitrary Shape," NACA-TR-411, 1931.

Proceedings of the ASME 2010 International Mechanical Engineering Congress & Exposition
IMECE2010
November 12-18, 2010, Vancouver, British Columbia, Canada

IMECE2010-38606

ELLIPTIC AIRFOIL STALL ANALYSIS WITH TRAILING EDGE SLOTS

Jonathan Kweder, Mary Ann Clarke, James E. Smith
Center for Industrial Research Applications
West Virginia University
Morgantown, WV, USA

ABSTRACT

Circulation control (CC) is a high-lift methodology that can be used on a variety of aerodynamic applications. This technology has been in the research and development phase for over sixty years primarily for fixed wing aircraft where the early models were referred to as "blown flaps." Circulation control works by increasing the near surface velocity of the airflow over the leading edge and/or trailing edge of a lifting surface This phenomenon keeps the boundary layer jet attached to the wing surface thus increasing the lift generated on the surface. The circulation control airflow adds energy to the lift force through conventional airfoil lift production and by altering the circulation of stream lines around the airfoil.

For this study, a 10:1 aspect ratio elliptical airfoil with a chord length of 11.8 inches and a span of 31.5 inches was inserted into the West Virginia University Closed Loop Wind Tunnel and was tested at varying wind speeds (80, 100, and 120 feet per second), angle of attack (zero to sixteen degrees), and blowing coefficients, ranging from 0.0006 to 0.0127 depending on plenum pressure. By comparing the non-circulation controlled wing with the active circulation control data, a trend was found as to the influence of circulation control on the stall characteristics of the wing for trailing edge active control. For this specific case, when the circulation control is in use on the 10:1 elliptical airfoil, the stall angle decreased, from eight degrees to six degrees, while providing a 70% increase in lift coefficient. It should be noted that due to the trailing edge location of the circulation control exit jet, a "virtual" camber is created with the free stream air adding length to the overall airfoil. Due to this phenomena, the actual stall angle measured increased from eight degrees on the un-augmented airfoil, to a maximum of twelve degrees.

INTRODUCTION

In recent years there has been a renewed interest in high lift technology, as well as lift augmentation devices. This interest has been generated mainly for use on both fixed and rotary wing vertical/short take-off and landing (V/STOL) aircraft. A rotor on a helicopter is similar to an airplane wing moving through the air in that it has a relative angle of attack with respect to the airflow. All aircraft are supported in the air by a lifting force, which is generated by the wings or rotor of the aircraft. As the air flows around the wings of the aircraft, these surfaces generate lift due to a pressure reduction above the wings. The lower pressure above the wing coupled with the higher relative pressure under the wing results in the total lift for the body. An immediate increase in lift can be obtained by increasing the angle of attack (AOA) which is defined as the angle between the wing chord and the relative airflow. But increasing the AOA also increases drag. Above a particular angle, the critical angle of attack, the airflow behind and above the lifting bodies becomes turbulent, and the wing stalls. This means that the wing could rapidly cease to provide enough lift to support the aircraft.

Another way of increasing the lift on an airfoil is to use circulation control technology. Circulation control is a form of lift enhancement that relies on the blowing of high velocity air around the leading and/or trailing edge of the lifting body (Ambrosiani, 1971, Abramson, 1975, Gibbs, 1975). Circulation control works by increasing the near surface velocity of the airflow over a specially designed aircraft wing. The wing has a rounded trailing edge, and ejects the air tangentially, through these slots, inducing the Coandă effect. This phenomenon keeps the boundary layer jet attached to the wing surface longer than a conventional wing and thus increases the lift generated on the wing surface due to the relaxation of the Kutta condition for the rounded trailing edge (Myer, 1972, Loth, 1976, Abramson, 2004). The circulation control airflow adds to the lift force through conventional airfoil lift production, by altering the circulation of stream lines around the airfoil.

Circulation can be described as the line integral around a closed path of the velocity component tangential to the path. Also, the Kutta-Joukowski Theorem states that the lift per unit span of an airfoil in moving air is equal to the product of the circulation, free stream velocity and the density of the fluid that the lifting body is moving through (Anderson, 2001, Keuth, 1967). This shows that the total lift force on the wing can be significantly controlled by restricting the circulation around it (Kind, 1968, Loth, 1984, Englar, 2005). As with changing the AOA of the wing to increase lift, it is necessary to analyze how CC affects the stall angles of the wing. It is the purpose of this study to identify and define the angles at which stall occurs without circulation control and compare them to the experimental stall angles using circulation control.

NOMENCLATURE

Symbol	Description	Units
A	Area	ft2
B	Span	ft
c	Chord	ft
C_L	Lift Coefficient	-
C_μ	Blowing Coefficient	-
h	Height	ft
L	Lift	lb
P	Pressure	atm or psi
r	Radial Location	ft
R	Ideal Gas Constant	ft-lb/slug-oR
S	Wing Area	ft2
t	Slot thickness	ft
T	Temperature	oR
V	Velocity	ft/s
V_j	Jet Velocity	ft/s
V_o	Downstream Velocity	ft/s
V_s	Upstream Velocity	ft/s
w	Error Term	-
z	Altitude	ft
μ	Dynamic Viscosity	slug-s/ft3
Δ	Change in	-
ρ	Density	lb/ft3
g	Specific Heat Ratio	-
Γ	Circulation	-

TESTING PROCEDURE

For this experiment, an airfoil was designed and constructed to have both a leading edge and trailing edge blowing slot, completely separate from one another. To achieve this, two air plenums were constructed inside the wing, between the ribs. These plenums were separated and sealed so that the air pressure could be regulated for the trailing edge separately from the leading edge. Figure 1 shows a computer aided drawing of the inside of the wing model. Only the trailing edge slot was utilized for this analysis.

Figure 1. Wing Model use for Experimentation (Pro-Engineer)

The center spar which runs along the span of the wing is sealed to both the upper and lower skins so that the two air plenums are separate from one another. There are also five porous walls situated before both the leading and trailing edges provided in an attempt to create a uniform exit jet velocity. The orientation of the wing model in the wind tunnel with respect to the free stream velocity is shown in Figure 2. A design aspect of this model, the small radius Coandă surface along the trailing edge is also shown.

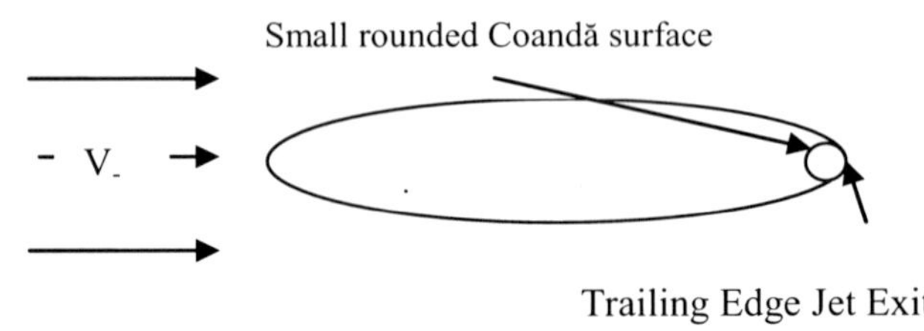

Figure 2. Airfoil and Slot Exit Orientation (not to scale)

The CC model used for this experiment was suspended in West Virginia University's Closed Loop Wind Tunnel via a support structure which reached around the top and bottom of the test section, but did not allow the model to touch the wind tunnel. This reduced the amount errors potentially caused by vibration that might be transferred to the model through the tunnel. The loading on the model was recorded using four linear load cells and because the model was also suspended from these load cells, it was necessary to record the tare values at each angle of attack.

BLOWING SLOT CONTROLS

A cylindrical cam was used to control the jet outflow at the blowing slots. The cam had the ability to be in three different positions, simply by turning them and was designed in order to seal one side of the wing while opening the other side. The "top open position" allows the circulation control jet to exit on the top side of the airfoil. Similarly, the "bottom open position" allows the airflow to exit over the bottom surface of the wing. The third position is the "closed position" which restricts any airflow from exiting both the top and bottom blowing slots. Figure 3 shows the cross-sectional area, as well as the dimensions, of the trailing edge cam.

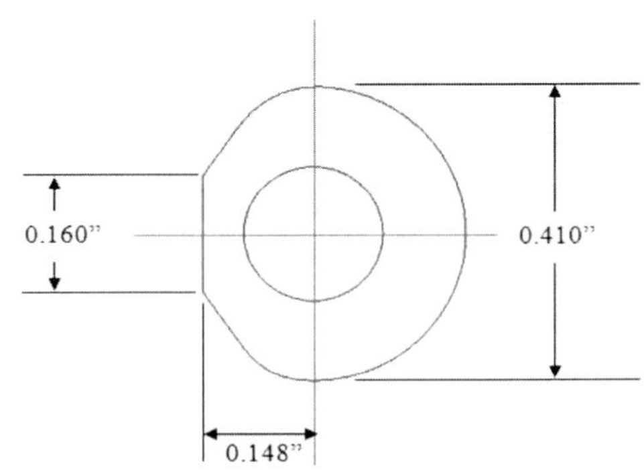

Figure 3. Trailing Edge Blowing Slot Control Cam

To ascertain where the stall of the airfoil occurred in a wind tunnel environment, the lifting force, as well as the drag force, needed to be measured. For this experimentation, four linear regression type load cells were attached to the tunnel stand holding the model. The load cells were positioned on-axis in relation to the lift and drag force directions to prevent having to find the force components based on the angle of attack for each data series of the experiment, as well as on both the top and bottom of the model to avoid saturating the load cells at high angles of attack. The rated loading for these gauges were 25 pounds. With one set at the top and one set at the bottom of the model, a total of 50 pounds could be measured without over stressing the load cells.

A linear calibration was required to mathematically change the read voltage from the strain gauges to a physical loading. This calibration needed to be done for each of the four load cells used for this experimentation. To get a suitable set of data points, known weights were hung in all 4 directions of loading, and the voltages were read by the data acquisition software. The loadings were applied and recorded in both the negative and positive directions of lift and drag.

The samples were taken at 200 Hertz for 3 seconds. After the samples were reduced, and averaged, it was shown that there was a drag force cross-loading when weight was applied to the lift direction only, as well as lifting force cross-loading when load was applied in the drag direction only. Consequently, several multi-loading data fields were recorded to add to the calibration. Several different parameters were varied to obtain a set of usable data for the circulation control wing model. One parameter that was varied was tunnel speed, from 80 fps to 100 fps, and finally to 120 fps. No slower speeds were tested because of wind tunnel limitations. At each of these three velocities, three different circulation control jet velocities were also examined. The three regulated plenum pressures tested on the leading and trailing edge slots were 10, 20, and 30 psi. Each of these tests was run at angles of attack from zero to sixteen degrees in two degree increments. Table 1 shows the testing schedule used for data acquisition in this experiment.

Table 1. Testing Schedule

Tunnel Velocity (fps)	80			100			120		
Plenum Pressure (psi)	10	20	30	10	20	30	10	20	30
Angle of Attack	0 – 16 Degrees								
# of Runs	3	3	3	3	3	3	3	3	3

STALL ANGLES WITHOUT USING CIRCULATION CONTROL

The first task in testing the model was to obtain the baseline data without the circulation control activated. This produced the expected values of lift and drag to be used as a reference when circulation control was turned on. Three freestream air velocities were tested, 80, 100, and 120 feet per second. Due to tunnel limitations, namely unsteady tunnel section velocity, lower velocities were not considered. The formula for the section lift coefficient (C_L) is given by Equation 1, (Anderson, 2001 and Bertin, 2002).

$$C_L = \frac{2L}{\rho V_\infty^2 S} \qquad \text{Equation 1}$$

This equation, in addition to depending on lift (L), also relies upon the density of the fluid (ρ), freestream velocity (V_∞), and the wing area (S). The values for these model parameters were measured over a series of testing runs. The data acquisition software was set to record the voltage readings of the load cells in a text file at a rate of 10,000 Hz for a three second duration. This sample rate gave 30,000 recorded points for each of the four load cells which were then averaged to get a voltage reading for the forces in each load cell direction. The calibration data taken for the load cells was then used to transfer the voltages readings over to a force value. Figure 4 shows the lift coefficient as a function of angle of attack of the model compared for each of the three velocities tested. The stall characteristics of this particular airfoil, seen in Figure 4, at the onset of the reduction of the lift coefficient, occurs between a four and six degree angle of attack for this blade section configuration.

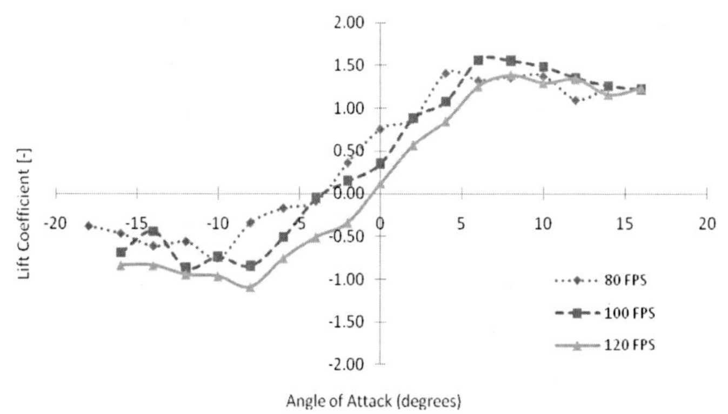

Figure 4. Section Lift Coefficient Under no Blowing Conditions

The resulting value for section lift was then compared with historical data obtained in previous research. Below, in Equations 2- Equation 5, the formulae for the computed uncertainty values, which were used to calculate for the instrumentation used in the experiment, are shown. Each of these equations is shown to be based on the partial derivatives of each of the measured quantities, as well as the instrumentation uncertainty. As an example, the uncertainty in the lift coefficient is based on the lift force uncertainty (w_L), the density uncertainty (w_ρ), freestream velocity uncertainty ($w_{v\infty}$), and the wing area uncertainty (w_A).

$$W_{C_L} = \left[\left(\frac{2w_L}{\rho V_\infty^2 S}\right)^2 + \left(\frac{2Lw_\rho}{\rho^2 V_\infty^2 S}\right)^2 + \left(\frac{4Lw_{V_\infty}}{\rho V_\infty^3 S}\right)^2 + \left(\frac{2Lw_A}{\rho V_\infty^2 S^2}\right)^2\right]^{0.5} \qquad \text{Equation 2}$$

$$W_\rho = \left[\left(\frac{w_{P_\infty}}{RT}\right)^2 + \left(\frac{P_\infty w_T}{RT^2}\right)^2\right]^{0.5} \qquad \text{Equation 3}$$

$$W_{V_\infty} = \left[\left(\frac{w_{\Delta P}}{2}\sqrt{\frac{2}{\Delta P * \rho}}\right)^2 + \left(\frac{w_\rho}{2}\sqrt{\frac{2\Delta P}{\rho^3}}\right)^2 \right]^{0.5} \qquad \text{Equation 4}$$

$$w_A = [(bw_c)^2 + (bw_b)^2]^{0.5} \qquad \text{Equation 5}$$

With instrumentation uncertainties calculated and plotted along with the tested and historical data, it is shown in Figure 5 that for the values of lift coefficient the tested model falls within the uncertainty values. The previous trial only experimented at angles of attack ranging from zero to six degrees. The results of this study were tested out to sixteen degrees angle of attack.

Also compared in Figure 5 are the results from wind tunnel work done on a NACA 0010-34 airfoil (Abbott and von Doenhoff, 1949). Because the airfoil tested in this experimentation does not have any historical work completed, the closest fit was chosen for comparison. This data does not fall within the error calculations for this experiment mainly because of the small radius Coandă surface on the trailing edge of the model. This alteration inhibits the Kutta condition at the trailing edge of the model while providing the elliptical airfoil with an increased lift coefficient by bending the streamlines around the trailing edge.

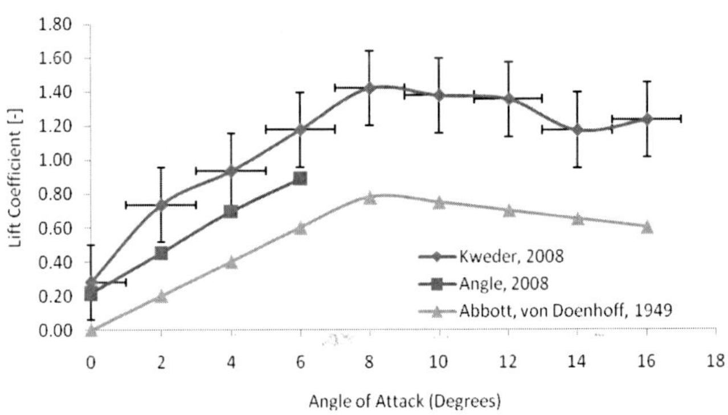

Figure 5. Lift Coefficient Comparison to Previous Experiment [Angle, 2008, Abbott and von Doenhoff, 1949]

CIRCULATION CONTROL STALL ANGLES

When investigating the effect of circulation control on an aerodynamic body, it is important to note that the results are based on a blowing coefficient (C_μ). In past experiments conducted in circulation control theory, the blowing coefficient has been defined in several different ways. The most common experiments define C_μ in terms of velocity (V_∞), slot thickness (t), chord length (c) and density (ρ) of the free stream and jet exit. The equation for this relationship is shown below in Equation 6 (Harness, 1970). For this experiment, the blowing coefficient ranged from 0.0006 to 0.0127 by varying the freestream velocity in the tunnel, as well as varying the air pressure in the blowing slot plenum.

$$C_\mu = \frac{\rho_j v_j^2 t}{\frac{1}{2}\rho_\infty V_\infty^2 c} \qquad \text{Equation 6}$$

Flow characterization of the velocity of the jet exiting the blowing slots of the airfoil was necessary. This data was taken using a water manometer and a pitot static pressure probe. A sample was taken one third of the distance between each of the five exit jets on the trailing edges. The three regulated jet pressures that were examined were 10, 20 and 30 psi. Jet velocity was then calculated through incompressible relations using Equation 7. In this equation, jet velocity (V_j) is dependent on, the gas constant (R), the ratio of freestream pressure (P_∞) to plenum exit pressure (P_d), specific heat ratio (γ = 1.4 for air) and plenum exit temperature (T_d).

$$V_j = \sqrt{\frac{2\gamma R T_d}{\gamma - 1}\left[1 - \frac{P_\infty}{P_d}^{\frac{\gamma-1}{\gamma}}\right]} \qquad \text{Equation 7}$$

The leading and trailing edge jet velocities were calculated and normalized according to the maximum across the entire exit slot. The results are found in Figure 6.

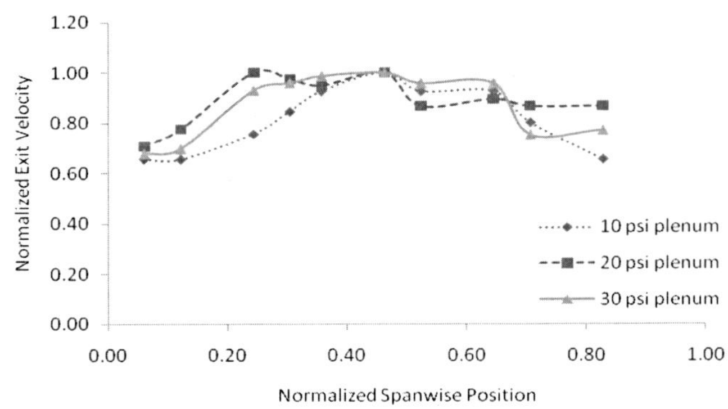

Figure 6.Trailing Edge Jet Velocity Profile Versus Spanwise Location

Following the completion of the test cases at angles of attack from zero to sixteen degrees, the section lift coefficient for the various tunnel speeds could be compared to the results of the airfoil without active circulation control to show the effect they had on the stall angles of the airfoil. As an example, Figure 7 shows the effect of trailing edge circulation control on the airfoil. The stall angle is decreased from eight degrees without CC to six degrees with the active circulation control. This trend continues as blowing coefficient (C_μ) varies from 0.0006 to 0.0011 and finally to 0.0019. The plot also shows as the blowing coefficient decreases, that is the tunnel velocity increases or the jet velocity increases, the circulation control becomes less effective. This plot also illustrates that the use of circulation control ejector slots on the trailing edge of an airfoil shows evidence of camber on the airfoil, so the model ceases to act as an elliptical airfoil under blowing conditions.

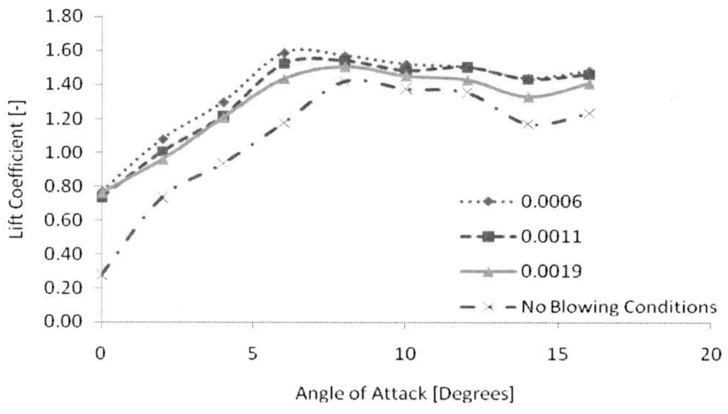

Figure 7. Section Lift Coefficient Versus Angle of Attack (Tunnel Speed 120 FPS)

At other tunnel speeds tested (80 and 100 FPS), the results were similar, with the stall angles occurring between six and seven degrees. The same trends also occurs in the lift coefficient, the maximum occuring when the blowing coefficient is lowest, suggesting that the higher the tunnel speed or exit jet velocity the less of an effect that CC has on the airfoil.

COMPUTATIONAL VERIFICATION

In order to check the accuracy of the experimental results, a model simulation was conducted using a computational fluid dynamics (CFD) software package. A CC model was constructed in Gambit, including a mesh with two separate sections, a close field and a far field, to add precision directly around the airfoil surface. The near field growth rate was 1.15 and had a maximum distance between mesh points of 0.2 inches. The far field dimensions used a growth rate of 2.5 and had a maximum distance between mesh points equal to 1 inch. This meshed model is shown in Figure 8, where the airfoil is at zero angle of attack.

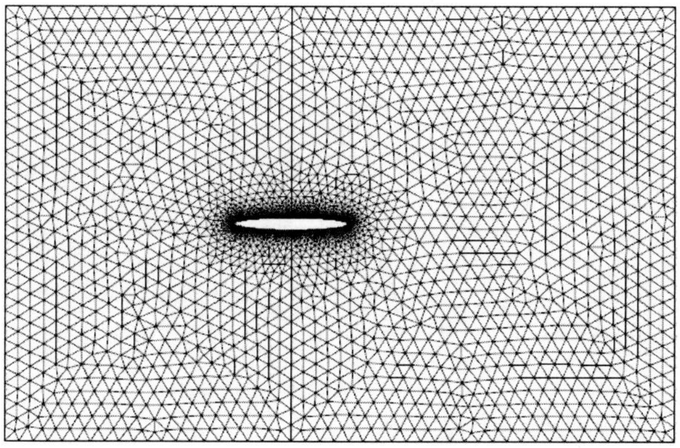

Figure 8. Meshed Circulation Control Model

A two-dimensional double precision k-ω turbulence simulation in the Fluent software package was used. Using the energy equation and the viscosity model solution settings, as well as a $1x10^{-3}$ continuity convergence criterion, angles of attack of six, eight, and ten were examined at a free stream velocity of 120 feet per second. The first sets of trials were run with no blowing conditions to examine how close a comparison the wind tunnel experiments were for the baseline model case. Three more case runs were tested with the top trailing edge jet velocities corresponding to 10, 20 and 30 psi plenum pressures. These slot activated tests were also run at six, eight, and ten degrees angle of attack, similar to the trials completed with no blowing conditions. Figure 9, 10, and 11 show the average lift coefficient CFD results plotted with the average experimental results versus the four different blowing conditions at tunnel wind speeds of 120 feet per second. These plots, particularly the 8° angle of attack, show a 2-3% difference between the experimental and computational results at blowing coefficients less than 0.001. The results for blowing coefficients larger than 0.001 show a difference of 6-9% between the CFD work and the experimental lift coefficients. As shown in Figure 12, stall is predicted to occur in the angle of attack range of 8-10° which is shown via the difference of almost 15% at low blowing coefficients. This was also congruent with the physical testing of the airfoil at the selected augmentation values.

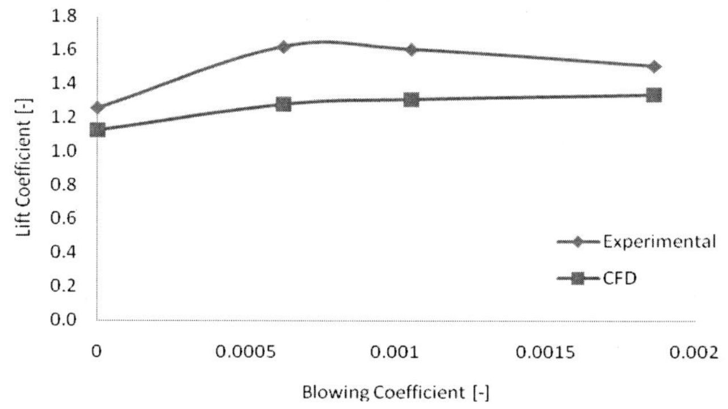

Figure 9. Lift Coefficient versus Blowing Coefficient at 6° Angle of Attack

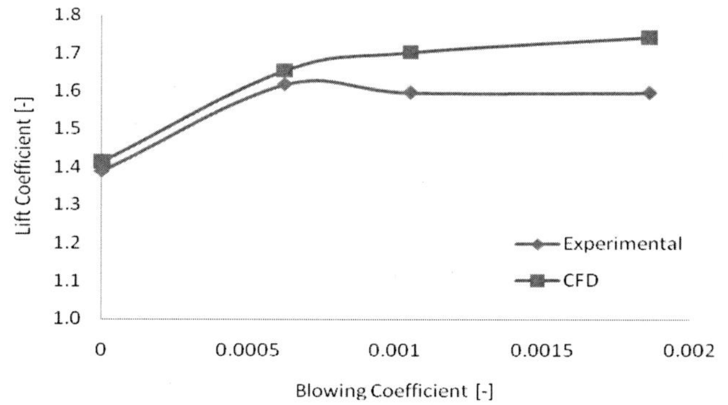

Figure 10. Lift Coefficient versus Blowing Coefficient at 8° Angle of Attack

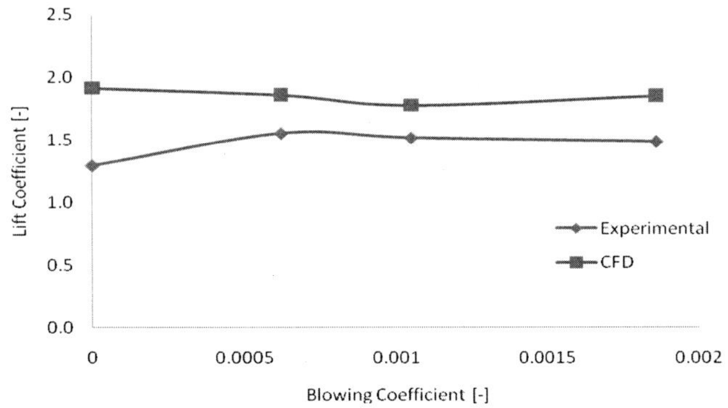

Figure 11. Lift Coefficient versus Blowing Coefficient at 10° Angle of Attack

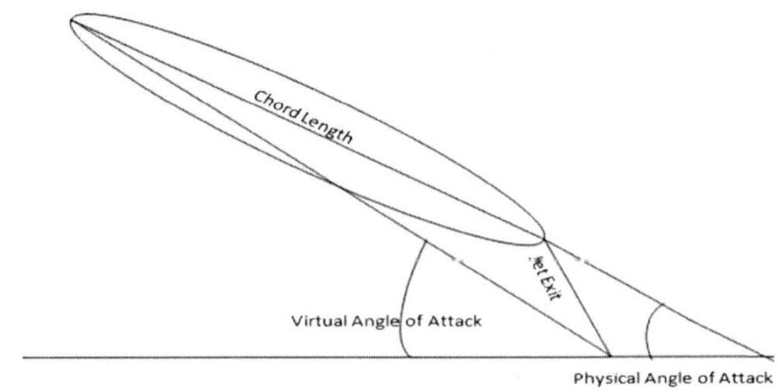

Figure 12. Physical Angle of Attack Compared to Virtual Angle of Attack

CONCLUSIONS

This research set out to study the effect that training edge circulation control had on the stall angles of a 10:1 elliptical airfoil. A variety of different test scenarios were completed which examined blowing coefficients varying between 0.0006 and 0.0127 by stepping the wind tunnel speeds between 80, 100 and 120 feet per second, as well as changing the jet exit plenum pressures between 10, 20, and 30 pounds per square inch. The result as to how the circulation control causes the stall angle of this particular airfoil to change was found to occur at six degrees for the trailing edge active blowing control.

The section lift coefficient of the airfoil was greatly increased from a maximum of 1.4 at stall without circulation control to a maximum value of 2.4 using CC, a 70% increase. This increase in lift and consequently, in drag as well, comes at a reduction of stall angle of two degrees.

In previous research, several experiments were conducted on a 20% elliptical airfoil with a 5% camber ratio, which showed an increase in stall angle from eight to fourteen degrees. Several conditions can be attributed to the fact that the data in this current work does not compare with that prior work, including,

- Wind tunnel blockage ratio changes when circulation control is activated on trailing edge slots, and therefore wall interference could become troublesome at high angles of attack,
- There is a force reaction on the wing when circulation control is activated, producing an artificial lifting force, and
- Physical angle of attack is different than the virtual angle of attack the freestream air flow sees, this phenomenon is shown in Figure 12.

To continue with future research in the area of circulation control, several aspects need to be studied. The first of these topics is low speed results. It has been shown that circulation control is less effective as the free-stream velocity of the air moving around the wing increases. Because of tunnel limitations, velocities lower than 80 feet per second were excluded from this experiment. Slower speeds must be examined to be able to use the circulation control wing in other applications, such low rotor speeds or for windmills.

REFERENCES

Abbott, I.H., Von Doenhoff, A. E., Theory of Wing Sections, Dover Publications, Inc, New York, NY, 1959.

Abramson, J. "Two-Dimensional Subsonic Wind Tunnel Evaluation of a 20-Percent-Thick Circulation Controlled Airfoil." David W. Taylor Naval Ship Research Center, Bathesda, MD, 1975.

Abramson, J. "Characteristics of a Cambered Circulation Control Airfoil Having Both Upper and Lower Surface Trailing Edge Slots." Naval Surface Warfare Center – Carderock Division, Bathesda, MD, 2004.

Ambrosiani, J.P., "Analysis of a Circulation Controlled Elliptical Airfoil," Ph.D. Dissertation, Department of Aerospace Engineering, West Virginia University, Morgantown, WV, 1971.

Anderson Jr, J.D., Fundamentals of Aerodynamics, McGraw Hill, Inc., New York, NY, 2001.

Angle II, G.M., "Aerodynamic Benefits of Near-Surface-Actuated Circulation Control Blowing Slots for Rotorcraft Use," Ph.D. Dissertation, Department of Aerospace Engineering, West Virginia University, Morgantown, WV, 2008.

Bertin, J.J., Aerodynamics for Engineers, Prentice –Hall, Inc., Upper Saddle River, NJ, 2002.

Englar, R.J., "Overview of Circulation Control Pneumatic Aerodynamics: Blown force and Moment Augmentation and Modification as Applied Primarily to Fixed Wing Aircraft," Proceedings of the 2004 NASA/ONR Circulation Control Workshop, NASA CP-2005-213509, Pages 23-66, 2005.

Gibbs, E.H., "Analysis of Circulation Controlled Airfoils," Ph.D. Dissertation, Department of Aerospace Engineering, West Virginia University, Morgantown, WV, 1975.

Harness, G.S., "An Experimental Investigation of a Circulation Controlled Cambered Elliptical Airfoil," Master's Thesis, Department of Aerospace Engineering, West Virginia University, Morgantown, WV, 1970.

Katz, J., Plotkin, A. Low Speed Aerodynamics, Cambridge University Press, Cambridge, NY, 2001.

Kind, R.J., Maul, D.J., "An Experimental Investigation of a Low-Speed Circulation-Controlled Aerofoil," The Aeronautical Quarterly, Vol. XIX, May, 1968.

Kueth, A.M., Schetzer, J.D., Foundations of Aerodynamics, John Wiley and Sons Inc., New York, NY, 1967.

Loth, J.L., Fanucci, J.B., Roberts, S.C., "Flight Performance of a Circulation Controlled STOL Aircraft." Journal of Aircraft, Vol. 13, No. 3, 1976.

Loth, J.L., Boasson, M. "Circulation Control STOL Wing Optimization," Journal of Aircraft, Vol. 21 No. 2, February, 1984.

Myer, D.P., "An Investigation of a Circulation Controlled Cambered Elliptical Airfoil with a Rounded Trailing Edge," Master's Thesis, Aerospace Engineering Department, West Virginia University, Morgantown, WV, 1972.

Proceedings of the ASME 2010 International Mechanical Engineering Congress & Exposition
IMECE 2010
November 12-18, 2010, Vancouver, British Columbia

IMECE2010-38816

EXPERIMENTAL TESTING OF A WIND TUNNEL MODEL FOR USE IN A VERTICAL AXIS WIND TURBINE

Chad C. Panther, Kenny A. Williams, Jay P. Wilhelm, James E. Smith
Center for Industrial Research Applications (CIRA)
Mechanical and Aerospace Engineering Department
West Virginia University, Morgantown, WV 26506, USA

ABSTRACT

Experimental testing was performed on a circulation controlled airfoil with upper and lower trailing edge blowing slots, controlled by span wise pneumatic valves. The augmented blade was designed for application to a circulation controlled vertical axis wind turbine. The design is based upon a conventional NACA0018 shape, replacing the sharp trailing edge with a rounded Coanda surface and blowing slots. A scale model with a chord of 8 inches and span of 16.5 inches was created using an ABS plastic rapid prototyping machine. In the past, circulation control wind tunnel models have been constructed with a separate blowing slot and trailing edge using conventional machining methods. The slot must be tediously aligned along the span for a consistent height which ultimately affects the uniformity and performance of the circulation control jet in combination with the flow rate. The rapid prototyping machine eased fabrication as a modular trailing edge section was printed which includes the Coanda surface, blowing slot, and diffuser all in one piece. Pressure taps were integrated by the prototyping machine into both the printed skin and trailing edge module. This method left additional space inside the model for circulation control valving components and eliminated the need for machining pressure ports. This paper will outline the model building procedures, wind tunnel test rig, and experimental results. Aerodynamic forces were determined by both load cells and surface pressure measurements; the agreement between the two methods will be analyzed and addressed. Test conditions include various angles of attack (±20°) at C_μ = 0, 0.02, 0.06, and 0.10; the test Reynolds number was kept constant at 300K. The results indicate that the blade performed at $\Delta C_l/C_\mu$ near 30 for C_μ = 0.02.

1. INTRODUCTION

The search for renewable energy solutions is growing worldwide. As technology continually progresses, wind energy conversion is gaining popularity as a viable option to replace non-renewable energy sources. Capturing and converting wind energy into electricity using a turbine can provide an inexpensive and renewable source of power. Wind turbines can be classified by the force which drives the rotation of the turbine blades. Hence, there are both lift and drag type wind turbines. Most modern wind turbines use aerodynamic lift to drive their blades through the air [1]. Drag devices use the force of the wind striking the turbine blade to push the rotor about its axis, generating torque. Lift devices are nearly two times more efficient at capturing the energy in the wind relative to the turbine's frontal area [1].

Wind turbines can also be classified into two orientations, based on the axis upon which the blades rotate. The most common orientation is the Horizontal Axis Wind Turbine (HAWT), which looks similar to a propeller. Currently in competition with the HAWT is the Vertical Axis Wind Turbine (VAWT); an example is shown in Figure 1. This orientation of a lift driven turbine has two distinct advantages over a HAWT. First, unlike conventional HAWT's, which need to be reoriented as the wind direction varies, VAWT's are operational in winds from all directions. Second, the VAWT places the entire generator, gearing, and other power equipment at the bottom of the structure, near the ground, making service repairs more convenient and safer. This advantage could also make the VAWT attractive to developing regions that do not have access to construction equipment necessary to assemble a typical HAWT.

Figure 1: Vertical Axis Wind Turbine [2]

Recent research and development efforts have been devoted to the application of circulation control to the turbine blades of VAWT's. To enhance an airfoil to include circulation control, the sharp trailing edge of a conventional wing is replaced with a rounded surface; blowing slots are included on both the upper and lower airfoil surfaces, located fore and adjacent to the rounded trailing edge. Pressurized air exiting the slots increases the momentum of the free stream air traveling over the airfoil. The jet naturally follows the rounded trailing edge via the Coanda effect. This form of circulation control acts as boundary layer control, preventing separation and deflecting the flow across the blade's top surface downward. As the flow is deflected downward, the apparent camber of the blade is increased, similar to the effect of a mechanical flap used on airplanes. This increase in the airfoil's apparent camber will provide significant increases in lift. The goal of applying circulation control to a VAWT's blades is to increase the wind speed operating range, ultimately increasing the power generated.

Circulation control (CC) is an attractive enhancement to turbine blades because it essentially increases the effective area of the wing without physically altering the airfoil, resulting in increased lift [3-9]. This concept was first proposed to improve the power performance of a wind turbine by Walters [3].

In order to evaluate the actual performance of a CC-VAWT, a prototype was constructed and tested. Before fabrication of the prototype, an analytical model was implemented to estimate the forces generated by the aerodynamics, shaft torque, and power generated. The aerodynamic forces measured during wind tunnel testing will be used as inputs to the analytical model. This model could prove to be invaluable, as it could lead to the configuration of optimal design parameters before construction and testing of a CC-VAWT prototype.

2. EXPERIMENTAL SET-UP

The tests were performed in West Virginia University's (WVU) subsonic closed loop wind tunnel; the tunnel is a single return, low turbulence type with variable speed capacity ($V_{\infty,max}$ = 250 ft/sec). The test section measures 32" x 45" and contains small openings fore and aft to maintain atmospheric pressure within. The tunnel has a turbulence factor (TF) of 1.31 at 100 ft/sec [12]. When comparing test results from different wind tunnels, the effective Reynolds number is used.

$$\mathrm{Re}_{eff} = TF \; x \; \mathrm{Re}_{test} = 1.31 \, \mathrm{Re}_{test} \qquad (1)$$

The effective Reynolds number is higher because it accounts for the additional turbulence in the wind tunnel caused by the propeller, the guide vanes, and the vibration of the tunnel walls [11]. The testing Reynolds number (Re_{test}), based on the airfoil chord, and was kept constant during all experiments at a value of 472K.

Endplates, each with a diameter of 11.875", were fitted to both ends of the model to minimize three-dimensional effects; the endplates were sized based on recommendations found in van den Berg [13]. A photograph of the model mounted in WVU's wind tunnel can be seen in Figure 3.

2.1. BLOWING AIR SUPPLY

The WVU Aerospace Engineering Laboratory's compressor and storage tank were used to supply the CC air to the wind tunnel model. The compressor, manufactured by the Worthington Corporation, is capable of producing 104.2 CFM displacement at a discharge pressure of 175 psi. Although the storage tank is capable of withstanding a maximum internal pressure of 200psi, in practice the automatic shutoff valve on the compressor was limited to 150 psi to protect the supply line components. A pressure regulator was located in the lines between the tank and measurement components to provide initial control and stability of the airflow. The approximate respective locations of the airflow system components can be seen in Figure 2. Plenum air was supplied to both sides of the model to minimize plenum air velocities and produce quasi stagnation conditions internally.

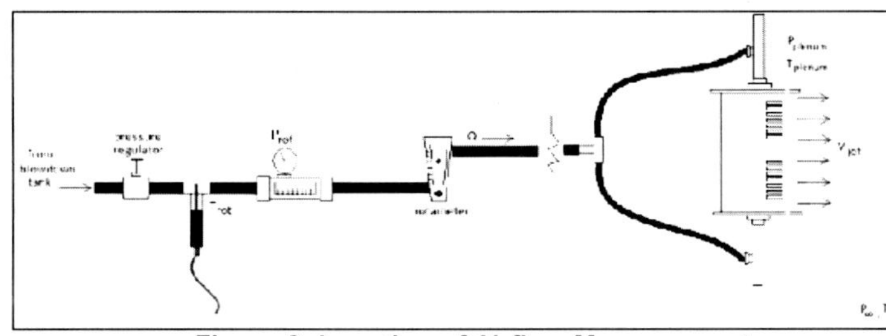

Figure 2: Location of Airflow Measurements

The primary control parameter for this experimental work dealing with circulation control is the momentum coefficient (C_μ), defined by:

$$C_\mu = \frac{(\dot{m}V)_{jet}}{q_\infty A} = \frac{(\dot{m}V)_{jet}}{\frac{1}{2}\rho_\infty V_\infty^2 A} = \frac{\left(\rho V^2\right)_{jet} h_{slot}}{\frac{1}{2}\left(\rho V^2\right)_\infty c} \qquad (2)$$

From Equation 2, the mass flow rate ($\dot{m}_{jet}$) was determined from volumetric flow rates (Q) recorded just upstream of the entrance to the model's plenum; Q was recorded with a King variable area rotameter (0-42 scfm). Pressure (P_{rot}) and Temperature (T_{rot}) measurements were recorded prior to the flow meter to calculate the density (ρ_{rot}) of the airflow entering the rotameter, via the ideal gas law. Equation 3 was then used to calculate the mass flow rate.

$$\dot{m} = \rho_{rot} Q \qquad (3)$$

It should be noted that the mass flow used in Equation 2 is per unit span, based on the total span of the blowing slot.

The velocity (V_{jet}) of the circulation control jet was calculated based on the pressure ratio of the plenum to the freestream static pressure; isentropic expansion of the jet was assumed (Equation 4).

$$V_{jet} = \left[\frac{2\gamma R T_{plenum}}{\gamma - 1}\left(1 - \left(\frac{P_\infty}{P_{plenum}}\right)^{\frac{\gamma-1}{\gamma}}\right)\right]^{\frac{1}{2}} \qquad (4)$$

Where the freestream air density (ρ_∞) and speed (V_∞), and model plan form area (A) are all included to determine the dimensionless blowing coefficient. The jet exit velocity is based on the pressure ratio of the plenum to freestream static pressure; isentropic expansion was assumed to calculate the speed of the jet exiting the blowing slot:

A static test (V_∞ = 0 ft/sec) of the Coanda jet can be seen in Figure **3**. The test was an initial check of the assembled model, with tufts for flow visualization. The upper slot is blown and the effective turning of the jet around the Coanda surface can be seen in Figure **3**. The skin and Coanda surfaces were sanded smooth with 2000 grit sandpaper and spray painted black and metallic silver, respectively.

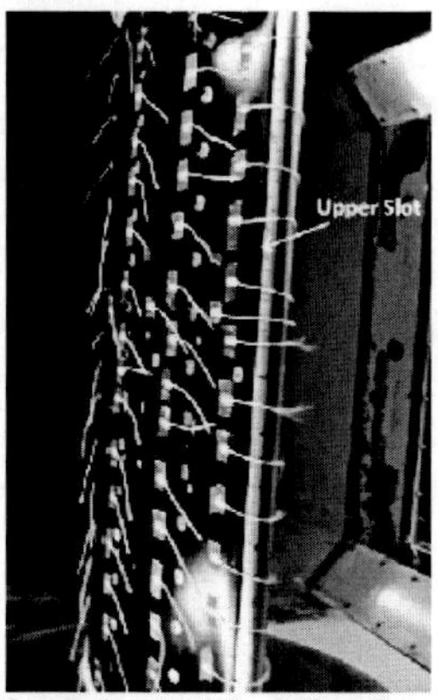

Figure 3: **Coanda Jet Visualization (C_μ = .10)**

2.2. AERODYNAMIC COEFFICIENTS

Aerodynamic forces were measured using both a force balance and surface pressure integrations. The force balance consisted of a rigid steel frame mounted around the wind tunnel test section. The frame was bolted to the lab floor and did not contact the wind tunnel walls at any point, allowing the force measurement equipment to be uninfluenced by tunnel vibrations (Figure 4).

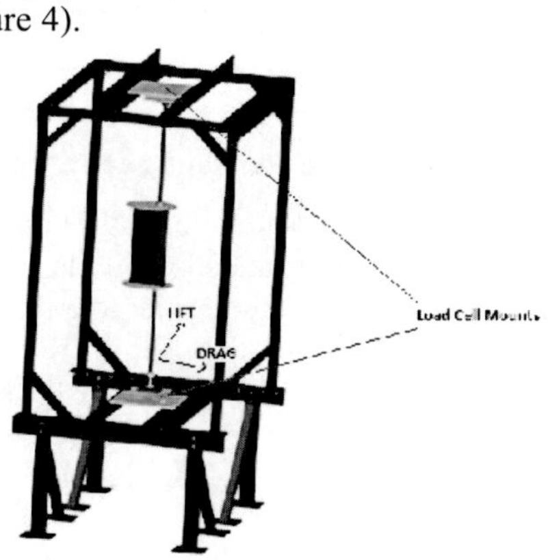

Figure 4: Steel Frame Surrounding Test Section

Full bridge thin beam force sensors were mounted above and below the test section. 20 pound capacity load cells were aligned perpendicular to the freestream air to monitor aerodynamic lift while 5 pound capacity strain gages were mounted parallel to tunnel flow to measure drag. To ensure accurate force measurements, each load cell was statically calibrated; after each load cell was mounted, the system was re-calibrated to check for cross-talk between gages due to misalignment. The experimental lift coefficient, based on force measurements, was calculated from Equation 5. Voltages were recorded by the load cells, applied to calibration curves to convert to a lift force (L) and used to determine the resulting lift coefficient.

$$C_l = \frac{L}{q_\infty} = \frac{L}{\frac{1}{2}\rho_\infty A V_\infty^2} \tag{5}$$

The model was also equipped with 2 separate rows of 33 surface pressure orifices, aligned along 0.4span and 0.6span.. The notation and location of each surface pressure measurement can be seen in Figure 5.

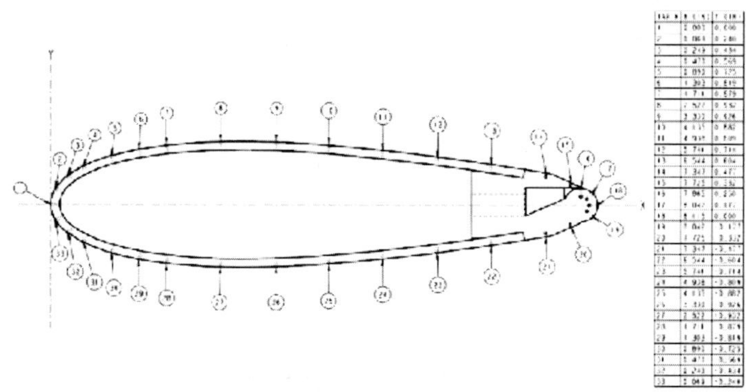

Figure 5: CCNACA0018 Pressure Tap Orientation

The pressure taps were connected to a Scanivalve pressure transducer measurement system and a computer program was used to numerically integrate the measured values to determine the sectional force coefficients.

The pressure coefficient (C_p) was first determined at each pressure tap, Equation 6.

$$C_P = \frac{P - P_\infty}{q_\infty} \tag{2}$$

The pressure distribution along the upper and lower airfoil surfaces was integrated to determine the axial (C_a) and normal (C_n) force coefficients (Equations 7 and 8).

$$C_a = \int_0^1 C_p \, d\left(\frac{y}{c}\right) \tag{7}$$

$$C_n = -\int_0^1 C_p \, d\left(\frac{x}{c}\right) \tag{8}$$

The axial and normal coefficients are then used to determine the sectional lift coefficient, based on angle of attack (α), as seen in Equation 9.

$$C_l = C_n \cos\alpha - C_a \sin\alpha \tag{9}$$

3. WIND TUNNEL MODEL

The design of the CC-VAWT blade was based on a conventional NACA 0018 airfoil with a 9 inch chord. The sharp trailing edge (TE) of the profile was removed and replaced by a rounded Coanda surface, reducing the augmented airfoil's chord to approximately 8 inches. The span was then limited to 16.5 inches, based on blockage considerations outlined by Barlow, Rae, and Pope [11]. With a chord of 8 inches, the chord-to-tunnel height ratio is 0.178, below the maximum of 0.25 suggested by van den Berg [13] and Wood [16].

The airfoil profile and trailing edge geometry were selected based on a preliminary computational study [14]. For reference, the conventional profile augmented to include circulation control capabilities, will be referred to as the Circulation Control NACA0018, or CCNACA0018. The specific dimensions of the augmented NACA profile can be seen in Table 1.

Table 1: TE Geometry

r_{TE}	0.25	[in.]
h_{slot}	0.015	[in.]
c	8.1	[in.]
$(x/c)_{slot}$	0.969	[-]
h_{slot}/r_{TE}	0.06	[-]
h_{slot}/c	0.00185	[-]
r_{TE}/c	0.031	[-]

A comparison of both a conventional and augmented NACA0018 (c = 8.12 in.) can be seen Figure 6.

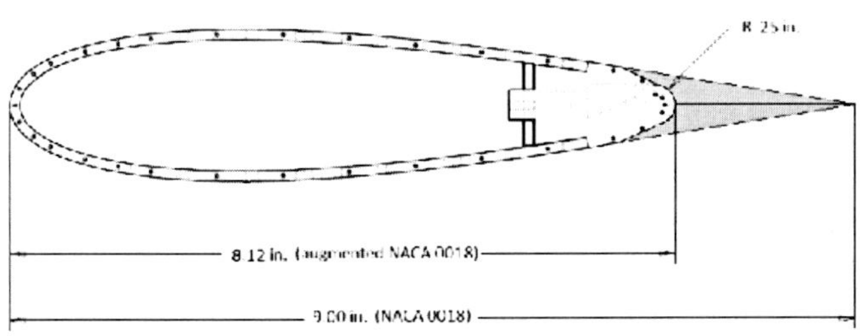

Figure 6: Conventional and Augmented NACA Profiles

3.1. RAPID PROTOTYPING MACHINE

The wind tunnel model was first designed using CAD software, and then the exported files were sent to a rapid prototyping (RP) plotter for fabrication. Several trial runs for various parts had to be completed, as the prototyping machine had improved accuracy and tolerances in the x-y plane versus y-z plane. The machine uses successive layers of 0.010" of ABS plastic to fabricate the 3D objects designed using ProE computer software.

The most critical aspect of this wind tunnel model is the TE diffuser design. It is of critical importance to have a fully convergent nozzle, with enough of a contraction angle to ensure that fabrication tolerances, hand finishing, and slot gap expansion under pressure will never result in a non-converging exit [17]. Possible errors associated with experimental determination of the momentum coefficient for small scale experiments are generally dominated by slot height measurement or by slot expansion under pressure [21]. A major problem experienced in previous models fabricated at WVU, also common to past experimental work completed elsewhere, is expansion of the skin at the blowing slot exit due to the high pressure jet exhausted from the plenum. As outlined in Equation 2, the slot height (h_{slot}) is an integral component used to calculate C_μ. Difficulties arise as the slot height becomes variable when trying to test various blowing conditions (i.e., C_μ = .01, .05, etc.). The plenum pressure of the model must be increased to amplify C_μ. As a result, the slot bulges, making the slot area dependent upon plenum pressure. Static testing is then necessary to measure the slot height at all desirable testing conditions; this can prove to be tedious, time consuming, and imprecise, depending on method of measurement. The ultimate result is an increased level of uncertainty in the value of C_μ, the most significant of all testing parameters. This makes validation by comparison of results to past experimental, computational, and analytical models pointless.

To alleviate this issue, a solid TE insert was designed and created by a Rapid Prototyping machine. The motivation was to provide a uniform slot exit velocity along the entire span to maximize effectiveness of the CC jet. The TE section included the Coanda surface, blowing slot, and a rail to connect the pneumatic valves, all in a single piece. A profile of the diffuser, making up approximately 0.15c of the TE, can be seen in Figure 6; a 3D rendering of the design created using ProE can be seen in Figure 7.

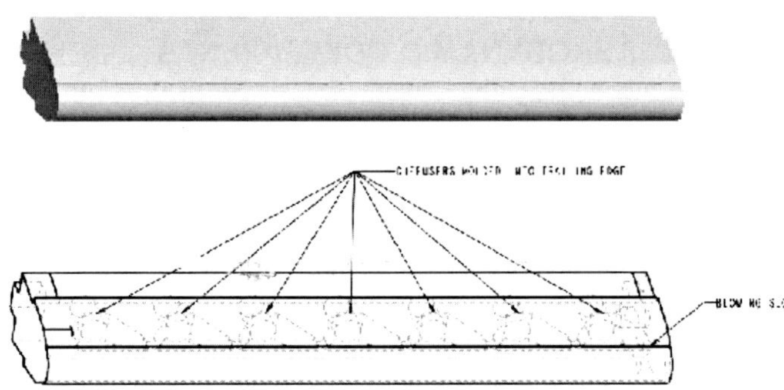

Figure 7: ProE Rendering of TE Diffuser Insert

The DYNAMCO pneumatic valves used had an orifice diameter of 0.008" that produced a circular jet; the diffuser was then used to transition the circular jet to a thin, uniform sheet of air, distributed evenly along the span of the model. The valves were located as far aft in the plenum as possible to minimize flow attenuation; similar procedures used by Jones and Cagle [26] were used as an outline. This left less than 2" of usable space for flow expansion and transformation.

A preliminary jet survey was completed under static conditions; the purpose was to first check the validity of the isentropic expansion assumption to calculate V_{jet}. The jet characterization was also completed to observe the spanwise uniformity of the jet to analyze the effectiveness of the diffuser.

A total pressure probe (0.010" OD, 0.006" ID) was inserted just inside the slot exit along with a J-type thermocouple to measure the conditions of the airflow. A simplified version of the Bernoulli equation was then used to calculate the measured velocity at the exit of the slot. The measurements are compared to theoretical calculations using Equation 4; the results can be seen in Figure 8.

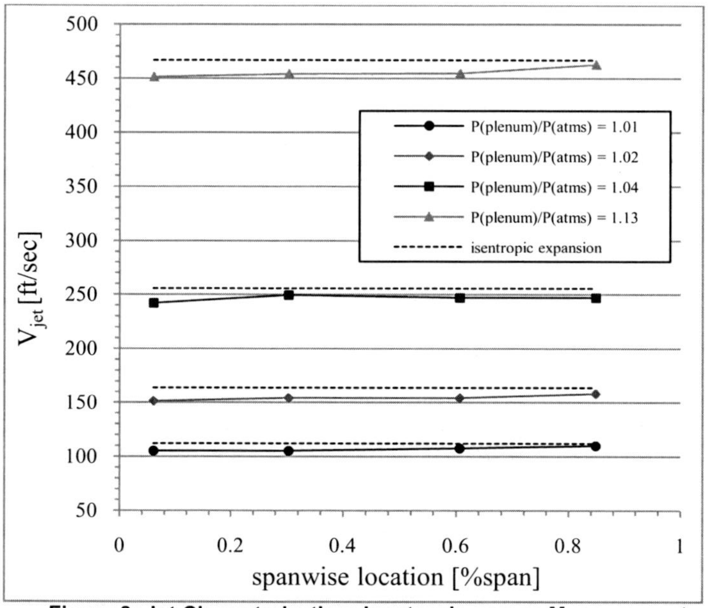

Figure 8: Jet Characterization: Isentropic versus Measurement

An increased discrepancy between experimental and isentropic theoretical methods is noticed as the plenum pressure is increased; the largest error was found to be 4%. The isentropic expansion assumption was thus deemed an accurate method to calculate jet velocity. It can also be seen that the jet is relatively uniform at random check points across the span.

Another unique advantage of using the RP plotter was the ability to fabricate the model's skin with pressure orifice channels integrated within. The skin was 0.125" thick, with 0.012" diameter orifices flush to the airfoil surface. Having the channels within the skin thickness allowed for more room in the model's plenum for valving, wiring, and the diffuser. The pressure tap channels can be seen at the end of the model profile in Figure 6.

4. EXPERIMENTAL RESULTS

Tests were conducted over a geometric incidence range of $-20° \leq \alpha \leq +20°$ and a momentum coefficient range of $0 \leq C_\mu \leq 0.10$; the test Reynolds number remained constant throughout all testing at a value of 472K.

4.1. DATA CORRECTIONS

Conventional corrections were applied to the measured free stream dynamic pressure (q_∞) to account for solid blockage due to model thickness [11], [15]. No wake blockage corrections were applied due to the varying effect of the tangential Coanda jet on wake size. After using the corrected value of (q_∞) to determine the lift and moment coefficients, which could be further corrected for tunnel wall constraint (induced camber, i.e., lift interference) as seen in Equations 12 and 13 [11][19]. The subscript 'u' denotes the uncorrected value measured during testing.

$$\alpha = \alpha_u + \frac{57.3\pi c^2}{96h_t^2}\left(C_{l_u} + 4C_{m_u}\right) \tag{12}$$

$$C_l = C_{l_u}\left(1 - \frac{\pi^2}{48}\frac{c^2}{h_t^2}\right) \tag{12}$$

The corrections outlined above offered negligible changes from measured to corrected values. Further induced AoA corrections, as outlined by [13][29][30] were .

4.2. RESULTS

The results from wind tunnel testing are included in the following figures. During preliminary testing, an unmodified NACA0018 airfoil (c = 8") was tested to validate the load cell force balance set-up. C_l was compared to thin airfoil theory as well as past experimental work completed by NACA [28] and Sandia National Laboratories/Wichita State University [10]. The lift curve slopes ($dC_l/d\alpha$) of the three experiments all coincide with a value of 0.10 per degree; this is slightly less than the thin airfoil theoretical value of 0.11 per degree.

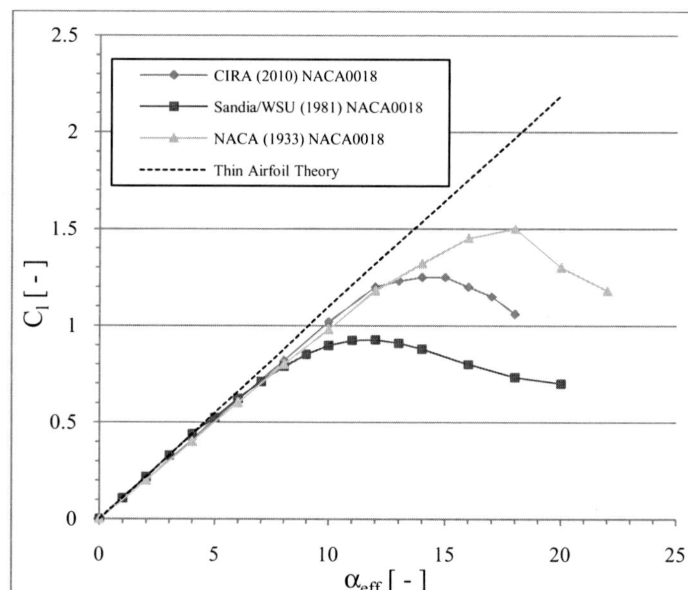

Figure 9: Load Cell Set-Up Validation: Unmodified NACA0018

Figure 10 compares the baseline ($C_\mu = 0$) lift curves from both the force balance and surface pressure integration. Close agreement between the two methods is show, with the largest error between the two methods being 13%. Both methods revealed α_{stall} to be 17°. A more gradual stall can be seen from the surface pressure integration due to the location of the pressure taps being near center span. When testing a two-dimensional circulation control airfoil, stall first occurs at the TE junction of the model to the endplates or sidewalls [16][21][24]. Thus, the lift loss due to stall on the wind tunnel had a larger impact on the load cell measurements which revealed a sharp and drastic stall. Similar results are seen in Figure 11 at a momentum coefficient of 0.10.

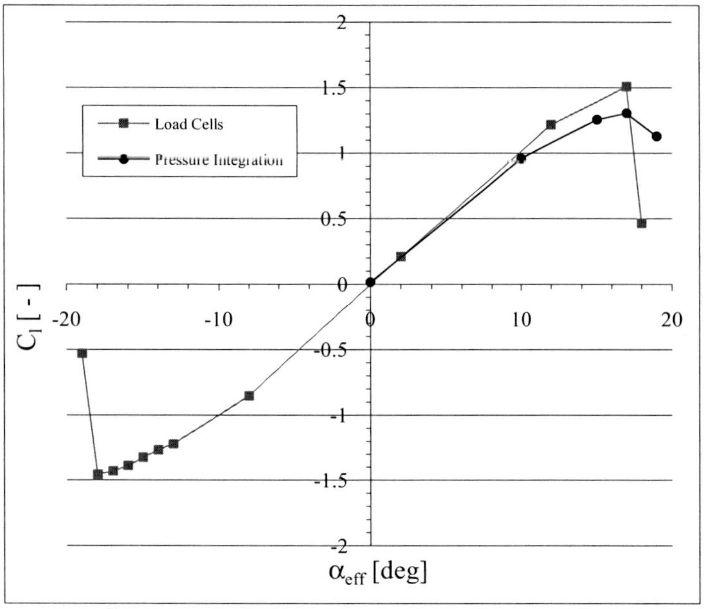

Figure 10: Measurement Method Comparison (C_μ = 0)

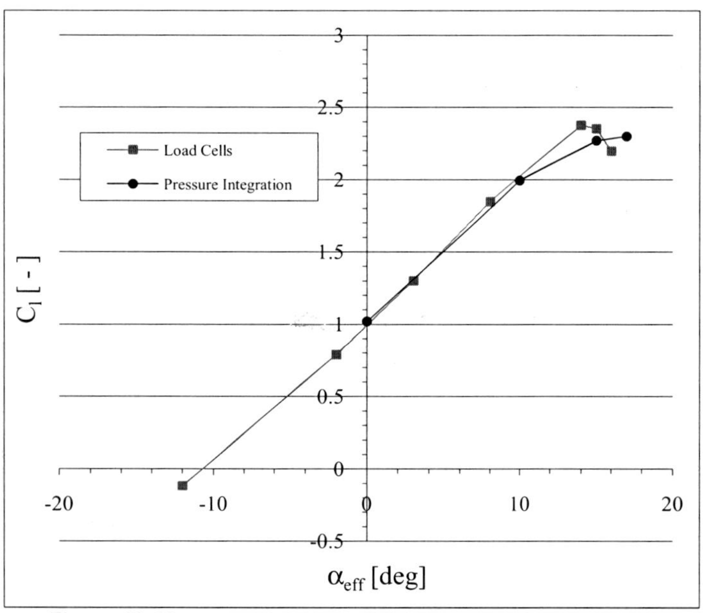

Figure 11: Measurement Method Comparison (C_μ = 0.10)

Figure 12 shows the effects of increased blowing on lift curves; sold lines represent positive angles of incidence while dashed lines denote those that are negative. At high positive α, an aft separation occurs, resulting in a large lift degradation as fully detailed by Englar and Williams [24]. As C_μ is further increased, the lift begins to restore itself; past test data has shown that this restoration is almost a step function in C_μ.

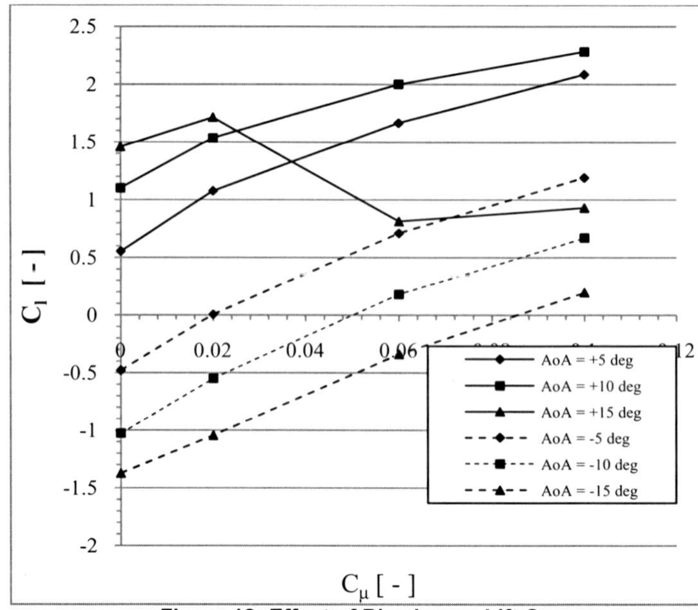

Figure 12: Effect of Blowing on Lift Curves

From Figure 9, it can be seen that boundary layer control was achieved during low levels of blowing (C_μ = 0.02); the lift is slightly augmented while α_{stall} is increased due to prolonged separation from circulation control. As C_μ is increased, the lift is further augmented but α_{stall} is reduced due to an increase virtual camber of the airfoil.

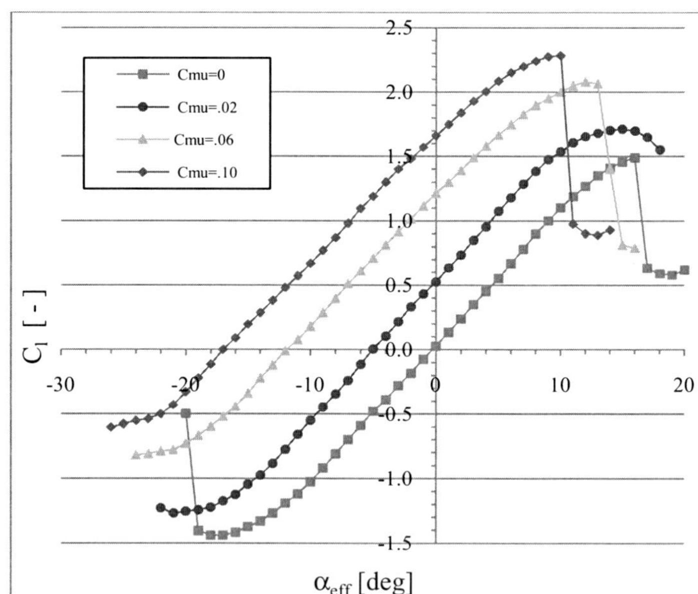

Figure 13: Effect of Blowing on Lift and Stall

Figure 14 shows the lift augmentation ($\Delta C_l/C_\mu$) versus blowing, where ΔCl was defined as the increase in lift coefficient from the unblown to the blown value for a given C_μ. For low rates of blowing, the lift augmentation was most effective at higher α.

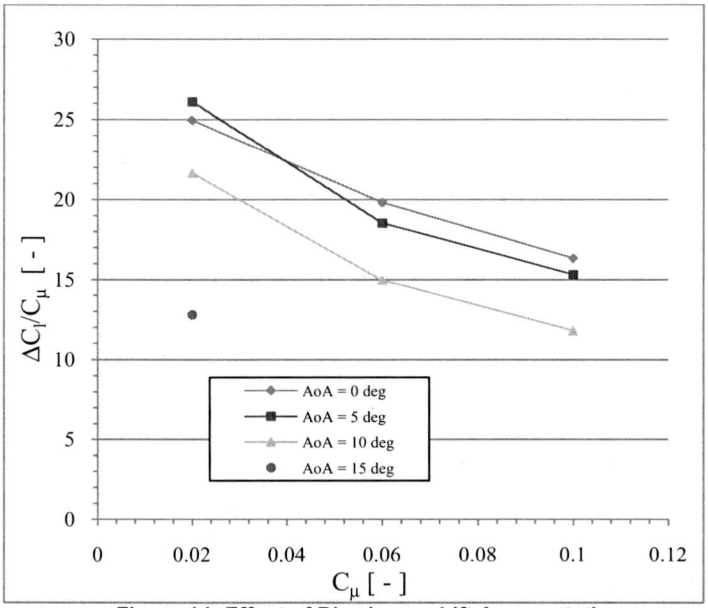

Figure 14: Effect of Blowing on Lift Augmentation

The same results are better seen in Figure 15; C_μ stall, defined by [30] as the AoA when the lift augmentation equals zero, is being approached as the momentum coefficient increases.

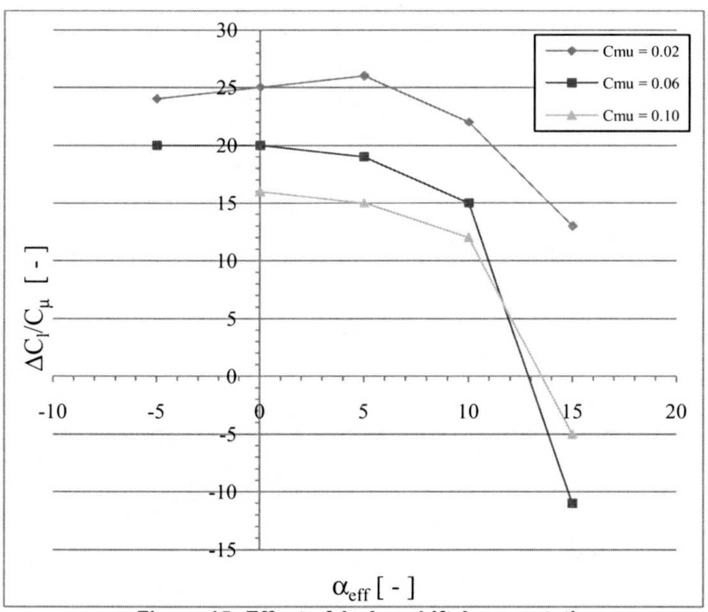

Figure 15: Effect of AoA on Lift Augmentation

5. CONCLUSION

Experimental testing of an augmented wind turbine blade was completed. The original goal of creating look-up tables of aerodynamic data for an augmented NACA airfoil was completed; a set of data tables was composed and will be used as input to the recently developed analytical vortex model to calculate performance predictions for a CC-VAWT. A baseline was established for steady circulation control for the newly developed turbine blade. A $C_{l,max}$ of 2.3 was achieved with a momentum coefficient of 0.10. When operating during low tip speed ratio conditions, where α ranges are largest, low levels of blowing (C_μ = 0.02) appear to offer the highest lift augmentation. These results are as expected due to the increase in apparent camber with increased blowing. At moderate incidence angles (± 10°), the lift is enhanced over the entire range of momentum coefficients tested ($0 \le C_\mu \le 0.10$). Lift augmentation factors in excess of 25 were found for this airfoil at a 2% blowing coefficient. As C_μ is increased, alpha stall is reduced, as expected due to an increase in virtual camber.

6. RECOMMENDATIONS

Future models should have internal pressure transducers and temperature instrumentation directly inside the plenum to allow air conditions to be monitored during real time testing. Additional surface pressure taps at various span wise positions should also be added to observe three dimensional effects at various testing conditions.

The current model only has a functional blowing slot on the upper surface, due to flow limitations of the pneumatic valves used. Alternative valving options are being investigated to allow for controllable dual surface blowing capabilities (upper and lower surface). Experimental investigations into dual blowing for drag reduction [26][27] and controllable jet deflections with varied upper and lower surface blowing [17] could be studied.

Finally, boundary layer control (BLC) should be added to the endplates at the junction of model and barriers, to reduce induced effects from extreme pressure gradients. [16][21][24][25]. Further details outlining BLC on walls, from past experimental efforts, can be found in [16], [20-24].

NOMENCLATURE

A	area; airfoil or blowing slot (ft^2)
c	chord (in.)
C_a	axial force coefficient (-)
C_l	corrected section lift coefficient (-)
$C_{l,u}$	uncorrected section lift coefficient (-)
C_m	corrected moment coefficient (-)
$C_{m,u}$	uncorrected moment coefficient (-)
C_n	normal force coefficient (-)
C_p	pressure coefficient (-)
C_μ	momentum coefficient (-)
h_t	height of test section (ft.)
h_{slot}	height of blowing slot (in.)
L	lift force (lbs.)
$\dot{m}_{jet}$	mass flow, per unit span (slug/ft-sec)
P_{jet}	power requirements to supply jet (W)
P_{plenum}	pressure inside plenum (lbs/in^2)
P_{rot}	pressure at rotameter inlet (lbs/in^2)
Q	volumetric flow rate (ft^3/min)
q_∞	freestream dynamic pressure (lb/ft^2)
R	gas constant (ft-lb$_f$/°R-lb-mol)
Re_{eff}	effective Reynolds number (-)
Re_{test}	testing Reynolds number (-)

T_{plenum}	temperature inside plenum (°R)
T_{rot}	temperature at rotameter inlet (°R)
V_{jet}	velocity of jet (ft/sec)
	Symbols
α_{eff}	effective angle of attack (deg)
α_{geom}	geometric angle of attack (deg)
α_u	uncorrected angle of attack (deg)
$\Delta C_l/C_\mu$	lift augmentation factor (-)
γ	ratio of specific heats (-)
∞	denotes freestream conditions (-)
	Acronyms
AoA	angle of attack
AR	aspect ratio ($span^2/A$)
CC	circulation control
HAWT	Horizontal Axis Wind Turbine
LE	leading edge
VAWT	Vertical Axis Wind Turbine
RP	rapid prototype
TE	trailing edge
TF	turbulence factor

REFERENCES

[1]. Gipe, P., *Wind Energy Comes of Age*. John Wiler and Sons, 1995, ISBN 047110924X, 9780471109242.

[2]. Wolfe, W., "Analysis of Test Results for the WVU Straight Bladed Darrieus Wind Turbine," *M.S. Thesis*, West Virginia University, Morgantown, WV, 1978.

[3]. Walters, R., "Innovative Wind Machines," West Virginia University, 1975.

[4]. Angle, G., "Aerodynamic Benefits of Near-Surface-Actuated Circulation Control Blowing Slots for Rotorcraft Use," *Ph.D. Dissertation*, West Virginia University, Morgantown, WV, 26506, 2008.

[5]. Abramson, J., "Two-Dimensional Subsonic Wind Tunnel Evaluation of a 20-Percent-Thick Circulation Control Airfoil," David W. Taylor *NSRDC-ASED 311*, 1975.

[6]. Abramson, J., "Characteristics of a Cambered Circulation Control Airfoil Having Both Upper and Lower Surface Trailing Edge Slots," *NSWCCD-50-TR-2004/030*, 2004.

[7]. Englar, R., "Experimental Investigation of the High Velocity Coanda Wall Jet Applied to Bluff Trailing Edge Circulation Control Airfoils," September 1975, Naval Research and Development Center *Report 4708*.

[8]. Fanucci, J., Walters, R., "Innovative Wind Machines: The Theoretical Performances of a Vertical Axis Wind Turbine," In Proc. Of the VAWT Technology Workshop, Sandia Lab. *Report SAND 76-5586*, 1976.

[9]. Loth, J., Chandra, Subrato, Gibbs, E., "Inlet and Internal Aerodynamics of VTOL Thrust Augmenter," Final Report, Office of Naval Research Contract *N00014-75-C-0324*, 1976.

[10]. Sheldahl, R. E. , Kilmas, P. C., "Characteristics of Seven Airfoil Sections Through 180 Degrees Angle of Attack for Use in Aerodynamic Analysis of Vertical Axis Wind Turbines," Alburquerque, NM: *SAND80-2114*, Sandia National Laboratories, 1981.

[11]. Barlow, J.B., Rae, W.H., and A. Pope, Low-Speed Wind Tunnel Testing (3rd Edition). John Wiley & Sons, New York, 1999.

[12]. Myer, D.P., "An Experimental Investigation of A Circulation Controlled Cambered Elliptical Airfoil with A Rounded Trailing Edge," *M.S. Thesis*, West Virginia University, 1972.

[13]. van den Berg, "Some Notes on Two-Dimensional High-Lift Tests in Wind Tunnels," from *AGARD Lecture Series No. 43 on Assesment of Lift Augmentation Devices*, Belgium, April 20-24, 1970.

[14]. Graham, H., Panther, Wilhelm, J., et al., "Airfoil Selection for A Straight Bladed Circulation Controlled Vertical Axis Wind Turbine," Proceedings from the ASME Energy Sustainability Conference: *ES2009-90343*, San Francisco, July 19-23, 2009.

[15]. Garner, H.C., E.W. Rogers, et al. "Subsonic Wind Tunnel Wall Corrections." Paris, Oct 1966 (*AGARDOGRAPH 109*), p. 29-74.

[16]. Wood, N.J., "The Aerodynamics of Circulation Control Aerofoils," *JIAA TR-41*, Stanford University, Stanford, CA, 1986.

[17]. Rogers, E.O., M.J. Donnelly, "Characteristics of a Dual-Slotted Circulation Control Wing of Low Aspect Ratio Intended for Naval Hydrodynamic Applications," AIAA 42nd Aero. Sciences Meeting, Reno, January 5-8, 2004.

[18]. Bradshaw, P. Experimental Fluid Mechanics. The Macmillan Company, New York, 1964.

[19]. Mangler, W., "The Lift Distribution of Wings with End Plates," *NACA TM No. 856*, Washington, April 1938.

[20]. Williams, R.M., "Some Research on Rotor Circulation Control," CAL/AVLABS Symposium Proceedings, Vol. 11, Buffalo, 1969.

[21]. Englar, R.J., G.S. Jones, et al., "2-D Circulation Control Benchmark Experiments Intended for CFD Code Validation," 47th AIAA Aerospace Sciences Meeting (*AIAA-2009-902*) Orlando, January 5-8, 2009.

[22]. Englar, R.J., "Two-Dimensional Subsonic Wind Tunnel Tests of Two 15-Percent Thick Circulation Control Airfoils," *NSRDC Technical Note AL-211*, Bethesda, August 1971.

[23]. Novak, C.J., R.K Roads, "Experimental Investigations of the Circular Wall Jet on a Circulation Control Airfoil," AIAA 25th Aerospace Sciences Meeting (*AIAA-87-0155*) Reno, January 12-15, 1987.

[24]. Englar, R.J., R.M. Williams, "Test Techniques for High-Lift, Two-Dimensional Airfoils with Boundary Layer and Circulation Control for Application to Rotary Wing Aircraft," *NSRDC Report 4645*, Bethesda, July 1975.

[25]. Rodman, L.C., N.J. Wood, "Verification of Performance Results for A Low-Speed 15% Elliptical Circulation Control Airfoil," *JIAA TR-56*, Stanford University, Stanford, CA, 1986.

[26]. Cagle, C.M., G.S. Jones, "A Wind Tunnel Model to Explore Unsteady Circulation Control for General Aviation Applications," 22nd AIAA Aerodynamic Measurement Technology and Ground Testing Conference (*AIAA-02-3240*), St Louis, June 24-26 2002.
[27]. Englar, R.J., R.M. Williams, "Design of A Circulation Control Stern Plane for Submarine Applications," *NSRDC Technical Note AL-200*, Bethesda, March 1971.
[28]. Jacobs, Ward, Pinkerton, "The Characteristics of 78 Related Airfoils from Tests in the Variable-Density Wind Tunnel," *NACA Report 460*, 1933.
[29]. Swanson, R.C., C.L. Rumsey, "Computation of Circulation Control Airfoils," *J. of Computers and Fluids*, Vol. 38, pp. 1925-1942, 2009.
[30]. Wood, N., J. Nielsen, "Circulation Control Airfoils: Past, Present, Future," (*AIAA-85-0204*), 1985.
[31]. Wood, N.J., E.O. Rogers, "An Estimation of the Wall Interference on A Two-Dimensional Circulation Control Airfoil," (*AIAA-86-0738*), 1986.

Proceedings of the ASME 2010 International Mechanical Engineering Congress & Exposition
IMECE2010
November 12-18, 2010, Vancouver, British Columbia, Canada

IMECE2010-38903

ELECTRO-THERMAL ICING MITIGATION SYSTEM FOR POLYMERIC COMPOSITE AIRFOIL

Maryam Mohseni, Pierre Mertiny, Alidad Amirfazli*
Department of Mechanical Engineering
University of Alberta
Edmonton, Alberta, Canada

ABSTRACT
Polymer composite materials are widely used in aircraft and wind turbine industries due to their high specific stiffness and strength. In this study the concept of an electro-thermal anti-icing system is developed using composite material. Controlling temperature distribution and minimizing energy consumption can be the main benefits of such an anti-icing system. The thermal behavior of a composite material, power consumption and the type of thermal system were characterized for designing the icing mitigation system. Constantan wires as thermal elements were embedded inside fiberglass/epoxy composite laminates, and the thermal characteristics of the system at specific power levels were investigated using thermocouples in a wind tunnel under room temperature and cold room conditions. When applying power, the surface temperature reached 50% of its final steady-state value in 20s. To obtain a relatively uniform temperature distribution and prevent the formation of cold spots on the surface, the maximum distance between the thermal elements was 1cm. The temperature of different points on the surface increased with increasing power, and its trend was the same in room temperature and cold room tests. In cold room tests, a surface temperature above 5°C was reached at the thermal elements and between them at 0.42 W/cm, which is sufficient to prevent ice formation in icing condition. This power did not cause thermal degradation of the composite material.

Keywords: Anti-icing; electro-thermal; polymeric composite; thermal characteristics; temperature map.

1 INTRODUCTION
Icing adversely affects the performance of aerodynamic structures such as aircraft wings and wind turbine blades. It is responsible for increasing the weight and decreasing the aerodynamic efficiency of the wing by decreasing the lift and increasing drag [1-3]. Over the years, a number of icing mitigation techniques have been developed and are categorized as anti-icing and de-icing systems [1, 4, 5]. De-icing is a process to remove ice from the surface after formation to provide a clean surface. There are various methods for de-icing e.g., pneumatic boots, thermal and chemical techniques using heat system and chemical fluids for ice removal [1, 4, 5].

Anti-icing systems prevent the formation of ice or frost for a limited period of time, and they are activated before entering icing conditions to prevent the adhesion of ice to the surface [1]. There are mainly three types of anti-icing systems, i.e., chemical systems, hot-air systems and electrical resistance heating (electro-thermal) systems [2]. There are also other systems such as super-hydrophobic coatings.

In chemical systems a chemical liquid with a low freezing temperature such as ethylene glycol is distributed on the ice prone surface [2, 4]. This system has been used successfully in some of aircrafts, but due to the weight load of the anti-icing liquid that the aircraft must carry, this system is generally not considered to be an effective anti-icing system.

In hot-air systems, which are the most prevalent thermal anti-icing systems, hot bleed air is ducted from the compressor of the engine to the wings through piccolo tubes or narrow gas passages to transfer thermal energy to the surface and thus heat the wing surface [2]. This is a highly reliable and well developed technology for conventional aircraft wings. But, due to the required volumetric flow rate of hot air, a relatively large amount of fuel is spent on heating and thrust is reduced which are important issues considering aircraft efficiency [2]. A system for polymer composite airfoils can be challenging since maximum operating temperature and thermal resistance of polymer composites pose a temperature limit for hot air [6, 7].

* Author to whom correspondence should be directed:
Phone: 780-492-6711; Fax: 780-492-2200;
Email: a.amirfazli@ualberta.ca

Ciardullo et al. [10] have developed a method for using hot-air as an anti-icing system for advanced composite structures. The concept involves conducting hot air through embedded glass capillary tubes on the surface of the composite structure rather than conducting heat through thick composite walls. Surface heating of composites employing this method was investigated, but its energy consumption and its effect on surface roughness, which can lead to increase the aerodynamic drag, has not been assessed. It was also not tested in icing conditions.

Electrical resistance heating systems are mostly used in the form of electro-thermal pads, applied to the surface of a structure, to prevent ice formation. In one version of such thermo-electric system, pads consist of a heating element, two layers of insulation and a layer of sheet metal as a shield. This system is very simple, but the power consumption can be very high [2]. Different types of electro-thermal pad systems with various electrical heating elements such as metal foils, carbon fibers, and graphite fiber/epoxy composites have been developed for different structures [3, 5, 7-9].

Hung et al. [3] reported on the use of a surface electro-thermal heater made from graphite composite material in the form of flat coupons. The heating element in this system is a graphite fiber/epoxy composite which is claimed to have high thermal and electrical conductivity. The protecting layers are made of glass fiber/epoxy composites. This system, which can be formed on irregular surfaces, was found to have acceptable heating performance in room temperature tests, but it has not been tested in wind tunnel icing experiments.

In this paper a new electro-thermal system is explored to be used as an anti-icing system for polymer composite structures. In this system, constantan wires which are used as electrical thermal elements are embedded in an aircraft grade fiber glass/epoxy composite in the form of flat coupons. The coupons were made and tested in room temperature and cold room conditions to investigate the feasibility of the concept. This system does not need any protection and sheet metal layers, and it can simply be incorporated during the manufacturing of a polymer composite structure such as airfoils. It is also applicable to irregular surfaces with different profiles. It provides a smooth surface and allows for the thermal elements to be placed close to the surface with direct heating, resulting in a nearly uniform heating of the surface. In principle this system can be customized so that the temperature of different locations on the surface can be controlled to reach a specific anti-icing temperature while minimizing energy consumption.

2 EXPERIMENTS

2.1 Materials and Manufacturing Procedure

In this study, polymer composite laminates were manufactured from a commercially available aerospace-grade fiberglass/epoxy prepreg, i.e., Necote E-765 epoxy/fiberglass prepreg from Park Electrochemical Corporation Advanced Material Technologies. Prepregs are pre-impregnated reinforcing fibers with a partially cured polymer matrix resin. The reinforcing phase in this prepreg is woven fiberglass (E-glass) fabric. Prepregs are in the form of sheets which can readily be applied to make the structure; hence, experimental uncertainties due to composite sample fabrication will be minimized. Since here polymer composites are subjected to relatively high temperatures, thermal stability of the prepreg in the anti-icing system is important. In polymer composites, thermal stability depends on the matrix system and it increases with an increasing glass transition temperature (Tg) [11]. Consequently, the prepreg was selected based on its Tg and maximum operating temperature. The specification of the selected prepreg is provided in Table 1.

Table 1: Physical properties of the prepreg as provided by the manufacturer

Prepreg	**Resin solids (%)**	**Volatile content (% max)**	**Tg (by DMA) (°C)**
7781 E-Glass	38	1.5	165
Gel time @275F (min)	**Void content (%)**	**Maximum wet service temperature (°C)**	**Maximum dry service temperature (°C)**
5	< 2%	82	137

Constantan wire was used as the thermal elements in the anti-icing system, because its resistivity is almost constant over a wide temperature range (0-100°C). Constantan wires with two diameters, i.e., 0.13 mm and 0.25 mm, were used from Omega Engineering Inc. Corresponding physical properties are presented in Table 2. Since both wires with different diameters are analyzed in this study, the wires will be termed ‚thin' and ‚thick' wire respectively.

The polymer composite laminates were fabricated by hand lay-up of six prepreg layers. The constantan wires as thermal elements were laminated in the middle of the composite sheet. The thermal wires were aligned straight inside the composite laminate using a specially designed jig. At first, three layers of prepreg were deposited, then the thermal wires were affixed on the third layer, and another three prepreg layers were laminated on top of the elements.

Table 2: Physical properties of the thermal elements as provided by the supplier

Type	Constantan T
Composition	45% Nickel 55% Copper
Size (mm)	0.13 0.25
Resistivity at 0°C & 20°C (μΩcm)	48.9
Temperature coefficient of resistance, 0-100°C (Ω/Ω/°C)	-0.1×10^{-4}
Thermal conductivity at 100°C (W/m°C)	21.115

The final dimension of the fabricated composite laminate was 15cm (L) × 16cm (W) × 1.4mm (H). The length of each constantan wire was 10cm and the distance between two adjacent wires was 1cm. Also, two wires were put near both edges of the sheet with a 2.5cm distance from the next wire to investigate edge effects. The laminate was then covered with vacuum bagging material and cured in the form of a flat plate for one hour at 82°C, followed by a two-hour cure at 135°C, in an oven applying a minimum vacuum pressure of 82kPa. Following curing, the thermal elements were connected to each other to form a serial circuit.

2.2 Measurements

Composite sheets with embedded heating elements were tested in an open circuit wind tunnel. This low speed wind tunnel is refrigerated and able to operate at subzero temperatures. The test section is a transparent, ice free cavity 26cm high, 30cm long and 38cm wide. Tests were conducted at room temperature and cold room conditions between -12°C and -10°C under a maximum air flow of 27m/s. Composite sheets were heated by connecting the thermal elements to a DC power supply. The power was adjusted in a range of 2W to 26W for room temperature tests and 24W to 40W in cold room experiments. The composite surface temperature at different power levels was recorded by eight fast response T-type thermocouples as a function of heating time. A schematic of the thermocouple positions on the sheet surface is shown in Fig.1.

Thermocouples were placed on a straight line on the surface 0.5cm apart from each other. In Fig.1, "W", "B" and "O" represent a location on a thermal element, the midpoint between two adjacent thermal elements, and a point outside the thermal element section respectively. Thermocouples were connected to a data acquisition system and surface temperatures were recorded using Labview software by National Instruments. The temperature as a function of heating time was recorded at different points on a straight line on the surface. In this study, test data were obtained from samples with seven thermal elements, of which five were put 1cm from each other and the remaining two, located near the sample edges, were 2.5cm far from the next thermal element.

3 RESULTS AND DISCUSSION

When the thermal elements were heated starting from room temperature with a DC current with constant electrical power, the temperature rise at selected points on the surface was monitored with time as shown in Fig.2. As can be seen in Fig.2, the surface temperatures on the thermal elements (TW) were higher than those of points between the elements (TB) and outside the thermal element section (TO). It was observed that the temperature rise for different points on the surface reached 50% of the final steady-state value in 20 seconds, which indicates a rapidly responding heating system.

The steady-state temperature at 250 seconds (T_{250} - $T_{ambient}$) as a function of their position relative to the sample edge at different normalized power amounts is shown in Fig.3. Power amounts were normalized with respect to the total length of the thermal elements in the composite sheets. It was found that the temperature distribution on the surface is relatively uniform for the points on the thermal elements and between them, except at the points where their distance from the thermal elements were more than 0.5cm (TB2, TB3), those with more than 0.5cm distance from one thermal element (TB1, TB4) and points located outside the thermal elements section (TO1, TO2). The lowest temperature was observed on TO1, TB2 and TB3 and their temperatures did not increased by increasing power. This indicates that temperature does not increase with power for the surface points with 1cm distance from the thermal elements.

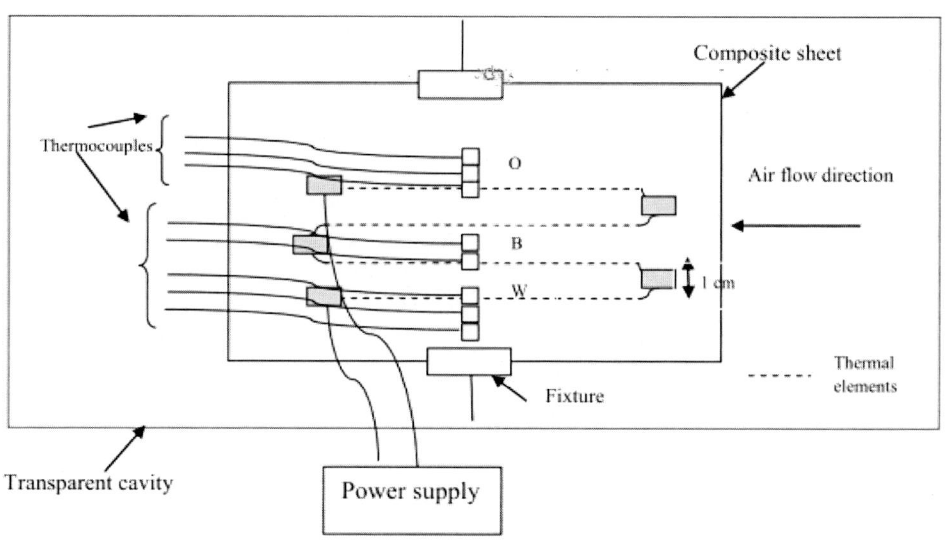

Figure 1: Top view schematic of thermocouple positions on the sheet surface in the transparent wind tunnel test section.

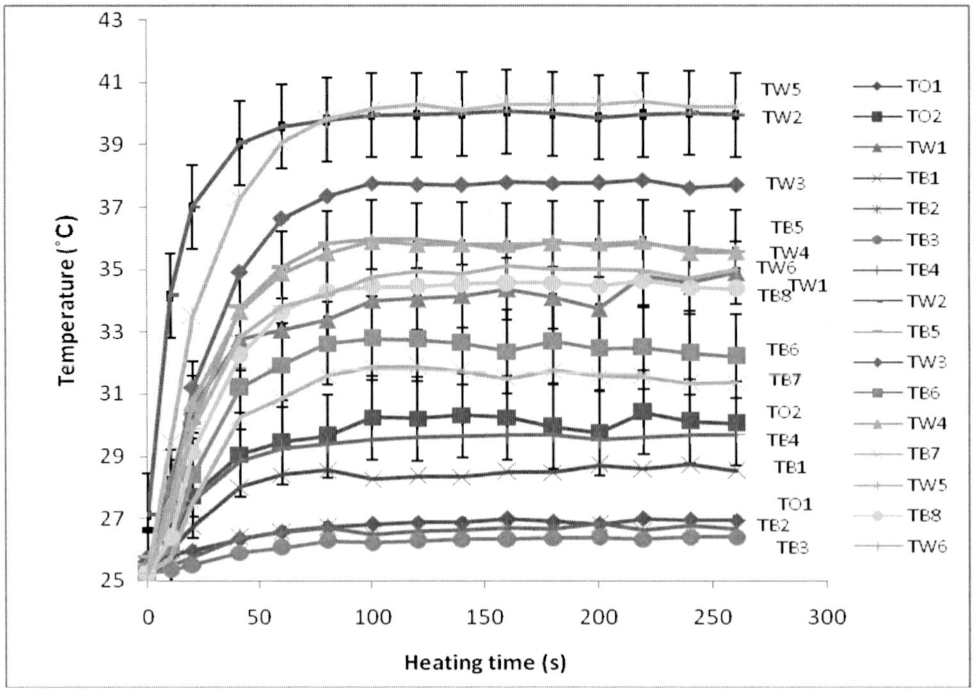

Figure 2: Temperature as a function of heating time for thick thermal element at 0.27W/cm (power per element length) at room temperature. (TW1: Temperature on the thermal element next to the sample edge; TW2: Temperature on the second thermal element 2.5cm away from W1; TW3, 4, 5, 6: Temperature on the thermal elements, each 1cm further from previous one; TO1: Temperature of the point 1cm from W1 and outside the thermal element section; TO2: Temperature of the point 0.5cm from W1 and outside the thermal element section; TB1: Temperature of the point 0.5cm from W1 and 2cm from W2; TB2: Temperature of the point 1cm from W1 and 1.5cm from W2; TB3: Temperature of the point 1.5cm from W1 and 1cm from W2; TB4: Temperature of the point 2cm from W1 and 0.5cm from W2; TB5: Temperature of the midpoint between W2 and W3; TB6: Temperature of the midpoint between W3 and W4; TB7: Temperature of the midpoint between W4 and W5; TB8: Temperature of the midpoint between W5 and W6.) (Note that lines connecting data points are shown to provide a guide for the eye of the reader.)

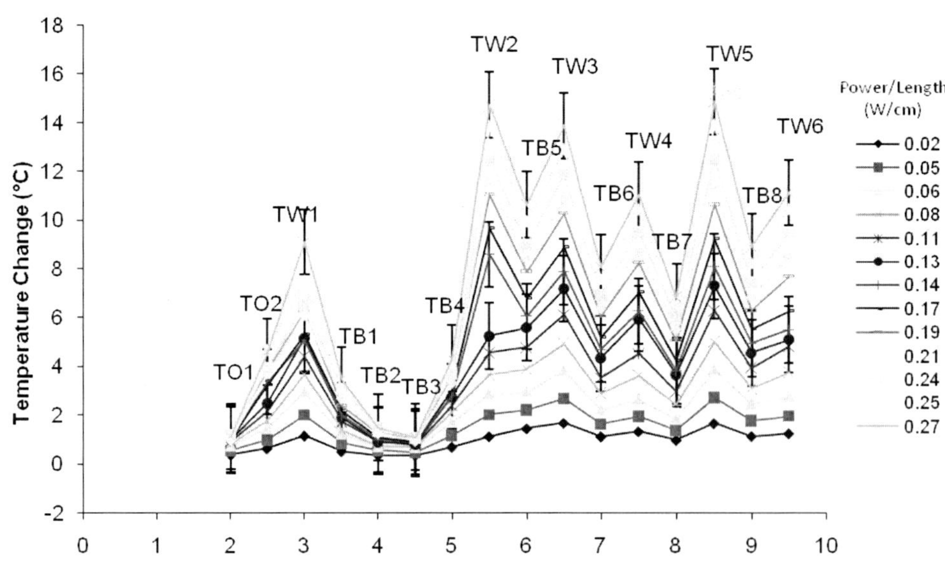

Figure 3: Steady-state temperature of different points as a function their position relative to the sample edge at various W/cm at room temperature for the thick thermal element. (Note that lines connecting data points are shown to provide a guide for the eye of the reader.)

As can be noticed from Fig.3, the temperature distribution on the surface is more uniform at lower power levels, and by increasing power, temperature differences between the various points on the surface increases. Considering the uncertainty of temperature measurements, the maximum temperature difference between the points on the wires (TW) and points between the wires (TB) was 5°C at 0.27W/cm. This can be explained by the low thermal conductivity of the composite material. It shows that the maximum distance between thermal elements should not be more than 1cm in order to ensure a relatively uniform temperature distribution on the surface. In icing condition, this pattern of thermal elements in the composite material, which can provide a nearly uniform temperature distribution, would be able to keep the entire surface warm enough for preventing ice formation. Furthermore, the required anti-icing surface temperature can be controlled by the spacing between thermal elements and electrical power input. Figures 3 and 4 show that the temperature of different points increases with increasing power; but its increase rate is higher for the surface points located on the thermal elements. The rate is lowest for the points located outside the thermal elements section and those which are not located between two thermal elements with 1cm distance. The maximum temperature change from room temperature at a power input of 0.27W/cm is about 13°C for the surface point on the thermal element, and the minimum is less than 1°C for TO1, TB3 and TB2 whose distance from the thermal elements is 1cm and more.

In the present application, polymer composite materials may potentially be subjected to high temperatures, and when being exposed to temperatures above or close to their upper service temperature, they will irreversibly be damaged. Hence, the surface temperature of the thermal elements must carefully be considered, i.e., it must be less than the maximum working temperature of the composite, which in this case is 82°C. In a separate study the present researchers determined that the surface temperature of the thermal elements was about 73°C, which is lower than the maximum service temperature of the composite. As a result, 0.27W/cm is a safe power level which can be applied without composite thermal degradation.

Constantan wires with two different diameters were used as thermal elements to investigate the difference between two types of thermal elements. As it can be seen from Fig.5 the temperature distribution for the thin wire is less uniform compared to thick wire (see Fig.3). Figure 5 shows that the temperatures on the wires are not necessarily the highest ones, and do not show a specific trend. This can be explained by the fiber orientation, the relative position of thermal elements with respect to the reinforcing fibers and resin matrix in the composite material and voids trapped during the manufacturing process. This behavior was verified through experimental results from other composite coupons with thin wire.

Steady-state temperatures of three specific points on the composite sheets with two types of thermal elements are provided as a function of power in Fig.6. This figure shows that for the selected points which are on the wire (TW), between two wires with 0.5cm from each wire (TB-0.5) and between two wires with 1cm from one wire and 1.5cm from other wire (TB-1), the temperature changes for thin and thick wires were the same at a specific power input. Since the diameter of the constantan wire did not affect its performance as a thermal element, and working with thick wire is generally easier than with a thin one, the thick wire was selected for further experiments at this stage.

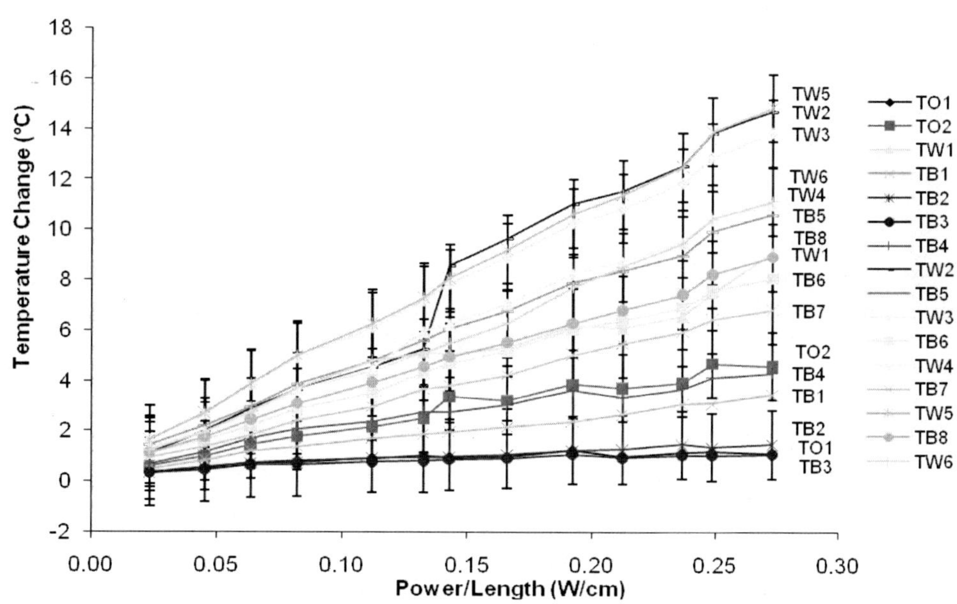

Figure 4: Steady-state temperature changes of different points of the surface as a function of power input at room temperature for the thick thermal element. (Note that lines connecting data points are shown to provide a guide for the eye of the reader.)

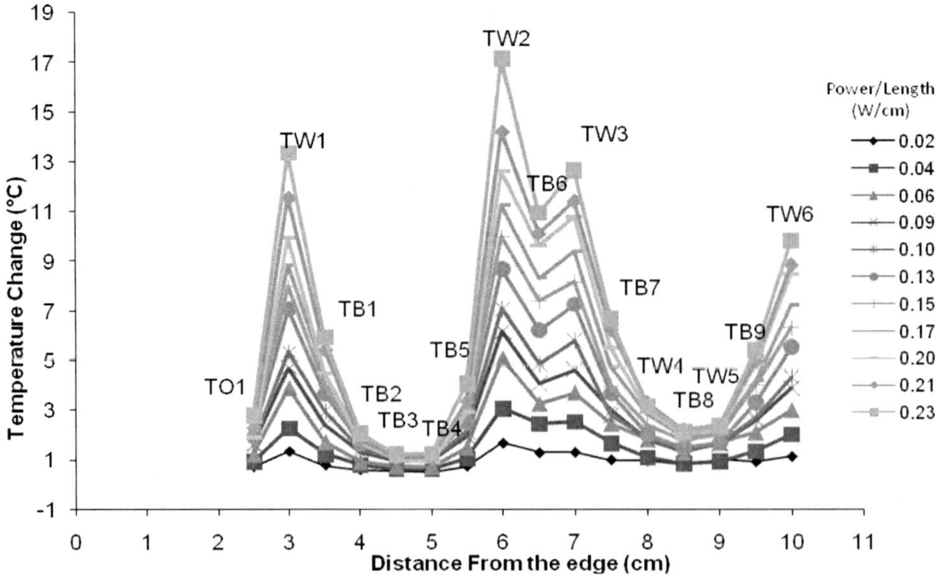

Figure 5: Steady-state temperature of different points as a function of their position relative to the sample edge at various W/cm at room temperature for the thin thermal element. (Note that lines connecting data points are shown to provide a guide for the eye of the reader.)

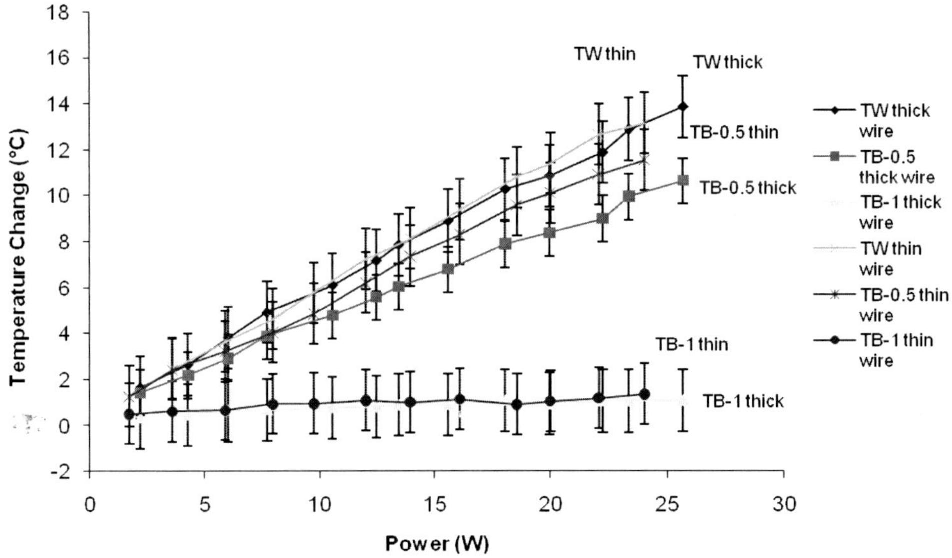

Figure 6: Comparison of surface temperature changes at three similar surface points for thermal elements with two different diameters as a function of power input. (Note that lines connecting data points are shown to provide a guide for the eye of the reader.)

Surface temperature measurements for the composite sample with the thick thermal elements in cold room conditions are shown in Fig.7. In this figure the steady-state temperature of the selected points at 250 seconds is presented as a function of their distance from the sample edge at different power inputs. It can be seen from Fig.7 that even at the highest power input of 0.42 W/cm, point TB3 (which is located 1.5cm from the edge thermal element and 1cm from the second one) and TB4 (located 2cm from the edge thermal element and 0.5cm from the second one) could not reach to above-zero temperatures. This means that under icing conditions, the points which are located farther than 0.5cm from one thermal element will be cold spots which could lead to the formation of ridges in cold regions. Similar to room temperature experiments, the surface points on the thermal elements showed the highest temperatures, and points between the wires had lower temperatures. All surface points, except TB3 and TB4, reached a temperature above 0°C at a power level of 0.32W/cm, and above 5°C at 0.42W/cm, which is high enough to prevent ice formation in icing conditions.

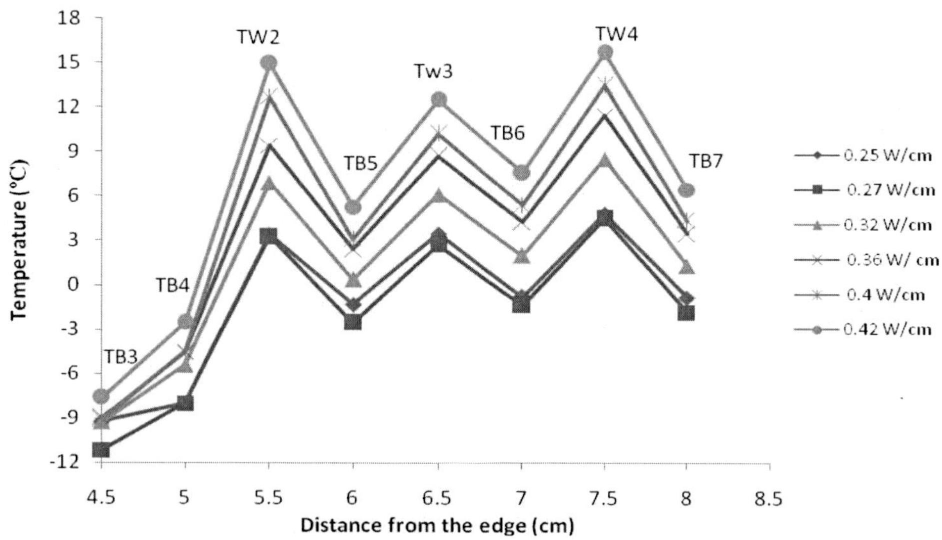

Figure 7: Steady-state temperature of different points as a function of their position relative to the sample edge at different W/cm for cold room tests with the thick thermal elements. (Note that lines connecting data points are shown to provide a guide for the eye of the reader.)

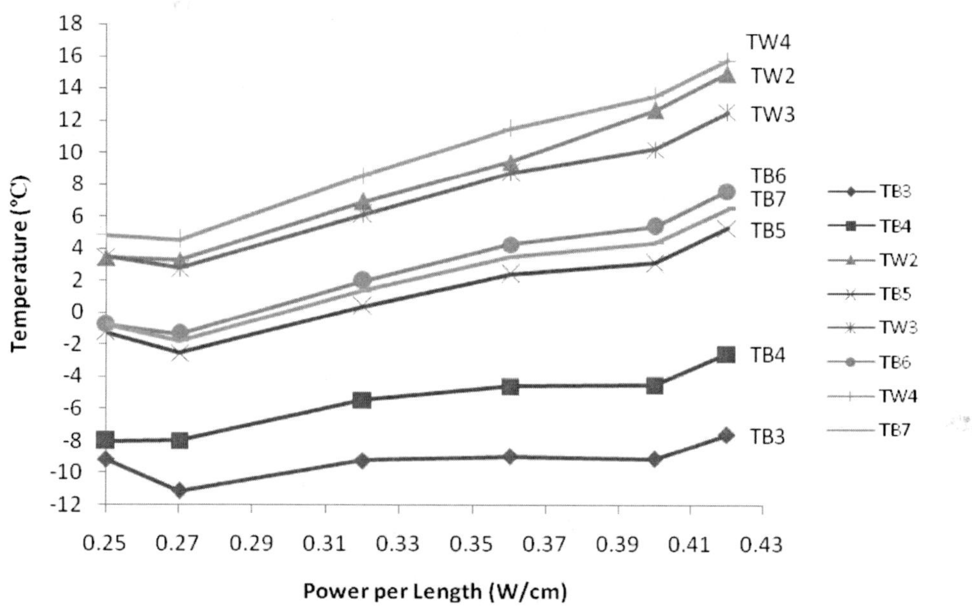

Figure 8: Steady-state temperature of different points as a function of power input for cold room test with the thick thermal elements. (Note that lines connecting data points are shown to provide a guide for the eye of the reader.)

Figure 8 shows the steady-state temperature of different points on the surface as a function of power-per-length for cold room conditions. The rate of temperature increase with power for different points, except for TB3 and TB4, is almost the same. As shown in Figs. 7 and 8, the surface temperature increases with increasing power input, and the surface temperature can be controlled with the amount of applied power.

Thermal analysis showed that the surface temperature of thermal elements at 0.42W/cm is about 61°C, which is less than the maximum service temperature of the polymer composite. Hence, this heating power can be applied to composite samples without causing thermal degradation. Also, this power level seems sufficient to prevent ice formation on the composite surface.

4 CONCLUSIONS

The concept of embedded thermal elements as an electro-thermal anti-icing system in polymer composite structures was developed. The feasibility of this system was investigated through making flat composite plates with constantan wires as thermal elements located in the middle of the composite laminate. Electrical power was applied to the heating system in wind tunnel tests at room temperature and cold room conditions. Temperature measurements of the surface showed that the maximum allowable distance between thermal elements resulting in a relatively uniform temperature on the surface was 1cm. The system also provided fast heating of the surface when power applied to the heating elements, and the surface temperature reached 50% of its steady-state temperature in 20 seconds. The surface temperature increased with increasing power input, and the maximum temperature rise at 0.27W/cm was 13°C at the thermal element locations. A similar trend of temperature rise with heating time and power input at similar points on the surface was observed for room temperature and cold room tests. Therefore, the required anti-icing temperature on the composite surface can be adjusted and controlled by the spacing of the thermal elements and the amount of applied electrical power. In cold room test at -10°C, the temperature of the surface at the thermal elements and between them were above 5°C at 0.42W/cm, which is sufficient high to prevent ice formation in icing conditions. This power level is also safe for the polymer composite material and can be applied without causing thermal degradation of the composite.

ACKNOWLEDGEMENTS

The authors would like to acknowledge the help of Kevin Breton with sample preparation and Siobhan O'Connor for assisting with experiments. The authors also wish to acknowledge the Canada School of Energy for their financial support and Park Electrochemical Corporation Advanced Material Technologies for providing composite prepregs, and Ron Duquette from technical sales of Park Electrochemical Corp.

REFERENCES

[1] Aircraft icing handbook, Civil aviation authority, 2000, March 19, 2010 URL http://www.caa.govt/

[2] Scott K. Thomas, Robert P. Cassoni, 1996, "Aircraft ant-icing and de- icing techniques and modeling", J. Aircraft **33**(5), pp. 841-854.

[3] Ching-Cheh Hung, M. E.Dillehay, M. Stahl, 1987, "A heater made from graphite composite Material for potential deicing application", J. Aircraft **24**(10), pp. 725-730.

[4] A.H. Tabrizi, W.S. Johnson, 1998, "Surface-blowing anti-icing technique for aircraft surfaces", J. Aircraft **26**(4), pp. 354-359.

[5] N. Dalili, A. Edrisy, R. Carriveau, 2009, "A Review of surface engineering issues critical to wind turbine performance", Renewable and Sustainable Energy Reviews **13**, pp. 428-438.

[6] Parent, O., Ilinca, A, 2010, "Anti- icing and de- icing techniques for wind turbines: Critical review", Cold Reg. Sci. Technol., doi:10.1016/j.coldregions.2010.01.005.

[7] Expert group study on "Wind Energy Projects in Cold Climates", edition 2005, March 19, 2010 URL http://virtual.vtt.fi/virtual/arcticwind/reports/recommendations.pdf

[8] Guy Fortin, Christine Mayer, Jean Perron, Sep. 2008, "Icing wind tunnel study of a wind turbine blade deicing system", Sea Technol. **49**(9), pp. 41-44.

[9] Seifert, H., 2003, "Technical requirements for rotor blades operating in cold climate", BOREAS 4 FMI, Pyhatunturi, Finland, pp.13.

[10] S.W. Ciardullo, S. C. Mitchell, R. D. Zerkle, Jan., 1987, "Evaluation of capillary reinforced composites for anti-icing", AIAA 25th Aerospace Sciences Meeting.

[11] George Lubin, Stanley T. Peters, "Handbook of composites", Chapman & Hall, 2th ed., 1998, Chap. 3.

Proceedings of the ASME 2010 International Mechanical Engineering Congress & Exposition
IMECE2010
November 12-18, 2010, Vancouver, British Columbia, Canada

IMECE2010-38931

AN AIRSHIP DESIGN METHODOLOGY BASED ON AVAILABLE SOLAR ENERGY IN LOW STRATOSPHERE

Antonio Dumas
Universita' di Modena e Reggio Emilia
Reggio Emilia, RE, ITALY

Michele Trancossi
Universita' di Modena e R.E. – I.T.I.S. Nobili
Reggio Emilia, RE, ITALY

Stefano Anzillotti
Universita' di Modena e Reggio Emilia
Reggio Emilia, RE, ITALY

ABSTRACT

The actual applicative research concerning airships and their use as HAP (High Altitude Platforms for telecommunications and military use) presents new applicative hypothesis of these systems, also concerning energetic high quote production. Authors present the energetic balance of a high quote photovoltaic platform with capability of static hovering realized by electric powered propellers. This is the first step trough the design of the P.S.I.C.H.E. (Photovoltaic Space Island for Conversion of Hydrogen as Energy vector) airship concept: a stratospheric airship which could be considered a platform for hydrogen and oxygen production by photovoltaic. It investigates the behaviour of a similar platform operating at altitudes between 10 and 20 km, positioned at 45° latitude north [1, 2].

This paper analyses the design process for a High Altitude Platform based on photovoltaic energy caption, but the process could be generalized in order to be applied to any airship project. It is considers airship shapes equipped with large PV array that covers energy request during the day. Surplus in power supplies electrolyser equipments for hydrogen and oxygen production from water, which could be captured by atmospheric humidity or brought by an auxiliary airship. Hydrogen and Oxygen are compressed and stored in gas cylinders. With insufficient solar irradiance, with severe wind conditions and during the night, a fuel cell system fed by hydrogen and oxygen tanks supplies power requirements.

The Standard Atmosphere Model is used to evaluate PV performance at various operative altitudes. A propulsion system with electric motors grants airship manoeuvrability and hovering. Energy balance of PV-hydrogen energy supply system has been analyzed for three airship shapes with equal volume with concern of overabundant hydrogen and oxygen production. Total weight and payload are calculated in relation to altitude. Storage tanks dimensions and products ground transportation frequency has been estimated. Hydrogen annual production for PV square meter has been evaluated in relation to ground production at the same latitude.

INTRODUCTION

A new interest of airships has begun about 15 years ago and it is witnessed by futuristic projects witnessed by futuristic and fascinating projects. But nowadays it seems that some projects are going to become reality, despite the economic crisis.

Specialized aerospace magazines presented the project JHL40 teamed by Boeing with the Canadian company SkyHook. It is a heavy-lift aircraft that combines helicopter rotor systems with a neutrally buoyant airship. It is considered the first heavy transport hybrid which is going to be operative. Some high altitude platforms are currently in pre-design phase for civil purposes, essentially telecommunications and digital broadcasting, and military systems such as air defense and surveillance.

Due to their heavy operative tasks both transport airships and high altitude platforms need large energy supply for very long enduring mission. A concrete possibility is constituted by photovoltaic energy production associated to a fuel cell system for hydrogen and oxygen. It is a closed mass loop cycle based on renewable solar energy, except a fractional dissipation of water in the hydrolytic process, which produces hydrogen and oxygen dissociation from water by electrolysis, coupled with a fuel cell system for night-time energy supply fed by O2 and H2 tanks.

This paper investigates hydrogen and oxygen production by stratospheric lighter-than-air platform with large hull photovoltaic surface, at operative altitudes above 12 km, which is defined in order to avoid dangerous meteorological phenomena like summer cumulonimbus and clouds shadowing in order to maximize solar energy intercepted in a year.

SHAPE DESIGN

Lighter-Than-Air history presents a large variety of specialized configurations for different uses. The most common shape is the conventional streamlined designs (ellipsoidal), but many projects use unconventional and innovative shapes [5] even specialized for special tasks. It is possible to observe different shapes: spherical, lenticular, toroidal, deltoid, flat-body, multi balloon, etc., and to analyze the characteristics of every kind of shape:

- spherical shape:
  - provides the optimum lift efficiency
  - it is not so efficient in aerodynamics;
- ellipsoidal (conventional) shape approximates a body of revolution about a longitudinal axis creating a streamlined body:
  - has a good aerodynamic performance;
  - presents a low efficiency in terms of lift and a high sensitivity to crosswinds and it could present problems at low speeds and during mooring.
- lenticular shape or "flying saucer" is an envelope geometry generated by the revolution of a flat shape about a vertical axis:
  - the aerodynamics of this shape has a reduced affection by horizontal wind direction and presents a large upper surface for PV array;
  - it could present some problems of stability in the case of not perfectly horizontal winds.

Table I - Configurations of the tested shapes.

Conf.	Volume	Front surface	Wet surface	Base diameter
	V	S_f	S_w	D
	$[m^3]$	$[m^2]$	$[m^2]$	[m]
(A)	2.224E+06	17168	89057	194.92
(B)	2.224E+06	12887	127223	269.20
(C)	2.224E+06	13383	110964	242.52

Configuration (A) presents a nearly hemispherical shape with the best surface-volume ratio while configuration (B) has been constructed starting from a segment of sphere and has the largest upper surface. Configuration (C) represents an intermediate surface between (A) and (B).

SHAPE DESIGN PRELIMINAR ANALYSIS

The history of recent large blimp projects has been started by aerodynamic performances and lift capacity only. Most projects have got in trouble by the energetic problem. This kind of problems has become more severe approaching photovoltaic powered systems.

This paper evaluates the energetic definition of HAP for energetic production and TLC, but results could be applied in any case. In order to solve this kind of problems it has been decided to use an energetic design approach.

The first objective limit is connected to the volume of the system. This is a completely defined parameter by structural weight and service ceiling. In a first pre-design phase we have defined three different layouts, by analyzing the main weight of structure, photovoltaic and energy conversion equipments, electrical engines and related accessories, etc.

By this component analysis and considering a service ceiling of 20 km, and applying a reasonable safety coefficient, we have determined an average volume of about 2.2 106 m2. This weight has been determined using a PV array on the top roof of the airship, with an area about 70% of the total roof surface.

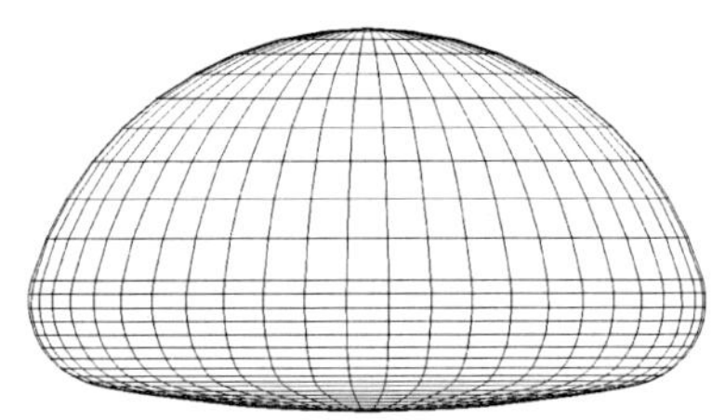

Figure 1 – Shape of configuration (A)

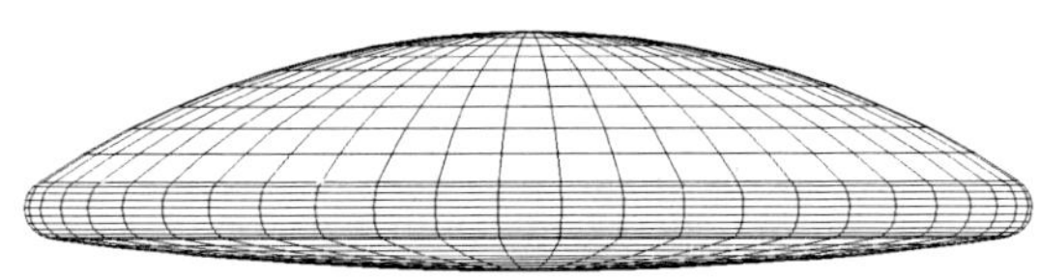

Figure 2 – Shape configuration (B)

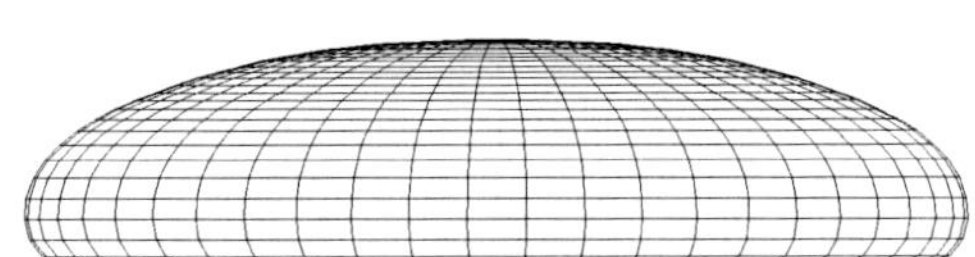

Figure 3 - Configuration (C)

Considering these prescriptions it has been evaluated an average system mass about 2*105 kg, adopting a safety factor 1.25. By this value, it is possible to evaluate the influence of the shape on energy system performance and payload capacity in the case of different shapes with the same volume (table I). The shapes represented in the figures above (Figure 1, 2, 3) has been analysed

AERODYNAMIC CONSIDERATIONS

To realize our research we have obtained and used meteorological and atmospheric data at the location of meteorological station of San Pietro Capofiume, located latitude 44.65, longitude 11.63 and elevation 6 meters.

This data have been compared and joined with the Standard Atmosphere model. It has been produced a preliminary analysis of meteorological conditions in a vertical plain Po.

It has been calculated mean velocity, density, and temperature profiles in the atmosphere up to 25 km, (Fig. 4) based on disposable measures of the last 20 years. Using the same data we have calculated average Mach number during different months (Fig 5). Reynolds number profiles has been calculated basing on V1/3 as reference length and show low speed subsonic and turbulent flow around the body (Figure 6).

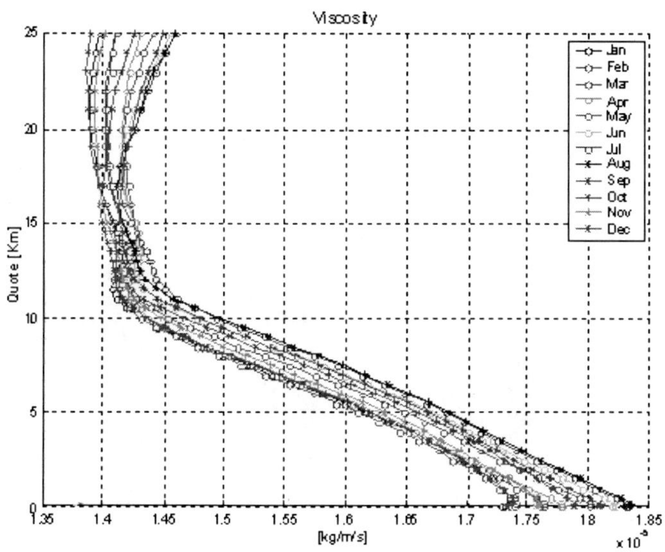

Figure 4 - Monthly average velocity profiles

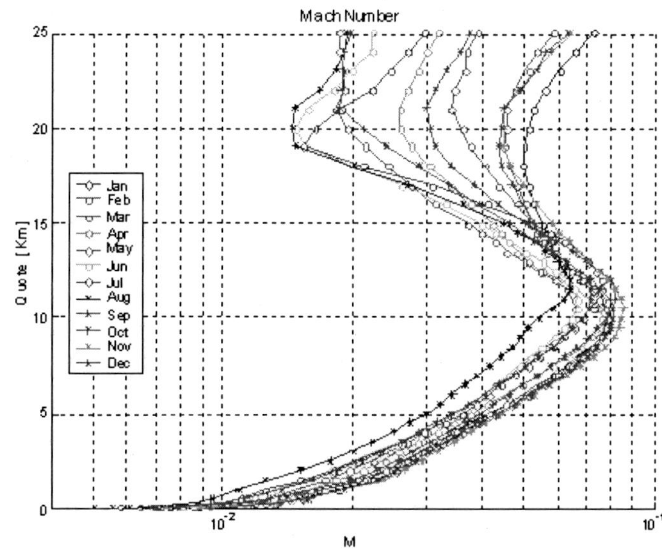

Figure 5 - Monthly Mach number evaluation

The platform needs power only for hovering, maneuvering, and supplying board systems and equipments. Mean monthly wind speed is the basis for calculation of total necessary energy on a year for hovering. The drag coefficient has been calculated for the three considered shapes by CFD simulation realized by the Fluent code.

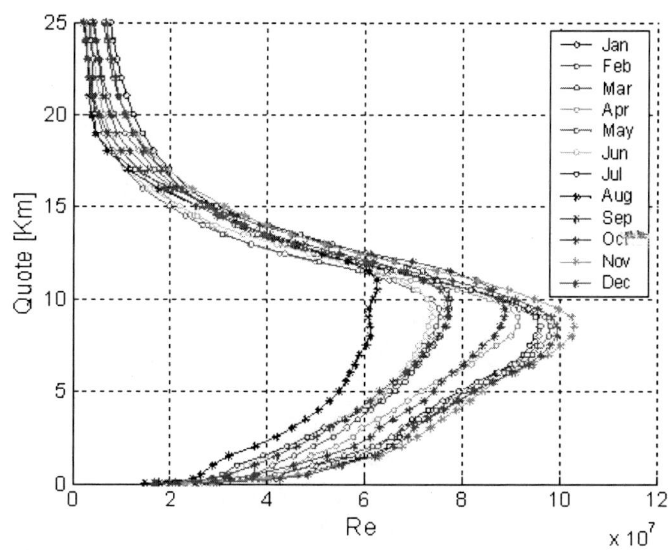

Figure 6 - Reynolds number based on monthly average wind speed

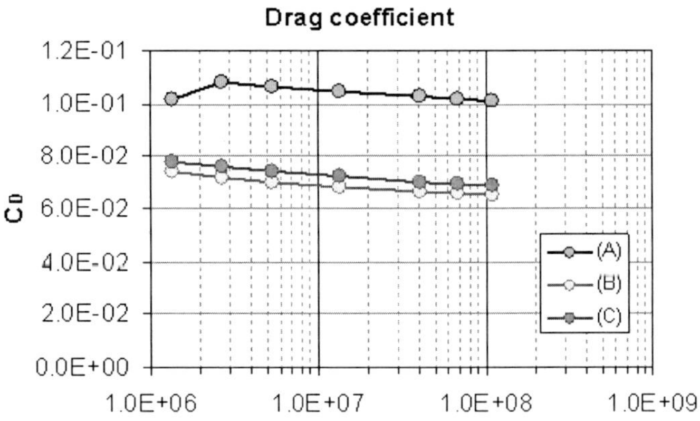

Figure 7 - Drag coefficient CD as a function of Re

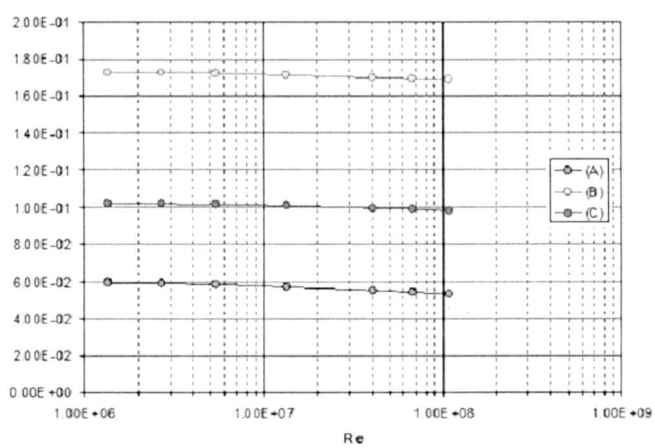

Figure 8 - Lift coefficient as a function of Re

Average aerodynamic parameters have been determined by simulations with CFD Fluent code of the three shapes. Simulations are realized using monthly data based on the locations of San Pietro Capofiume. Drag and Lift coefficients are calculated by velocity v_∞ and density ρ_∞ by the following expressions:

$$C_L = \frac{2 \cdot L}{\rho_\infty \cdot v_\infty^2 \cdot A} \text{ and } C_D = \frac{2 \cdot D}{\rho_\infty \cdot v_\infty^2 \cdot A} \qquad (1)$$

Data are represented before (Figure 7, 8). Profiles show a maximum in annual energy requirements around 10÷12 km, due to jet streams. At higher altitude it reduces considerably. Results show an energetic convenience for hovering below 3 km or over 15 km. The comparison between different shapes at in every altitude condition for the different shapes shows that (B) and (C) shapes presents a superior energetic efficiency. A comparison charts is produced (Figure 7). The characteristic peak correspondent to jet stream altitude is common to all the quantities related with wind speed. This analysis shows an energetic convenience for hovering only below 3 km or over 15 km (Figure 8).

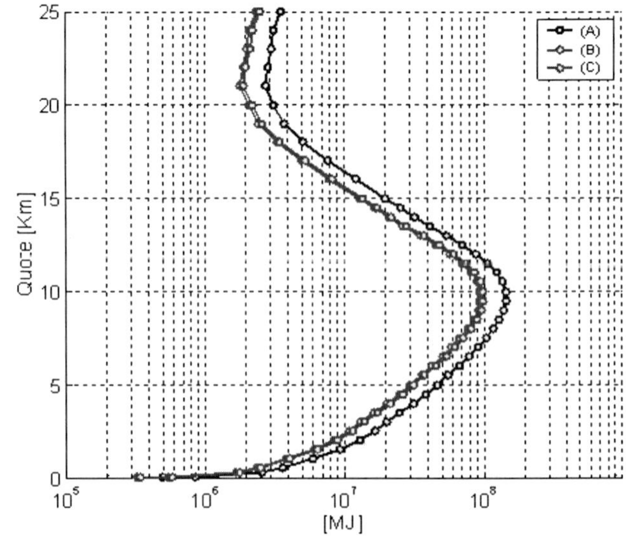

Figure 9 – Required annual energy for hovering

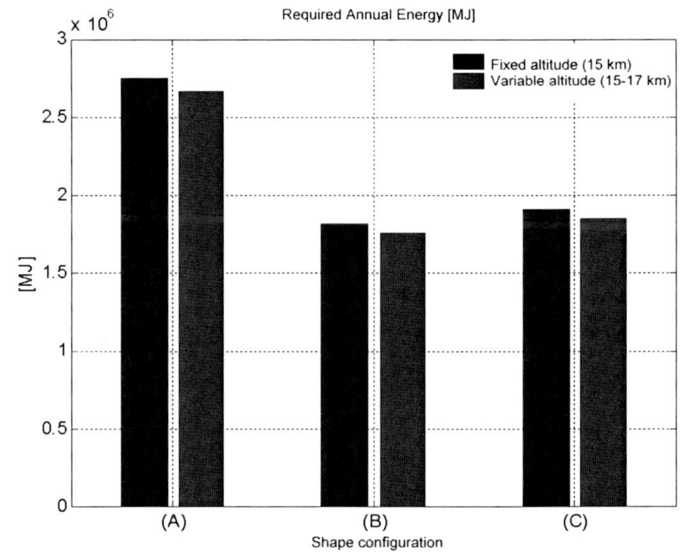

Figure 10 - Energy required for hovering: fixed quote of 15 km and variable quote in the range between 15 and 17 km

SOLAR IRRADIANCE EVALUATION

Direct normal irradiance increases with atmospheric height. Horizontal diffuse irradiance is negligible at high altitude. It has been calculated solar irradiation and irradiance for different azimuth and inclination from 10 to 25 km using the common Technical Standards ASTM G-173-03 and EN 60904-3.

In the calculation, in order to obtain a better approximation of the PV roof of the airship envelope has been reduced to an envelope of planar surfaces. The calculation, by quarter of hour steps, has involved direct and reflected irradiance on the entire shape. It has been considered the PV energy production which could be realized in operative conditions at high altitude, using panels with an overall efficiency at sea level of about 14%.

Photovoltaic efficiency improves by altitude because of physical conditions: reduction of atmospheric filtration, temperature and pressure and increase of wind speed.

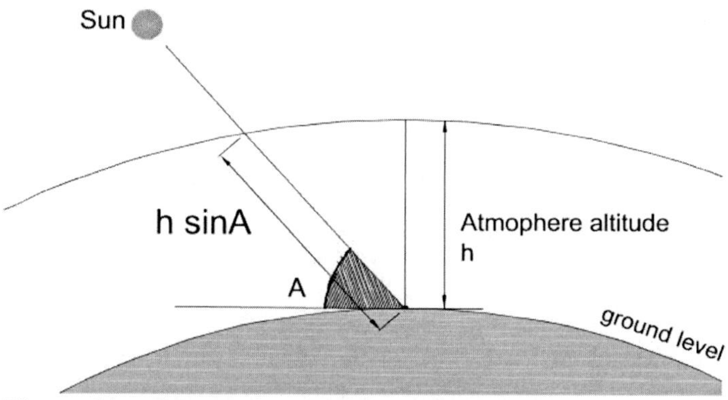

Figure 11 – Geometric model for calculations

Reduction of atmospheric filtration is computed by usual optic techniques using the following traditional calculation schema (Fig. 11), which is used also for ground based installations. Direct solar radiation increases with altitude, because it has a shorter way in the atmosphere. It reaches a value which is similar to the extra-atmospheric one about an altitude of 30 km. Different models has been evaluated and compared.

DRUMMOND AND LAUE MODEL - For calculation purposes one of the most significant equations is the equation of solar radiation modified by Drummond, Laue and others [6, 7, 8] . This equation expresses solar irradiation as a function of altitude and could be expressed as:

$$I_n(A,h) = SF(1-a\cdot h)\cdot e^{-c[1/\sin(A)]^S} + a\cdot h\cdot SF \quad (3)$$

with $a=0.14$.

OKI AND SHIINA MODEL – A more recent study by M. Oki and H. Shiina [9] based on experimental analysis based on different localities all over the world. They have obtained the following equation which is function of A and h, expressed in meters:

$$I = \left[650\sin\left(\frac{\pi}{180}\left(\frac{4A}{3}-30\right)\right)+325\right]\left(8\cdot 10^{-5}h+1\right) \quad (4)$$

Both the presented models are studied in order to represent the irradiation conditions in high quote localities and loose significance over 6 km altitude. A comparison of the considered equation results has been plotted (Fig.16).

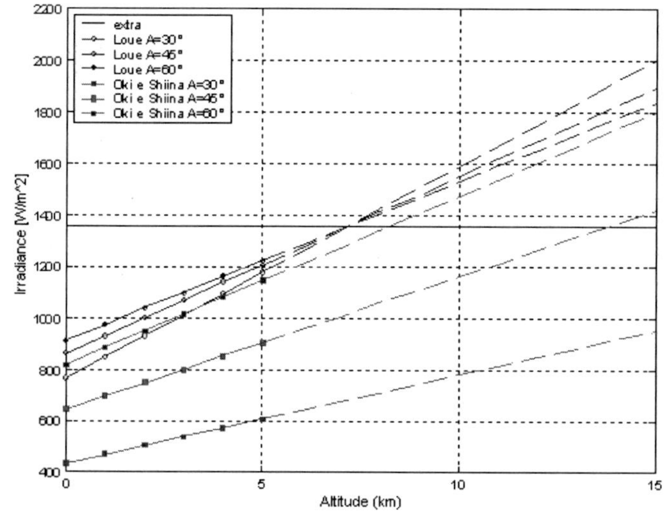

Figure 12 - Solar irradiance as a function of quote (comparison between Drummond-Laue Model and Oki Shiina model)

ASHRAE CALCULATON METHOD - An interesting alternative formulation is the one adopted by ASHRAE [10], which is the most used in engineering calculations. Direct normal solar radiation assumes the following expression:

$$I_d(A, p/p_0) = (C_n)SF\cdot e^{-\frac{B}{\sin A}\left(\frac{p}{p_0}\right)} \quad (5)$$

Table II – Coefficient B, ASRAE model

Date	B W/m^2]
January, 21	0.142
February, 21	0.144
March, 21	0.156
April, 21	0.180
May, 21	0.196
June, 21	0.205
July, 21	0.207
August, 21	0.201
September, 21	0.177
October, 21	0.160
November, 21	0.149
December, 21	0.142

C_n is dimensionless, and it represent the relationship between normal radiation calculated using the average value of water vapour in a sunny day and the same quantity calculated by the tabulated value of the ASHRAE Handbook [10]. At considered altitudes (8÷20km), the atmospheric physical properties are conforming to the Standard Atmosphere model. By this hypothesis it is possible to assume $C_n=1$. The relation p/p_0 is the correction as function of the altitude. The B coefficient is the dimensionless coefficient of Threlked e Jordan, reported in ASHRAE calculation method.

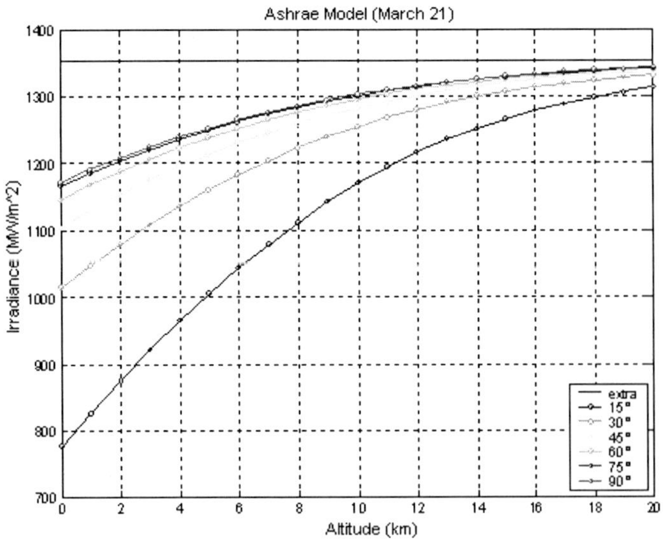

Figure 13 - Irradiance as function of the altitude (ASRAE Model)

W. Knaupp and E. Mundschau [10, 11] have analyzed the solar irradiance using the MODTRAN code, with a constant zenith angle. The code uses data of temperature, pressure and atmosphere composition of US 1976 Standard Atmosphere Model. They calculate phenomena of solar radiation absorption and dispersion of solar spectrum at various level of atmosphere by wave length and produce spectral and global values at various altitudes. They assume that diffuse irradiation over 7.5 km results lower than 40 W/m2 and is negligible.

It is necessary a more accurate evaluation about reflected component, in particular the one due to presence of clouds under the station. In these conditions the albedo coefficient assumes great importance (~0.9). Assuming the hypothesis of an isotropic radiation under the station and it is possible to obtain

$$I_r = \rho \cdot I_{t,h} \frac{1-\cos(\beta)}{2} \qquad (5);$$

Total radiation on an inclined surface, neglecting the diffusive component, is:

$$I_t = I_d + I_r = I_n \cos(\vartheta) + I_r \qquad (6)$$

The total radiant energy intercepted by the surface can be calculated using the following expression:

$$Q_{month} = \sum_{n=i_1}^{i_2} \int_{-\omega_s}^{\omega_s} I_t \cdot d\vartheta \qquad (7)$$

where s is the angle between sunrise and sunset is

$$\omega_s = \arccos\left[-\tan(\varphi)\tan(\delta)\right] \qquad (8)$$

Direct normal irradiance increases with atmospheric height. Horizontal diffuse irradiance is negligible at high altitude. It has been calculated solar irradiation and irradiance for different azimuth and inclination from 10 to 25 km. The calculation, in order to obtain a better approximation we have simplified the airship envelope by planar surfaces. This calculation, by quarter of hour steps, has involved irradiance and irradiation distribution on the entire platform.

It has been calculated daily values of irradiance of south oriented surface, for various tilt angles, in particular conditions and particular days. It has been represented the different average values of irradiance for unit of surface as a function of tilting angle at a quote of 15 km (example in Figure 14). The numerical integration of equation (7) gives the average monthly values of intercepted energy by a surface. The sum of those values is the total annual captured energy.

$$Q_{year} = \sum_{i=1}^{12} Q^{i}_{month} \qquad (9)$$

Annual values of radiant energy captured in fixed and tracking configurations are calculated, with a constant azimuth angle. Maximum values are obtained with an angle of departure from the south of up to 30 degrees and inclinations of between 30 and 50 degrees (Figure 14, 15, 16).

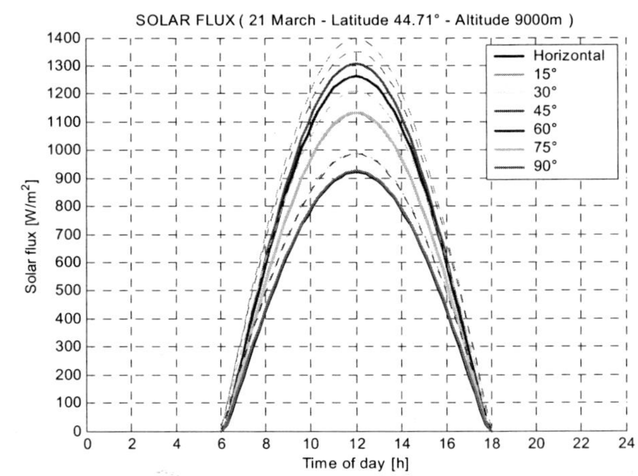

Figure 14 – Total irradiance (March, 21)

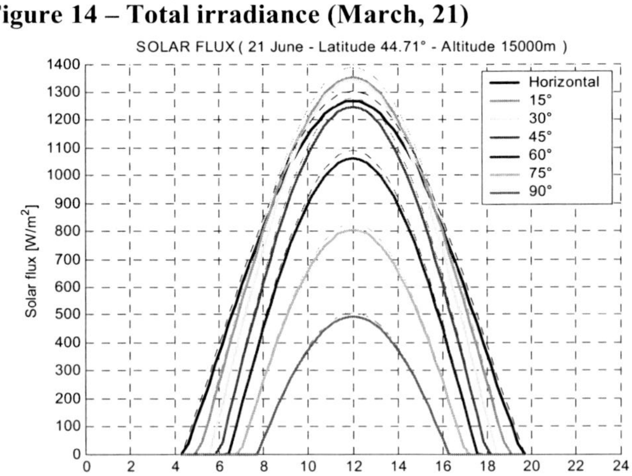

Figure 15 – Total Irradiance (June, 21).

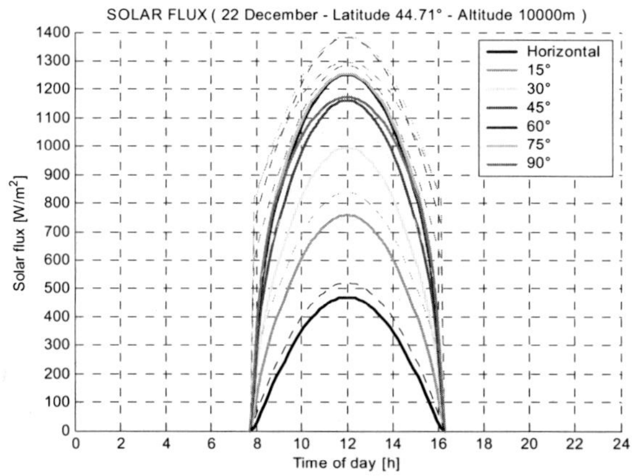

Figure 16 - Total Irradiance (June, 21)

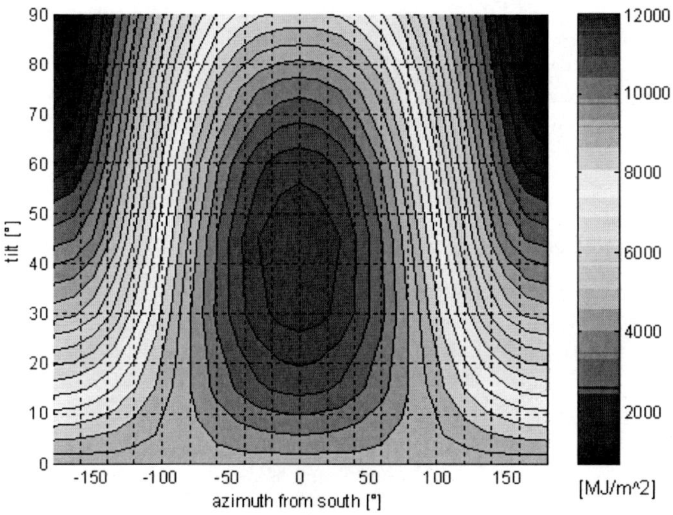

Figure 17 - Global irradiation on a flat plane in summer [MJ/m²]

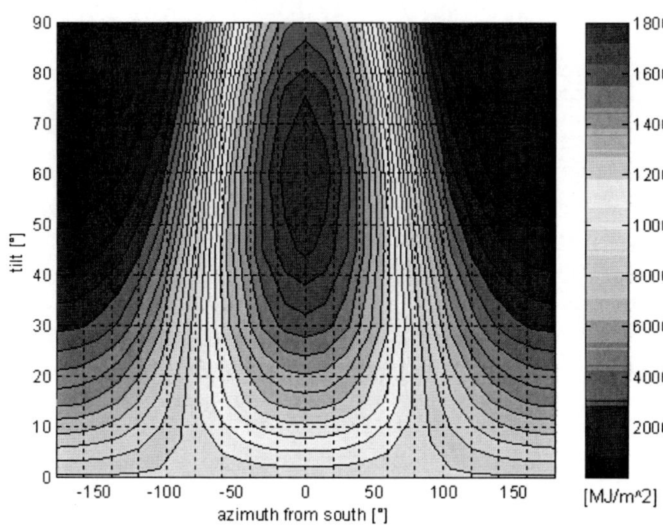

Figure 18 - Global irradiance on a flat plane summer [MJ/m^2]

Under these conditions we have verified that energy collection by PV panels is obviously symmetric. The symmetry plane is a vertical plane passing through sun and the centre of the platform itself. The two symmetric half-shapes defined by this plane intercept the same solar flux. This results shows that it is possible to have a correct and efficient electric configuration, which guarantee a good behaviour of the system.

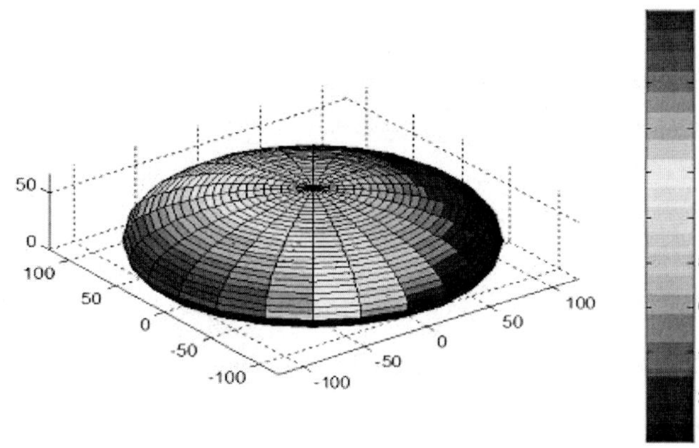

Figure 19 - Solar energy produced December 21, 12 a.m. shape (B)

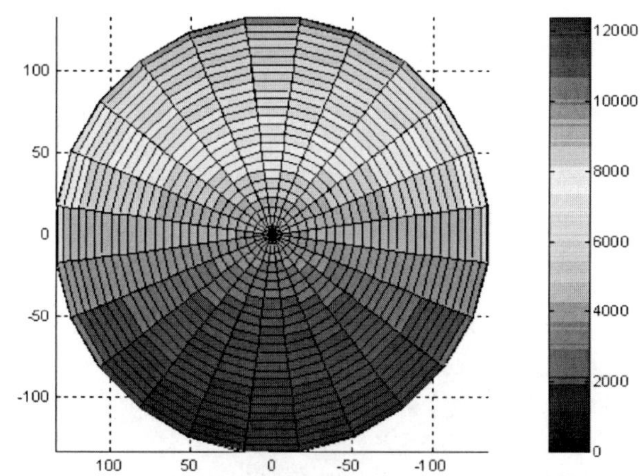

Figure 15 – Global Annual Irradiance configuration (B) at an altitude of 15 km [MJ/m²]

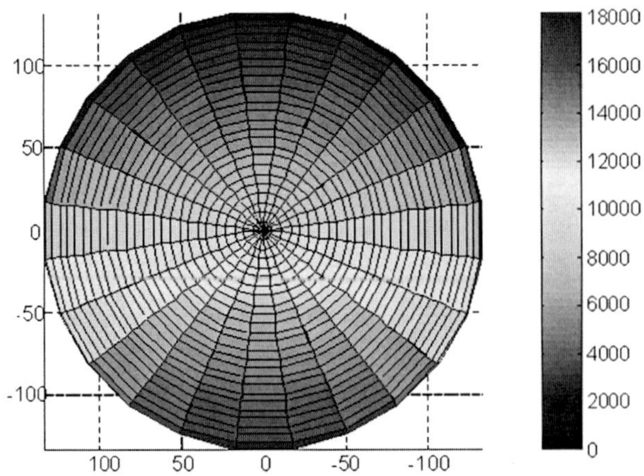

Figure 16 – Global Annual Irradiance configuration (B) tracking at an altitude of 15 km [MJ/m^2]

PV SYSTEM

PV energy production was carried out under consideration of operative conditions at high altitude. In particular PV efficiency is affected by high altitude atmosphere conditions such as low temperature and wind speed. Equilibrium temperature of PV cells has been calculated by the thermal balance equation suggested by Colozza [4], involving radiant, convective and conductive heat transfer:

$$\tau_{\alpha}\Phi \simeq \Phi_{cond} + \Phi_{conv} + \Phi_{rad} + \Phi_{el} = f\left(T_{cell}\right) \qquad (3)$$

the photovoltaic efficiency has been computed from equilibrium temperature and solar irradiation [3]. Photovoltaic power has been computed all year long by steps of 15 minutes. Energetic production over a year has been computed by numerical integration of these values.

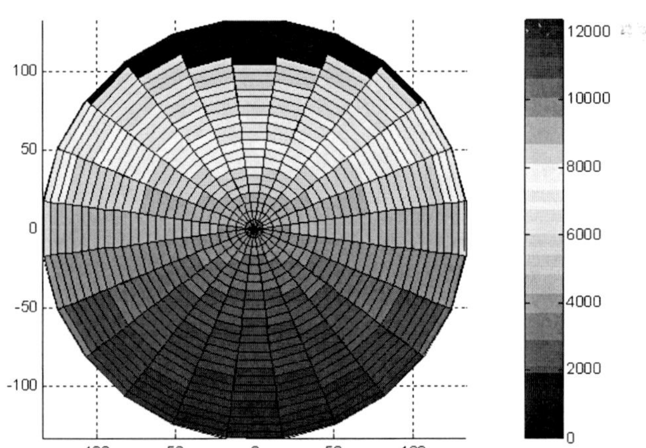

Figure 17 - Annual irradiance distribution with cut-off value of 5400 MJ m2 - 15 km, configuration (B) fixed [MJ/m2]

Setting a cut-off annual irradiance on a fixed surface value at ground level with best tilt angle and orientation (5400 MJ/m2) is possible to calculate envelope surface that grant a better irradiance value (Figure 17).

It is clear (Figure 17 and 18) that with tracking option is possible to produce the same energy on a year with an important reduction in PV surface in comparison with fixed configuration.

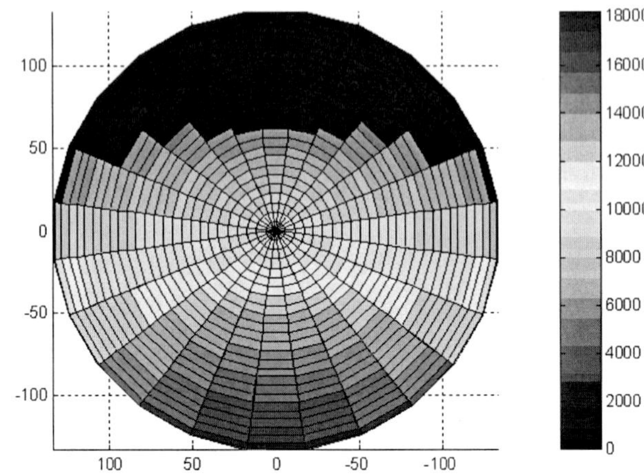

Figure 18 - Annual irradiance distribution with cut-off value of 5400 MJ m^2 - 15 km, configuration (B) fixed [MJ/m^2]

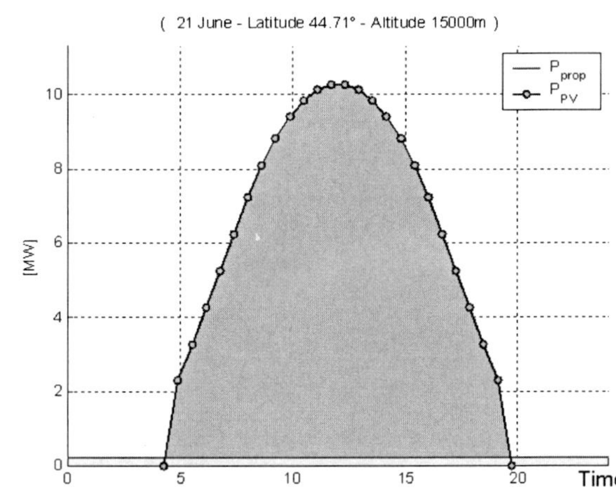

Figure 19 - PV power production (cyan) and mean necessary power (yellow), 21 June, configuration (B) - [MW]

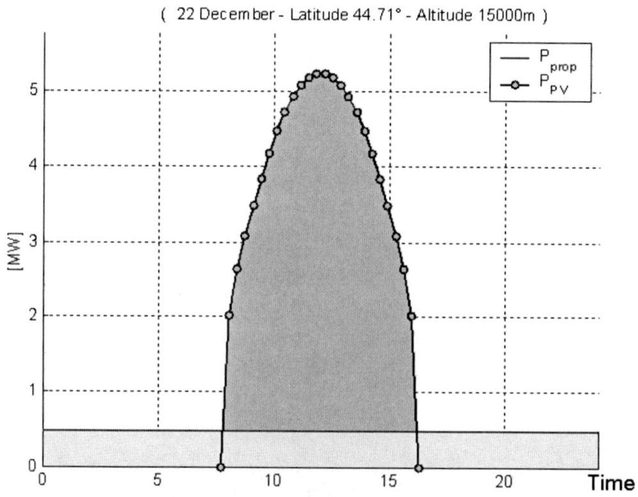

Figure 20 - PV power production (cyan) and mean necessary power (yellow), 21 December – configuration (B) - [MW]

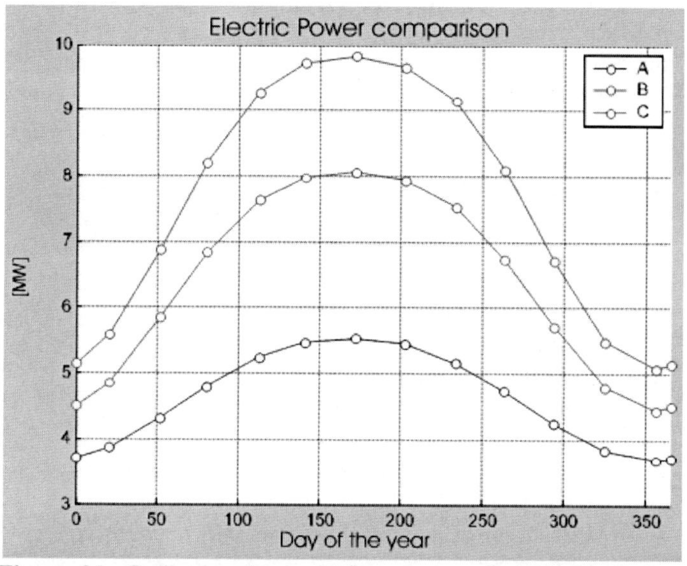

Figure 21 - Daily maximum PV power production during the year

PV power has been calculated for every hour (Fig. 14, 15) of the year and total energy production over a year was computed by numerical integration of these values.

Overabundant summer energy supplied by PV array (Figure 19) allows effective hydrogen and oxygen production by electrolysis and even during winter solstice (Figure 20) daily energy production is greater than energy requirements for hovering.

The maximum daily power delivered by PV system is valued for the three configurations (Figure 21). Configuration (B) reaches top production (almost 10 MW) on summer solstice.

ENERGETIC BALLANCE

The energetic problem has been defined as the most important problem. It has been analyzed the energy balance for hovering at various operative altitudes. An annual system energy balance is presented in Figure 22.

The yellow area represents the conditions in which the energy supply system is not able to meet energy requirements for hovering (negative balance over a year).

This balance gives different results for different designs (Fig. 7). It has been obtained different results depending on the design of the airship shape, varying from 11 km for lenticular configuration (B) to 13 km for a spheroid shape (A).

The configuration (C) is intermediate. All the tested designs present a maximum in annual energy production near the altitude of 18 km.

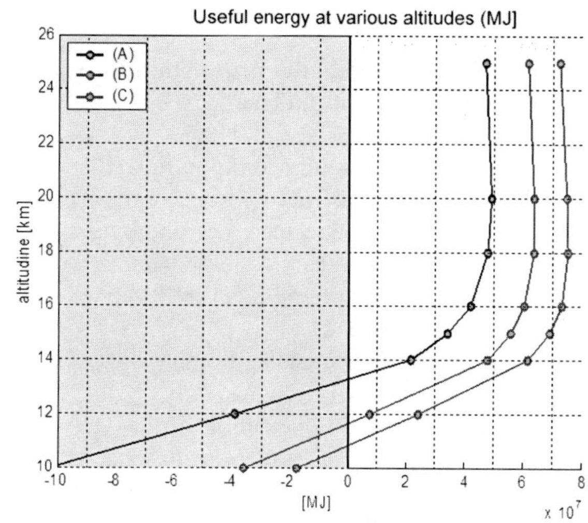

Figure 22 - Energy balance over a year [MJ]

The energy exceeding the immediate need of power by the propulsive system can supply the hydrolysis system and could realize hydrogen and oxygen production from water. A part of the products can be stored in gaseous form to satisfy nighttimes' engine consumption for static hovering [1, 15, 16, 17]. The main part of hydrogen and oxygen can be liquefied (or compressed) and stored in cylinders tanks waiting to be transported to the ground.

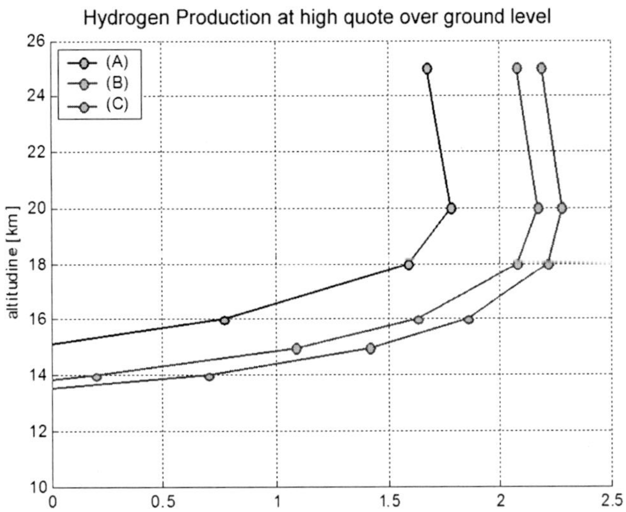

Figure 23 - Hydrogen annual net production for square meter (high quote vs. at ground level).

It has been verified that lenticular configuration (B) presents the best performance in terms of net hydrogen production. It can realize a production for square meter of photovoltaic array in high quote which is more than double, than the production which could be realized at ground level by PV fixed surface at best tilt and orientation (Fig. 24).
In this comparison we have considered both systems at the same latitude using climatic data of Bologna (Italy).
The energetic needs for the liquefaction of the 1 Nm3 gas are quite small, but not ignorable, respect to low heating value of hydrogen. The liquefaction process presents a global efficiency about 20% which could be increased considering stratospheric physical conditions. A comparison with the same process at the ground level shows an energetic gain of almost the 40%. Similar results could be determined for 1 Nm^3 Oxygen, with a lower energetic gain due to different physical properties of the substance.
Hydrogen net production has been calculated (Figure 23) and configuration (B) presents the best performance with H2 output near 350.000 kg/year.
Table III – Net annual surplus energy production - shape (B)

16 Km Altitude	Production	Unit
Electricity	$8\ 10^7$	MJ
Hydrogen	$4\ 10^6$	Nm^3
Oxygen	$2\ 10^6$	Nm^3

Hydrogen net production by high altitude platform for square meter of PV system is more than double, for configuration (B), than production at ground level by PV fixed surface at best tilt and orientation (fig. 27).

Total production basing on Low Heating Value of hydrogen (~120 MJ/kg) is 42 TJ/year that corresponds to 8.4 millions of kilograms of methane (~50 MJ/kg) or 1.17•106 Nm^3.

CONCLUSION

This paper presents a case of general energetic design for airships. The presented method has been developed in order to design a High Altitude Platform, but it could be easily applied to any photovoltaic powered airship. In the considered case it possible to have an annual energetic production as calculated in Table III. In particular we have verified that the considered shape (B), which presents the best energetic production, could be good for a large surplus energy production.

By this considerations take origin the project of an high altitude platform for energetic production, formerly named PSICHE (Photovoltaic Space Island for Conversion of Hydrogen as Energy vector) [2, 15, 16, 17], a large high altitude platform designed in order to carry out simultaneously more than one task: hydrogen and oxygen production, telecommunications and digital broadcasting, air surveillance and control ground monitoring, and low temperature - low pressure high technology laboratory.

NOMENCLATURE

Symbols

A	solar altitude (m);
h	altitude of the site (m);
I_d	Direct normal radiation (W / m^2);
I_r	Reflected radiation (W/m^2);
ρ	albedo coefficient;
$I_{t,h}$	total radiation on the horizontal plain (W/m^2);
i_1	first day of the month, expressed as progressive day of the year;
i_2	last day of the month, expressed as progressive day of the year;
ω_s	angle between sunrise and sunset;
φ	sunrise angle;
δ	sunset angle;
C_n	normal radiation calculated by the average value of water vapor in a sunny day vs. normal radiation calculated by tabulated values of ASHRAE Handbook;
Q_{month}	Intercepted solar radiation in one month (J/m^2);
Q_{year}	Intercepted solar radiation in one year (J/m^2);
Φ	Thermal Flux [W/m^2]

Acronyms

PV	photovoltaic;
HAP	High Altitude Platform;
PSICHE	Photovoltaic Space Isle for Conversion in Hydrogen as Energy Vector

REFERENCES

1) Colozza, A.Initial Feasibility Assessment of a High Altitude Long Endurance Airship. NASA/CR 2003-212724.
2) Knaupp, W., Mundschau, E. Solar electric energy supply at high altitude. Aerosp. Science and Tech. 8, 245–254, (2204)
3) Knaupp, W., Mundschau, E.. Photovoltaic-hydrogen energy system for stratospheric platforms. 3rd World Conf. on PV Energy Conv. (2004)
4) Drummond A.J, Hickey J.R, Scholes W.J. Laue E.G. New Value for the Solar Constant of Radiation. Nature 218, 259 - 261 (20 April 1968); doi:10.1038/218259b0
5) Oki M., Shiina H. Preliminary Study On An Estimation Method For Annual Solar Irradiance At Various Geographical Altitudes Eighth International IBPSA Conference, Eindhoven, Netherlands, August 11-14, 2003
6) Anzillotti, S., Una piattaforma stratosferica per la produzione di idrogeno mediante energia fotovoltaica. Tesi di Dottorato in Ingegneria della Gestione Industriale ed Integrazione tra imprese, Università di Modena e Reggio Emilia, XVIII ciclo, (2006). Rel. A. Dumas
7) Khoury, G.A., Gillet, J. D. Airship technology. Cambridge University Press
8) Dumas A., Anzillotti S.: "PSICHE: A Stratospheric Platform Producing Hydrogen and Oxygen" 5th International Conference on Sustainable Energy Technologies. Vicenza, Italy (2006)
9) Anzillotti, S. "Una piattaforma stratosferica per la produzione di idrogeno mediante energia fotovoltaica." Tesi di Dottorato in Ingegneria della Gestione Industriale ed Integrazione tra imprese, Università di Modena e Reggio Emilia, XVIII ciclo, (2006). Rel. A. Dumas
10) Dumas A., Anzillotti S., Trancossi M., Zumbo F. "P.S.I.C.H.E.: The Concept of a Stratospheric Airship for Energy Production, Telecommunications and Territorial Surveillance", SAE 2008 Wichita Air Congress, Relation by Michele Trancossi, Wichita (KS) – USA August 20-22 (2008)
11) Dumas A., Anzillotti S., Trancossi M. "Available Solar Energy in Low Stratosphere", II Congresso Nazionale AIGE, Pisa 4-5 Settembre (2008)
12) Dumas A., Anzillotti S., Trancossi M. "Available Solar Energy in Low Stratosphere", II Congresso Nazionale AIGE, Pisa 4-5 Settembre (2008)

Proceedings of the ASME 2010 International Mechanical Engineering Congress & Exposition
IMECE2010
November 12-18, 2010, Vancouver, British Columbia, Canada

IMECE2010-40018

METAL CUTTING THEORY APPLIED TO THERMAL MAPPING OF THE FRICTION STIR WELDING PROCESS

Lewis N. Payton[1], Ph.D
Director, Design and Manufacturing Laboratory
270 Ross Hall
Auburn University, Alabama 36849
[1]Contact Author (Payton@Auburn.edu)

Vishnuvardhan Chandrasekaran
Masters, Mechanical Engineering
270 Ross Hall
Auburn University, Alabama 36849

KEYWORDS
Friction Stir Welding, Thermal Mapping,

ABSTRACT
Friction Stir Welding is a solid state "green" welding method developed by The Welding Institute (UK). An internal thermal mapping instrument has been developed which allows for symmetrical mapping of the thermal fields developed by a Friction Stir Welding tool as it passes through the material being welded. This symmetrical mapping conclusively documents statistically the asymmetrical nature of the heat sources within the friction stir welding process. The various models in the literature are compared against these results. A model developed by the authors using classic metal cutting theory predicts the observed thermal fields. A successful predictive model will facilitate tool optimization and welding schedules, while optimizing the mechanical properties of the weld.

INTRODUCTION
Friction Stir Welding (FSW) is a relatively new industrial process that was invented at The Welding Institute (TWI, United Kingdom) and patented in 1992 under research funded in part by the National Aeronautics and Space Administration (NASA). Often quoted advantages of the process include good strength and ductility along with minimization of residual stress and distortion. It produces superior weld products in difficult-to-weld materials without producing any toxic fumes or solid waste that must be controlled as hazardous waste. Noise pollution in the work place is greatly reduced. The process is robust and environmentally friendly.

Although there have been several flow models and heat source models proposed in the refereed literature of Friction Stir Welding, there is a shortage of real world thermal data against which to validate the proposed models. This is largely due to the contained nature of the actual Friction Stir Welding process. The heat sources of deformation are located within the base metal being welded and underneath the rotating tool shoulder. As such, surface pyrometry measures are ineffective. The limited amount of thermal data in the reported literature is usually coincidental to the primary goal of the experimental design and non-symmetrical.

An internal thermal mapping protocol/device has been developed which allows for symmetrical mapping of the thermal fields developed by the tool as it passes through the material being welded. This symmetrical mapping clearly documents the asymmetrical nature of the heat sources within the friction stir welding process. One side of the pin is always without fail hotter than the other, with similar results for the two sides of the shoulder.

BASIC PROCESS
In order to visualize the fundamentals of the process, consider Figure 1. The two materials to be welded are placed in contact via either an overlapping or, in this case, a butt joint fashion. A broad tool with a narrower pin on the end is fabricated. The tool is then inserted while rotating at a high speed into the material until the wider "shoulder" of the tool makes contact with the material being welded. At this point, the tool begins a traverse of the weld seam, deforming the material in its passage, leaving behind a formed weld. The material does not melt during this solid-state deformation. Personnel with shop experience will recognize similarities with a type of end-milling referred to as slot-milling. For the travel direction depicted in Figure 1, the left hand side of the weld is

referred to as the advancing side of the tool and the right hand side is referred to as the retreating side of the tool.

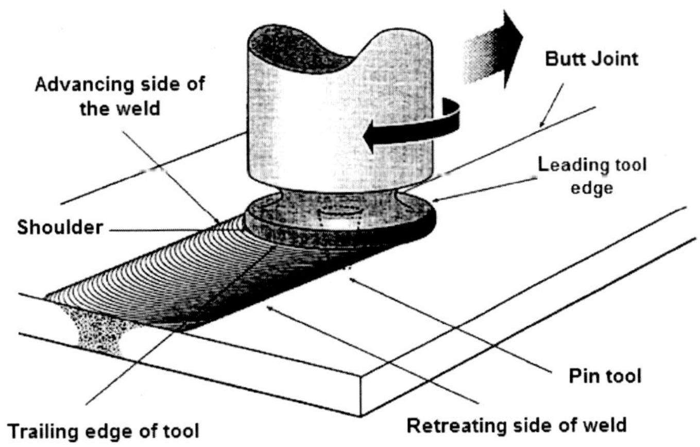

Figure 1: Basic Friction Stir Welding Process [1]

REVIEW OF PREVIOUS THERMAL STUDIES

The majority of the published literature in Friction Stir Welding consists of post-mortem material property studies and/or attempts to visualize the flow of material around the tool. The mechanistic studies have been analytical, post mortem, tracer studies of various types (e.g. [2-10]). Various flow models have been discussed with varying degrees of success and acceptance [11-22]. Parameterization of the welding process has been couched in terms most frequently used by welders [19, 23, 24]. Few modelers have incorporated classic metal deformation theory into their models [24-26]. Arbegast [25] based his modeling upon studies of flow stress in a high temperature Gleeble testing apparatus. Payton's [24, 26] approach is to treat the process as a specialized form of slot milling, in an attempt to design tools that will avoid the classic defects defined in the literature.

Although temperature and pressure within the weld drive the dynamic recrystallization process exactly as they do in classic forging, extrusion and burnishing operations, there has never been a systematic, symmetrical study of the temperature profile within the weld as the tool passes. Temperature measurements have been conducted as an afterthought or corollary to another major Friction Stir welding experiment.

Single sided observations have dominated these thermocouple recordings. Chao et al [13] conducted thermal measurements on one side of the weld in aluminum alloy 2195 while simultaneously measuring the bulk temperature of the tool with a slip ring assembly. Chen et al [27] measured the temperature with thermocouples on the retreating side only aluminum alloy 6061-T6 at different distances from the centerline. McClure, Tang et al [28] recorded similar readings on the retreating side of aluminum at various heights and distances from the weld centerline. Tang et al [29] later expanded on McClure's work, but the readings were again limited to one side. Song and Kovacevic [30] recorded temperatures in aluminum 6061-T6 on the retreating side.

A limited number of researchers have placed thermocouples on both the advancing and retreating side of the weld, but without a symmetrical (and comparative) alignment. Lienert et al [31] conducted studies in a mild steel on both sides but at different locations and moments along the weld. Lambrakos et al [32] placed thermocouples at a variety of asymmetric heights within the welded material at different distances from the centerline. In perhaps the most detailed "symmetrical" study to date, Xu et al [33] place two pairs symmetrically along the weld line while simultaneously recording the external thermal fields with a thermal camera. Unfortunately, the study did not establish the sensitivity or repeatability of the experimental setup.

The lack of a large database of thermal data serves as a handicap to the Friction Stir Welding process modeling community. The objective of this study is to develop an instrument capable of repeatedly measuring the thermal fields around a tool passing through the standard test geometry. Changing samples should take minutes, not hours or days of preparation in order to make the collection of large amounts of data possible quickly and accurately.

DESIGN OF THE INSTRUMENT

The experimental objective of the instrument is to simultaneously and symmetrically measure the thermal field around a tool as it passes a given vertical plane. Ideally, this would capture the temperature under the shoulder at its maximum radius, just under the pin as it passes overhead and below the midpoint side of the pin. This "nominal" configuration requires a pairing of 5 thermocouples to completely capture the transverse plane as the tool passes by the observation point. Multiple pairings of matched thermocouples allow the tool to pass through a "minefield" of temperature thermocouples.

Embedding thermocouples in such a manner for every sample is a very time consuming effort. Controlling the variability of thermocouple fabrication and placement is an extremely daunting prospect for even a skilled technician. Each of the earlier studies was conducted on a limited number of carefully prepared samples. It became apparent that it would be best to construct a standard holder with embedded thermocouples. Instead of changing the thermocouples, the experimentalist changes the sample. This allows the collection of a large body of statistical data rapidly. Figure 2 provides a schematic view. Figure 3 is a photograph of the instrument. The first two sets of thermocouples are separated from the final five sets to allow for the dwell in temperature during tool insertion. The last five sets or pairings all view the approaching tool as a steady state, if transient, heat source.

The overall length of the data run is 22 inches measured from the tool entry point until the tool exit point. Seven thermocouple pairs are arranged along the length of the unit in groups of five for a total of 35 thermocouples. Within each

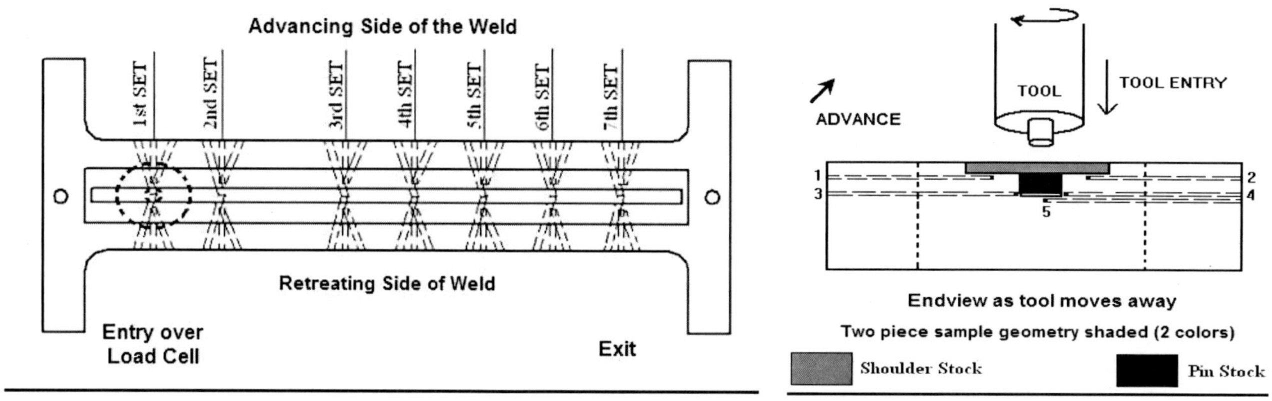

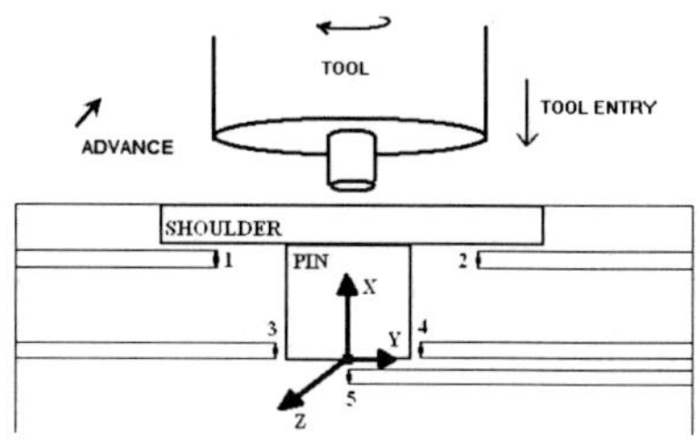

		Co-ordinates (inch)	
Number	Representations	X	Y
1	Advancing Shoulder	-0.45	0.485
2	Retreating Shoulder	0.45	0.485
3	Advancing Pin	-0.3	0.1
4	Retreating Pin	0.3	0.1
5	Under the pin	0	-0.15

Figure 2. Rough schematic of the instrument above. The thermocouple wells must be drilled at convergent angles in order to place all 5 thermocouple data points in the same plane as the tool moves through the plane. A load cell is located underneath the entry point on the left. The advancing side of the weld is located on the side away from the operator as he faces the machine. The tool enters from the left. Normal rotation is clockwise. The distance between the thermocouples and the stock (Shoulder or pin) are 0.05 inches. Representation of thermocouples: 1-Advancing shoulder, 2-Retreating shoulder, 3-Advancing pin, 4-Retreating pin, 5-Under the pin. Coordinate location of the thermocouple wells is illustrated in the lower half of figure.

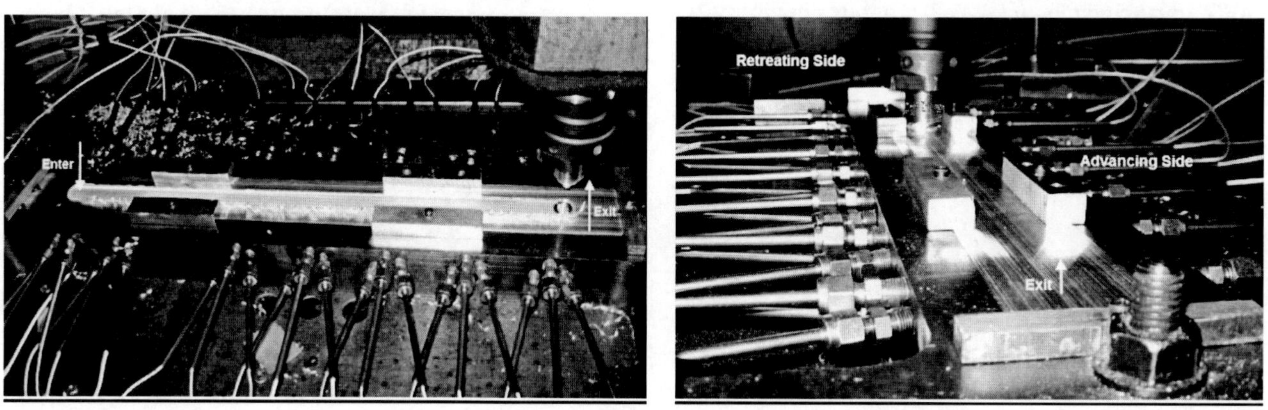

Figure 3: Side and end view of the thermal profile instrument. Tool enters on the left hand side and traverses to the right along a 22 inch long path. Exit is marked in each view in white. The run is complete in the view on your left (tool up). A run is in progress in the right hand photograph (tool down, moving towards the front).

planar group, there are two shoulder thermocouples (advancing and retreating side), two pin thermocouples (advancing and retreating sides), and one under the pin thermocouple. The first two of seven pairs are separated by a small offset to the left hand side from the final 5 paired sets. This allows for observation of the tool entry and initial traverse independently of the routine weld passage observed through the 5 pairs on the right hand side. Downward thrust force is also recorded at this time using a small force plate underneath the entry point. The instrument can therefore analyze "initial entry" as a separate event/goal from the routine passage of a welding tool along a weld.

The design objective was to achieve an instrument with a statistical power of 90%. The unit is currently capable of capturing a +/- 2 degree Celsius temperature difference 95% of the time with only 5 replicates of each factor level. The long term goal is to achieve 95% confidence with 3 replicates of each factor level. Many samples per hour are possible since only a slight cooling off period is needed between samples. The detailed collection of data on a new alloy can be rapidly conducted using very standard, inexpensive stock sizes.

EXPERIMENTAL RESULTS

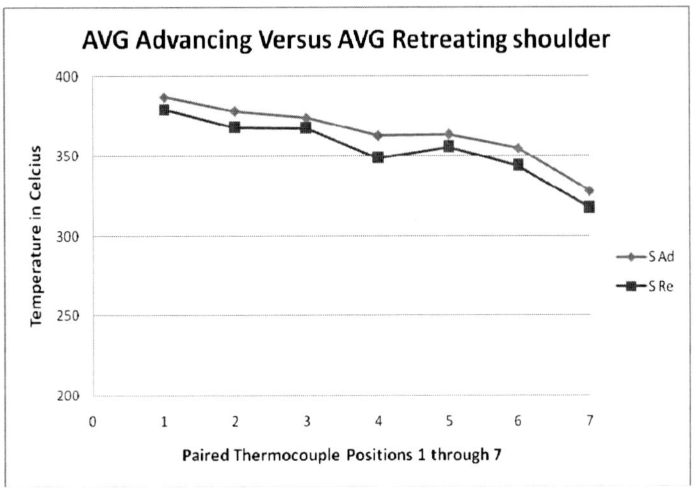

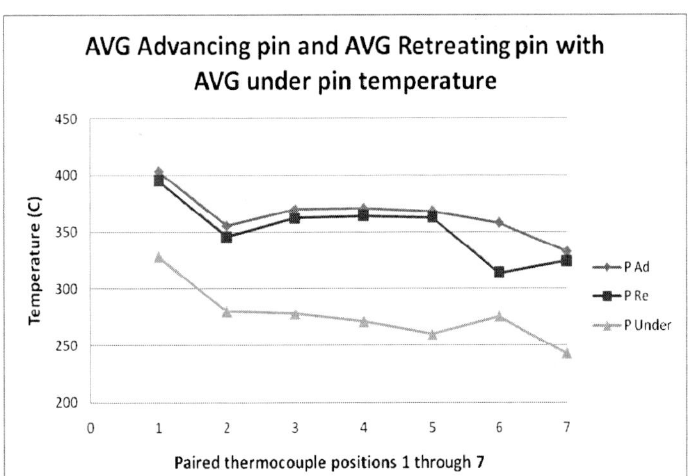

Figure 4. Average temperature at the advancing shoulder (S ADV C, in Celsius) in the left plot is always measurably higher than the average temperature of the shoulder on the retreating side (S RET C, in Celsius) during trials conducted in aluminum alloy 6061-T6. The same relationship has been observed between the advancing and retreating sides of the pin underneath the shoulder.

The trial runs used to establish the reliability, repeatability and sensitivity of the instrument have statistically proven with 99.99 percent certainty that there is always a measureable temperature difference in aluminum alloy 6061-T6 between the advancing and the retreating sides of the shoulders, as well as the advancing and retreating sides of the pin. Furthermore, there is a strong statistical difference between the pin temperatures and the shoulder temperatures on the same side. Figure 4 (in the left hand column) illustrates typical results with a 1 inch shoulder and a 0.250 inch pin with an RPM of 400 and a traverse speed of 4 inches per minute (4 IPM). This geometry allows direct comparison to earlier experiments in the literature review.

A PREDICTIVE MODEL

None of the earlier thermocouple measurements by researchers in the literature review observed a well behaved difference between the advancing and retreating sides. Perhaps as a consequence of this, only one of the researchers has modeled the FSW process as an asymmetric thermal process. Modelers need data in order to formulate realistic models. Simulators need large amounts of data in order to validate their simulations of the various models.

Only one model in the literature review predicted the asymmetric result observed during data collection above. Payton [24] equates the advancing side and retreating sides of FSW with what classical machining theory in general refers to as up milling and down milling. On the leading (up-milling side), as the tool enters the work, tool tip velocity (V) assumes a maximum value, which decreases as the tool progresses along the tool path. On the trailing side, velocity (V) has a minimum value when the tooth leaves the work and a somewhat higher value when the tool enters the work. The tool edge traces out a looped tracheiod path through the material. Incorporating the eccentric geometry and travel features of the tool, and generating the proper coefficients to use the data in ASM Handbook of Metals (9th Edition), Payton generated the following useful expressions for the FSW tool designer in aluminum alloy 6061-T6 which predict a hot advancing side and a colder retreating side using the following equations.

$$T_{cutting\_edge} = [16,148] HP_s \sqrt{\frac{V*t}{k*(pc)}} \quad (1)$$

HP_s is the specific horsepower for the alloy or workpiece, k is the coefficient of thermal conductivity for the workpiece, (pc) is the volume specific heat of the work material, V is the cutting speed at the tip of the tool. This is a geometric consideration which varies between left and right side (equation 2).

$$V = 2\pi n\sqrt{(R \pm r)^2 \pm 2ry} \quad (2)$$

One must use a (+) sign in equation (2) for the up-milling side and a (-) sign for the down-milling side of the geometry. The variable t is the uncut chip thickness, generally given by the following equation:

$$t_{avg} = \frac{F_t * d}{R\cos^{-1}\left(1-\frac{d}{R}\right)+\frac{F_t * n}{\pi * D}\sqrt{D*d-d^2}} \quad (3)$$

where R is the Radius of the cutting edge (specific to shoulder or pin in this application), d is the depth of the cut (specific to the design of tool), D is the diameter of the cutter, n is the number of teeth, and F_t is the feed per tooth. These equations predict the results of thermocouple measurements in the laboratory to date very exactly, since the velocity term V captures the difference between the advancing and retreating sides. The availability of a large amount of data over different machining values and modern multiphysics based finite element analysis using equations 1 through 3 permit simulation of the results accurately. Table 1 summarizes the material and machining parameters associated with equations (1-3).

Symbol	Nomenclature	Constant	Unit
HP_s	Specific horse power	0.34	hp/in^3/min
k	thermal conductivity of W/P	92	BTU/ft.h.F
pc	Volume specific heat of W/P	0.215	BTU/lb.F
R	Radius of shoulder	D/2	inch
D	Diameter of shoulder	.9,.7,.5	inch
t	Average chip thickness	Eq (3)	inch
F_t	Feed per tooth for shoulder (1 tooth)	V/n	inch
n	RPM	350,450	RPM
r	Radius of pin	0.15	inch
V	Cutting speed	4,7	inch/sec

Material and machining variables used in equations (1) through (3). English units are used throughout because of the predominance of those units in the applicable literature sources on milling and metal cutting. Table 1.

PHYSICS BASED SIMULATION

Equations (1) through (3) provide the heat input at both the pin and shoulder surfaces. As Song summarizes in his various papers [30, 34, 35], most simulation efforts only attempt to model the heat input from the tool shoulder, which is a very great simplification. Modern physics based software such as COMSOL Multiphysics allows the application of physics based expressions such as equations (1) through (3) directly to the contact surfaces of the pin and tool.

The equations developed above were utilized in COMSOL multiphysics FEA after the examples of Song using the geometry of the thermal collection unit [36] in Figure 5. The choice of a moving coordinate system analysis and the metal cutting equations results in a highly accurate match of the observed data to the model and the simulation.

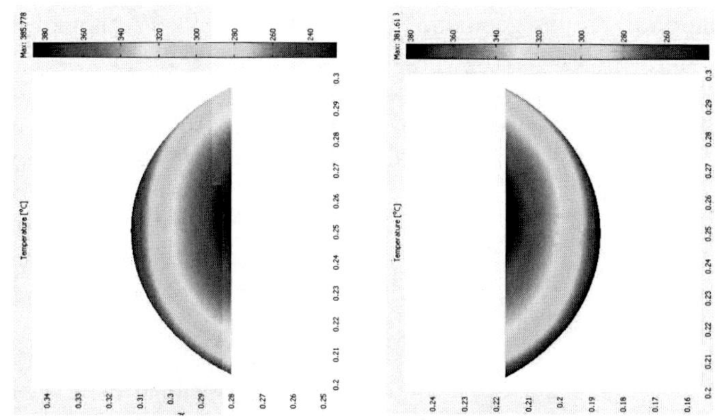

Figure 5. Advancing side (left) versus retreating side (right) of a Friction Stir Welding tools using Equations (1) through (3) to model heat source at each boundary of the tool. Advancing is hotter.

Since the early attempts using COMSOL, more advanced graphics models have been developed locally using the multiphysics modules of ANSYS which recently became available for use at the university. Figure 6 shows the asymmetrical nature of the FSW process quite well.

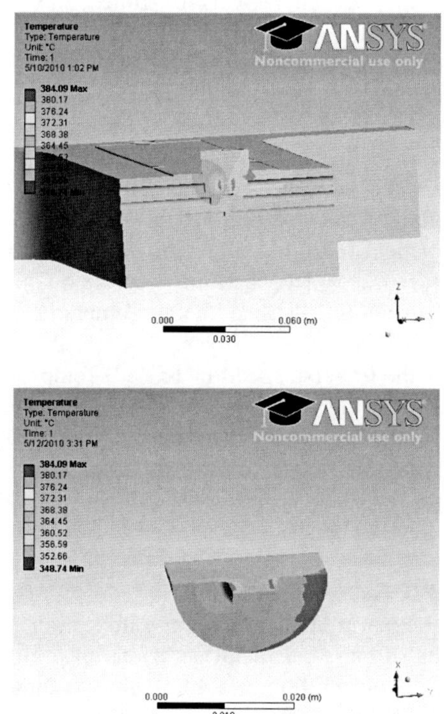

Figure 6. ANSYS Multiphysics simulation showing asymmetrical nature of the tool heating as it passes through the material being welded.

PARAMETERIZATION USING TOOL DESIGN

The thermal model described within this paper incorporates the material properties of the alloys being welded and tool designs geometric properties and path. Although the major experiment conducted by Kandaswaamy [36] utilized the very well known results for aluminum alloy 6061-T6 extensively, the researcher validated the sensitivity of the instrument to both material and tool changes using copper alloys and aluminum together along with various tool geometries, RPM's and feed rates across the welds.

The model and simulation capture the "dwell" of the tools motion as it passes through the material since the shoulder dominates the final passage over the welded junction. It correctly captures the transitions during insertion and initial should contact. Most importantly, the asymmetrical thermal difference on the two sides of the tool, and the variation of the temperature from the shoulder down along the pin are accurately predicted by the simulation of this model presented.

A model such as this lets the operator choose from the classic milling conventions (feed per tooth, RPM, velocity) in order to "target" a sweet spot or temperature range within the alloy of concern. Although the data collected during the initial experiments involved the simple geometry of a "straight" pin with no shoulder ridges or flutes, it is only a straightforward application of conventional milling theory to incorporate the cooling effects of flutes and scrolls in the design of the tool.

CONCLUSION

The advancing side of the tool is always hotter than the retreating side of the tool, with a statistical power of at least 90% for all the experiments and factor level combinations done. The rate of traverse dominated temperature responses by a large statistical margin. Shoulder diameter was the second largest impact factor. RPM, to the surprise of the investigator, was a very distant third impact factor. Temperature rose linearly as RPM and shoulder diameter increased and decreased inversely as the traverse speed increased. Temperature changed with material being welded with all other things constant. Classic metal cutting theory from the 1950s and 1960s has been successfully applied to a joining process. This suggests that classic extrusion theory and software such as DEFORM 3D might also be very beneficially applied to Friction Stir Welding.

FUTURE WORK

Several experiments are currently underway examining the relationships of classical metal deformation principles such as forging, extrusion, milling and burnishing as they apply to the Friction Stir Welding process. Various designed experiments investigating the factorial combinations suggested by equations (1) through (3) are continuing as funding allows.

ACKNOWLEDGEMENTS

The principal author would like to thank Dr. Arthur C. Nunes, Jr., as well as Kirby Lawless, Jeff Ding, Bob Carter, and Carolyn Russell at NASA's Marshall Space Flight Center and the National Space Science and Technology Center (NSSTC) which is funding further research with Auburn University. Additionally, the authors wish to express their appreciation to the Center for Advanced Vehicular Electronics (CAVE) who funded Dr. Kandaswaamy in part during his studies at Auburn University. Dr. Kandaswaamy would also like to extend his personal thank you to Dr. Alice Smith, Dr. Robert Thomas and the entire faculty of the Industrial and Systems Engineering department at Auburn University.

REFERENCES

1. Thomas, W.M. and E.D. Nicholas, *Friction Stir Welding for the Transportation Industries.* Materials & Design, 1997. 18(4-6): p. 269-273.
2. Murr, L.E., et al., *Friction-Stir Welding: Microstructural Characterization.* Materials Research Innovations, 1998. 1(4): p. 211-223.
3. Murr, L.E., et al., *Fundamental Issues and Industrial Applications of Friction-Stir Welding.* Materials Technology, 2000. 15(1): p. 37-48.
4. Murr, L.E., et al., *Microstructures in Friction-Stir Welded Metals.* Journal of Materials Processing & Manufacturing Science, 1999. 7(2): p. 145-161.
5. Murr, L.E., et al., *Intercalation Vortices and Related Microstructural Features in the Friction-Stir Welding of Dissimilar Metals.* Materials Research Innovations, 1998. 2(3): p. 150-163.
6. Murr, L.E., G. Liu, and J.C. McClure, *Dynamic Recrystallization in Friction-Stir Welding of Aluminium Alloy 1100.* Journal of Materials Science Letters, 1997. 16(22): p. 1801-1803.
7. Reynolds, A.P. and F. Duvall, *Digital Image Correlation for Determination of Weld and Base Metal Constitutive Behavior.* Welding Journal, 1999. 78(10): p. 355S-360S.
8. Esparza, J.A., W.C. Davis, and L.E. Murr, *Microstructure-Property Studies in Friction-Stir Welded, Thixomolded Magnesium Alloy Am60.* Journal of Materials Science, 2003. 38(5): p. 941-952.
9. Field, D.P., et al., *Heterogeneity of Crystallographic Texture in Friction Stir Welds of Aluminum.* Metallurgical and Materials Transactions a-Physical Metallurgy and Materials Science, 2001. 32(11): p. 2869-2877.
10. Field, D.P. and T.W. Nelson, *Tool Geometry Dependence of Local Texture in Friction Stir Welds of 7050 Aluminum Plate*, in *Textures of Materials, Pts 1 and 2.* 2002, TRANS TECH PUBLICATIONS LTD: Zurich-Uetikon. p. 1507-1512.
11. Li, Y., L.E. Murr, and J.C. McClure, *Flow Visualization and Residual Microstructures Associated with the Friction-Stir Welding of 2024 Aluminum to 6061 Aluminum.* Materials Science and Engineering a-Structural Materials Properties Microstructure and Processing, 1999. 271(1-2): p. 213-223.

12. Li, Y., L.E. Murr, and J.C. McClure, *Solid-State Flow Visualization in the Friction-Stir Welding of 2024 Al to 6061 Al.* Scripta Materialia, 1999. 40(9): p. 1041-1046.
13. Chao, Y.J., X. Qi, and W. Tang, *Heat Transfer in Friction Stir Welding - Experimental and Numerical Studies.* Journal of Manufacturing Science and Engineering-Transactions of the Asme, 2003. 125(1): p. 138-145.
14. Colligan, K., *Material Flow Behavior During Friction Stir Welding of Aluminum.* Welding Journal, 1999. 78(7): p. 229S-237S.
15. Dong, P., et al., *Coupled Thermomechanical Analysis of Friction Stir Welding Process Using Simplified Models.* Science and Technology of Welding and Joining, 2001. 6(5): p. 281-287.
16. Heurtier, P., C. Desrayaud, and F. Montheillet, *A Thermomechanical Analysis of the Friction Stir Welding Process*, in *Aluminum Alloys 2002: Their Physical and Mechanical Properties Pts 1-3*. 2002, TRANS TECH PUBLICATIONS LTD: Zurich-Uetikon. p. 1537-1542.
17. Khandkar, M.Z.H. and J.A. Khan, *Thermal Modeling of Overlap Friction Stir Welding for Al-Alloys.* Journal of Materials Processing & Manufacturing Science, 2001. 10(2): p. 91-105.
18. Khandkar, M.Z.H., J.A. Khan, and A.P. Reynolds, *Prediction of Temperature Distribution and Thermal History During Friction Stir Welding: Input Torque Based Model.* Science and Technology of Welding and Joining, 2003. 8(3): p. 165-174.
19. Nunes, A.C. *Wiping Metal Transfer in Friction Stir Welding.* in *TMS Annual Automotive Alloys and Joining Aluminum Symposia*. 2001. Dearborn: The Materials Society.
20. Reynolds, A.P., et al., *Utility of Relatively Simple Models for Understanding Process Parameter Effects on Fsw*, in *Thermec'2003, Pts 1-5*. 2003, TRANS TECH PUBLICATIONS LTD: Zurich-Uetikon. p. 2959-2964.
21. Reynolds, A.P., T.U. Seidel, and M. Simosen. *Visualisation of Material Flow in an Autogenous Friction Stir Weld* in *1st International Symposium on Friction Stir Welding* 1999. Thousand Oaks, CA: Abington, Cambridge CB1 6AH, UK; Abington Publishing.
22. Reynolds, A.P., et al., *Structure, Properties, and Residual Stress of 304l Stainless Steel Friction Stir Welds.* Scripta Materialia, 2003. 48(9): p. 1289-1294.
23. Reynolds, A.P., *Visualisation of Material Flow in Autogenous Friction Stir Welds.* Science and Technology of Welding and Joining, 2000. 5(2): p. 120-124.
24. Payton, L.N., *Metal Cutting Theory and Friction Stir Welding*, in *TMS Welding Conference*, T.J. Lienert, Editor. 2005, The Materials Society: Pine Mountain, Georgia.
25. Arbegast, W.J., *Modeling Friction Stir Welding as a Metal Working Process*, in *Hot Deformation of Aluminum Alloys*. 2003, The Materials Society.
26. Payton, L.N. and A.C. Nunes, *Metal Cutting Theory and Friction Stir Welding Tool Design*. 2003, Marshal Space Flight Center, NASA: Huntsville, Alabama. p. 1-5.
27. Chen, C. and R. Kovacevic, *Thermomechanical Modelling and Force Analysis of Friction Stir Welding by the Finite Element Method.* Proceedings of the Institution of Mechanical Engineers Part C-Journal of Mechanical Engineering Science, 2004. 218(5): p. 509-519.
28. McClure, J.C., et al. *A Thermal Model of Friction Stir Welding.* in *5th International Trends in Welding Research.* 1998. Pine Mountain, Georgia: The Materials Society.
29. Tang, W., et al., *Heat Input and Temperature Distribution in Friction Stir Welding.* Journal of Materials Processing & Manufacturing Science, 1999. 7(2): p. 163-172.
30. Song, M. and R. Kovacevic, *Heat Transfer Modelling for Both Workpiece and Tool in the Friction Stir Welding Process: A Coupled Model.* Proceedings of the Institution of Mechanical Engineers Part B-Journal of Engineering Manufacture, 2004. 218(1): p. 17-33.
31. Lienert, T.J., W.L. Stellwag, and L.R. Lehman. *Heat Inputs, Peak Temperatures, and Process Efficiencies for Fsw.* in *4th International Friction Stir Welding Symposium*. 2003. Park City, Utah: TWI.
32. Lambrakos, S.G., et al., *Analysis of Friction Stir Welds Using Thermocouple Measurements.* Science and Technology of Welding and Joining, 2003. 8(5): p. 385-390.
33. Xu, J., et al. *Experimental and Numerical Study of Thermal Process in Friction Stir Welding.* in *Joining of Advanced and Specialty Materials VI*. 2003. Pittsburg, Pa: ASM International.
34. Song, M. and R. Kovacevic, *Thermal Modeling of Friction Stir Welding in a Moving Coordinate System and Its Validation.* International Journal of Machine Tools & Manufacture, 2003. 43(6): p. 605-615.
35. Song, M. and R. Kovacevic, *Numerical and Experimental Study of the Heat Transfer Process in Friction Stir Welding.* Proceedings of the Institution of Mechanical Engineers Part B-Journal of Engineering Manufacture, 2003. 217(1): p. 73-85.
36. Kandaswaamy, S., *Internal Thermal Mapping of the Friction Stir Welding Process*, in *Industrial and Systems Engineering*. 2009, Auburn University: Auburn.

Proceedings of the ASME 2010 International Mechanical Engineering Congress & Exposition
IMECE2010
November 12-18, 2010, Vancouver, British Columbia, Canada

IMECE2010-40382

PITCH STABILITY OF AN UNPOWERED GROUND EFFECT VEHICLE

Meagan Hubbell[1], Steven Hard[2], Matthew Boots[3], Mary Ann Clarke[4], James E. Smith[5]

Center for Industrial Research Applications
West Virginia University
Morgantown, WV, USA

ABSTRACT

The ground effect regime was first utilized in the early 1900's with the advent of transatlantic flight. Aircraft such as the Dornier DO-X would fly close to the surface of the water in order to increase its payload and range. Since that time, research has been periodic with the largest resurgence of ground effect interest in the 1960's. The Russian government became involved in developing aircraft designed solely for ground effect flight. The design of these aircraft was difficult due to the inherent problems that exist within ground effect. There are natural instabilities that occur, especially in the longitudinal direction that are antagonized by shifting payload weights. Past researchers have handled the unique design requirements of ground effect through the usage of high-tail devices which operate outside of ground effect and power augmented ground effect which artificially generates the lift force through the use of thrust vectoring. The Center for Industrial Research Applications (CIRA) has developed a single passenger, unpowered, subsonic aircraft that relies on gravitational forces for momentum. AirRay combines the benefits of ground effect i.e. the increased lift and decreased induced drag, with a unique approach to maintaining stability.

The design of AirRay faced many challenges as a result of flying in the ground effect regime, similar to those found in the prior efforts. These include natural instabilities, primarily in the longitudinal direction, that cause the glider to want to pitch up. In addition the size requirements for a single rider to maximize maneuverability, as well as the potential for updrafts on a downhill slope are added constraints to the design of the ground effect vehicle. These issues, and others, are the subject of current study. This paper has focused on the most important aspect of the design, longitudinal stability.

This research has shown positive results with respect to the effectiveness of slots, on passively controlling the movement of the center-of-pressure at varying angles of attack. The 40 degree slot located at 20% of the chord line was most advantageous in stabilizing movement. These results indicate a craft can be designed that can be stable and function in the majority of the flight conditions that have been specified.

INTRODUCTION

Ground effect was a concept that was first introduced in the 1920's and was discovered by accident as a result of its influence on low flying Dornier DO-X seaplanes. These planes would fly close to the surface of the water and would experience additional lift capabilities which gave them the ability to conserve fuel. This could, also, be translated into an ability to carry heavier loads. [12]

Ground effect is defined as an increase in the lift-to-drag ratio on a lifting surface, such as a wing, when moving in close proximity to the ground. Ground effect is categorized by the direction of its influence, span dominated or chord dominated. Span dominated ground effect is utilized with wings that have a large aspect ratio such that the chord and ground clearance of the wing are significantly less than the span. As a result of span dominated ground effect, the induced drag is reduced. Chord

[1] Graduate Student, Center for Industrial Research Applications, Mechanical and Aerospace Engineering
[2] Graduate Student, Center for Industrial Research Applications, Mechanical and Aerospace Engineering
[3] Undergraduate Student, Center for Industrial Research Applications, Mechanical and Aerospace Engineering
[4] Assistant Professor, Mathematics
[5] Professor and Director, Center for Industrial Research Applications, Mechanical and Aerospace Engineering

dominated ground effect is utilized with smaller aspect ratios and results in a stagnation of airflow underneath the wing. This leads to an increase in the pressure under the moving body while in close proximity to the ground. [14,15]

One of the fundamental problems that has been found over the years when working within the ground effect regime has been an inherent instability in the longitudinal direction. As a result of this many efforts have been made to correct this problem in such manners as using ram-wing technology or relocating the tail section outside of the ground effect regime. [4,7,8,11] These options were not viable for AirRay and therefore a new method of pitch stability needed to be developed. As a result, ducted airfoils were examined for their ability to reduce the air flow over the wing at high angles of attack, effectively reducing the wings ability to pitch-up. [2]

NOMENCLATURE

CAD	Computer Aided Design
CFD	Computational Fluid Dynamics
AoA	Angle of Attack
Fr	Froude number
Re	Reynolds number
L	Characteristic length
V	Velocity
Xcp	Center of pressure
Xcg	Center of gravity
d	Angle of slot w.r.t. chord line
x	Distance of slot w.r.t. leading edge as percentage of chord line
w	Width of slot as percentage of chord line
ρ	Density
μ	Viscosity

BACKGROUND

AirRay was initially conceptualized during observations of low flying sea birds that would curve their wings when flying close to the surface of the water as a way of capturing the updrafts at the wave's edge. As a result of mimicking the biology of these animals, a conceptual design of AirRay was development as shown in Figure 1.

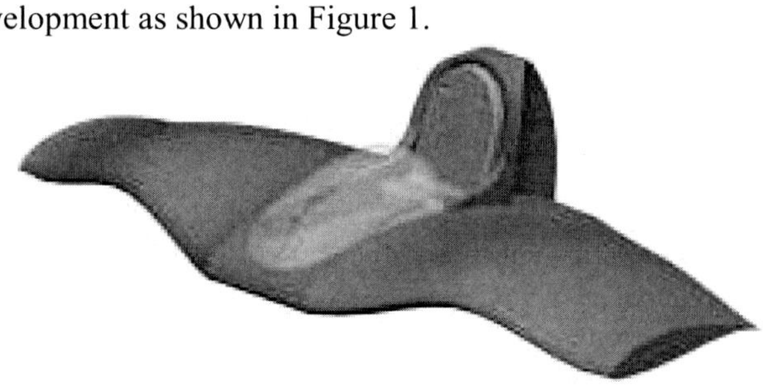

Figure 1: Concept Design of AirRay [10]

This aircraft would be a small, unpowered, extremely low speed, single passenger, recreational vehicle which would operate on downhill slopes and fly in ground effect. Historically, ground effect aircraft are powered, capable of carrying anywhere from four to a couple hundred people and fly at velocities upwards of 60 mph. AirRay is substantially different from historical ground effect vehicles and the concepts and techniques developed for them.

DESIGN WORK

An initial base design was developed that has the capability of flight without the assistance of specialty aerodynamic features such as twist, dihedral, wingtips, etc. This resulted in a straight wing, tapered design with moderate sweep to adjust for static stability. Figure 2 shows this design for which ducted wing concepts could be implemented and tested for their effects.

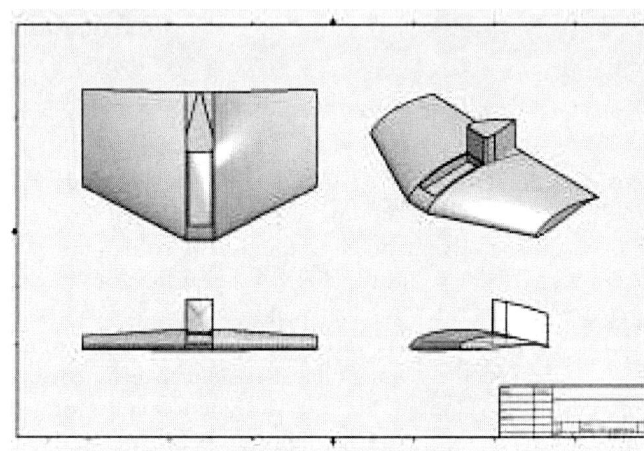

Figure 2: Straight Wing Design of AirRay

Pitch stability is vital to the flight of an aircraft, especially in close proximity to the ground where the craft could potentially pitch into the ground, aerodynamically stall, or pitch backwards. In an effort to minimize the hazards of longitudinal instability, it was necessary to develop a method to minimize the travel of the center-of-pressure for varying angles of attack. Pressure equalization slots will be used as a means to passively control this center-of-pressure movement which could potentially be placed at varying positions along the chord of the airfoil in order to redirect or interrupt the airflow at higher angles of attack.

The slots will be computationally evaluated two-dimensionally based on two main parameters: the angle of the slot with respect to the normal of the chord line (d = 20 °, 30 °, and 40 °) and the position of the slot along the chord line from the leading edge (x = 15%, 20%, and 25%) as shown in Figure 3. In addition, these airfoils will be tested at three velocities: 8, 12, and 18 m/s. These are representative of the average and maximum design speeds of the glider under normal operating conditions and the projected maximum speed for the final glider design.

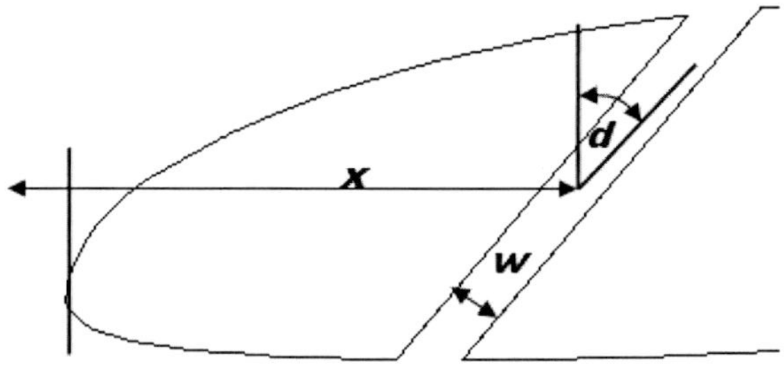

Figure 3: Representation of Slot Characteristics

These models were computationally evaluated with a computational fluid dynamics package CFD and compared against baseline testing of the Wortmann FX 63-137 which has been both theoretically and computationally modeled in previous research [9].

METHODOLOGY

This section discusses the initial setup and generation of the two-dimensional Wortmann FX 63-137 airfoils, both modified with slots and the baseline. In addition, it discusses the simulation setup and validation. [3,6]

Grid Setup of Airfoil

The two-dimensional figures for the FX 63-137 airfoil were setup and meshed in the CAD software Gambit 2.3.16. This software was chosen because it directly interfaces with the CFD software, Fluent 6.2.16. For this analysis, two main types of geometries were created, the baseline model, which simulated a scaled, unmodified FX 63-137 airfoil and a modified model which incorporated the varying slot concepts. The baseline model for an angle of attack of zero is shown in Figure 4. An example of a slotted model at 10 degrees angle of attack is shown in Figure 5. [1]

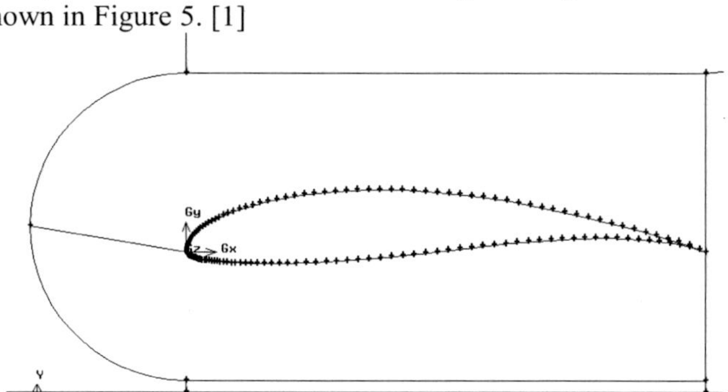

Figure 4: Wortmann FX 63-137 Baseline Model

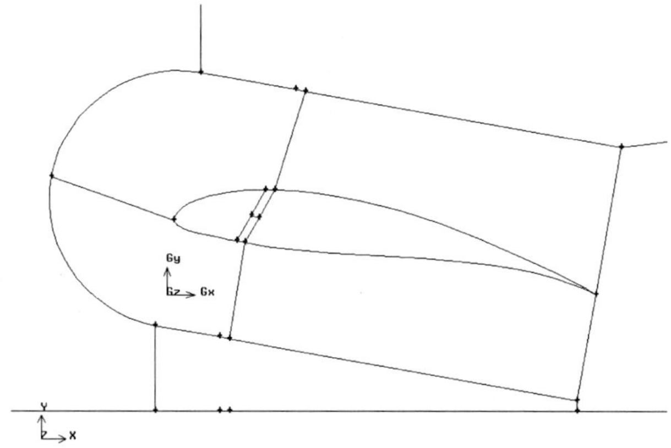

Figure 5: Wortmann FX 63-137 Slotted Model

Each model was scaled from the full version to have a chord length of one meter in order to simplify the modeling process and Reynolds number calculations. In addition both types of models were placed at 25% of the chord length above the ground from the lowest point on the airfoil. This location was chosen because it is within the most advantageous range of ground clearances for optimal ground effect. [14]

The grid of the airfoil was constructed with a structured mesh, concentrating nodes at the leading and trailing edge for the baseline, and for the modified airfoil, also around the slot entrance and exit. A close-up view of the slotted airfoil is shown in Figure 6 which illustrates the enhanced mesh immediately surrounding the airfoil and the concentration of cells within the slot.

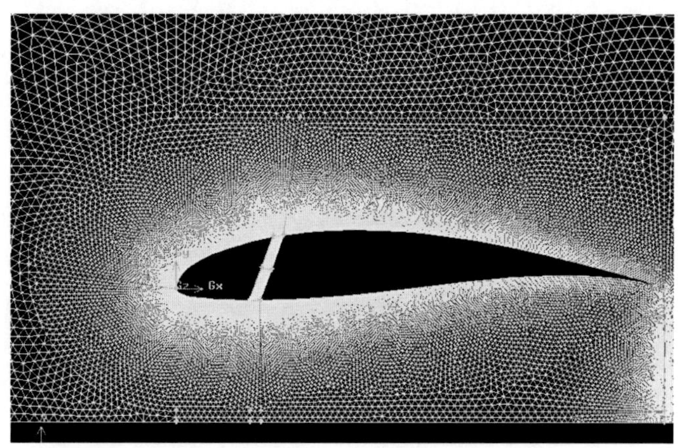

Figure 6: Close-up of Slotted Model with Mesh

Once the structured meshes were generated, they were used to create unstructured triangular cells which grew at a rate of 1.05 away from the airfoil. The specific meshes are consistent for all the slotted airfoils with a similar mesh being created for the baseline except without the focus on the slot. The maximum that the cells were allowed to grow to were 0.3 for the area acting as a boundary immediately around the airfoil and 0.25 for outside of the airfoil boundary and plume area.

To imitate real flow, a velocity inlet was placed nine chord lengths upstream with a pressure outlet located an equal distance downstream. In addition, a stationary wall was placed ten chord lengths above the airfoil, where it acts as a boundary to the simulation but does not restrict the movement of the flow, as well as a stationary wall a quarter chord beneath the airfoil. For positive angles of attack the trailing edge was used as the point of rotation whereas for negative angles of attack the leading edge was used. This procedure is to maintain the quarter chord distance from the ground plane. After generation, the baseline model had approximately 60,000 nodes and the slotted models had approximately 85,000.

Computational Setup of Airfoil

For the CFD analysis, Fluent 6.2.16 was chosen because of its ability to efficiently model two-dimensional flow and interface with the Gambit software. Within Fluent the stationary ground wall can be changed to a moving wall which will most accurately reflect the moving ground plane of the actual environment. Both the moving ground plane and the velocity inlet are set to the same velocity during simulation to mimic ground effect without the addition of environmental effects.

Since the airfoil has been scaled to a chord length of 1 meter from the original 8 foot mean chord, the associated wind speeds will also have to be scaled. For the prototype, the velocities of 12 and 18 m/s are encountered during flight and for the commercial craft, a maximum design velocity of 28 m/s was used. The scaled model will therefore encounter speeds of approximately 7.5, 11.5 and 18 m/s. This is found through relating the chord length and velocity through the Froude equation found in Equation 1. The Froude number is non-dimensional group which is used for flow with the free surface.

$$Fr = \frac{V}{\sqrt{gl}} \qquad \text{Equation 1}$$

In order to determine the type of flow around the airfoil, the Reynolds number was calculated using Equation 2. The density was taken at sea level to be 1.225 kg/m^3 with a viscosity of 1.789 x 10^5 kg/m-s.

$$\mathrm{Re} = \frac{\rho V L}{\mu} \qquad \text{Equation 2}$$

At 18 m/s the Reynolds number was calculated to be 1.9 x 10^6, which indicates turbulent flow. Turbulent flow occurs at Reynolds numbers of 1.5 x 10^6 and higher. The Reynolds number at 11.5 m/s was determined to be 7.9 x 10^5, which indicates transitional flow. This region is indicated by Reynolds numbers larger than 3 x 10^5 and less than 1.5 x 10^6. At 7.5 m/s the Reynolds number was calculated to be 5.2 x 10^5, which also falls within the transient region. For the slotted models, there is an acceleration of velocity within the slot itself which can trigger turbulent flow; therefore even with the transient region indicated in the 7.5 m/s flow, all models will be simulated with turbulent flows. [1]

For the calculations, a two-dimensional double precision simulation with a k-epsilon solver was used. The double precision will more accurately predict the flow field, while the k-epsilon solver is robust and reasonably accurate for a wide range of turbulent flows. [13] The additional Fluent solver settings are shown in Table 1.

Table 1: Fluent Settings

Pressure Solver	Presto!
Momentum Equation Solver	First Order Upwind Discretization
Energy Equation Solver	First Order Upwind Discretization
Turbulent Kinetic Energy Solver	First Order Upwind Discretization
Turbulent Dissipation Rate	First Order Upwind Discretization

The first order upwind discretizations were used in the solvers to assist in convergence of the flow. The Presto! solver is useful in that it improves accuracy with surfaces that are strongly curved. An non-equilibrium wall treatment was applied in order to accommodate for the increased pressurization at the walls. This was a result of the nearness of the airfoil to the ground. This particular treatment is useful for complex flows involving separation, reattachment and impingement [13].

There were two types of convergence criteria used for determining the solutions to the simulations. The first type of convergence checked the continuity, x-velocity, y-velocity, energy, k, and epsilon residuals for a convergence of at least 1 x 10-3. At this point, values were recorded for the forces and aerodynamics coefficients to comparatively analyze optimal performance with respect to one another in order to reduce the number of test cases. Once the optimal test cases were selected, the previous residuals were reassessed for a convergence of 1 x 10-5 in addition to the convergence of the residuals of the lift and drag coefficients.

Validation of Fluent Results

An initial grid of 82,000 nodes and 165,000 cells was generated for each airfoil in Gambit. This resulted from approximately 2000 grid points located on the surface of the airfoil. The generated surrounding triangular mesh contained reasonably negligible amounts of skewed cells and was assumed to be significantly fine for the initial grid independence check. Two types of adaptations were attempted in order to compare simulation time and productivity of modeling.

The first technique used Gambit to increase the number of nodes on the model to an amount one order higher than the original. This lead to 480,000 nodes on the baseline model once the adaptation took place. The model was simulated with an 18 m/s flow which required 36 hours to converge.

The second technique used the adapt tool in Fluent to make modifications to the existing mesh. Both boundary and region adaptations were used to increase the size of the mesh. These tools were used to refine the grid until a factor of at least 1.3 was achieved [5]. Boundary adaptation is primarily used to refine cells located within a certain number of cells of the boundary zones [13]. In this case within one cell of the selected zones was used.

The secondary type of adaptation used was the region function. It focuses on user selected areas of the mesh to improve meshes that grow as the distance increases from the boundaries [13]. In this case, the entire mesh area was selected for improvement. The ensuing simulation time for this technique was approximately two hours.

Considering the simulation time required for each technique, the adaptation method provided by Fluent was used to verify grid independence. The first adaptation of the original mesh used both the boundary and then region functions once. The difference between the center-of-pressure positions was approximately 6%. Therefore the adaptation was repeated again, but with each function being used twice alternately. This resulted in a value for the center-of-pressure position being less than 1% away from the previous adaptation method results. The numerical uncertainty in a fine-grid solution is at a maximum of 2.2% [5]. This indicates that the first adaptation method was grid independent.

RESULTS

The study conducted on the Wortmann FX 63-137 focused primarily on evaluating the effect of different geometric slots on limiting the travel of the center-of-pressure. This is of such significance because the stability of the center-of-pressure allows for the glider to be able to maintain steady flight. If the travel of the center-of-pressure does not remain restricted, AirRay could move in and out of ground effect and either loses the aerodynamic influence of the ground or accidentally impact with the ground plane.

Analysis of Initial Convergence Data

In order for any comparison to be made, the baseline Wortmann FX 63-137 airfoil was simulated with the initial convergence criteria to determine the range that the center-of-pressure traveled. Figure 7 shows the maximum travel that occurred over a range of angle of attacks, at each velocity, as a percentage of the chord length.

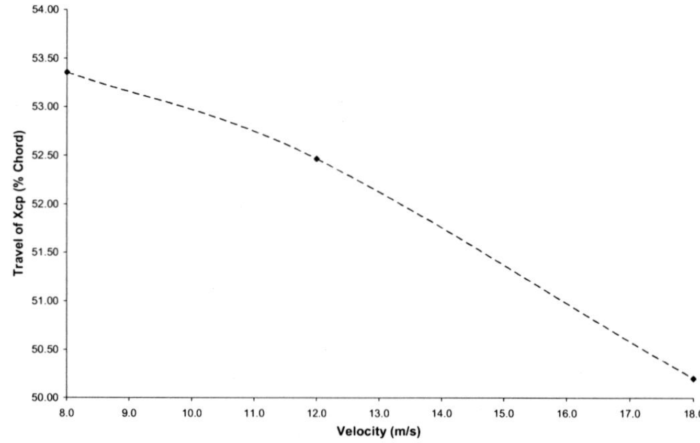

Figure 7: Max Travel of Baseline Xcp across AOA as % of Chord

The general trend that appears is the tendency for the center-of-pressure to travel less as the velocity increases. This indicates that control will be required the most at low speeds. Test cases were run for the three chords and vertical angle positions. The results from the CFD simulations using the initial convergence criteria are shown in Figure 8 and include the baseline data found in Figure 7.

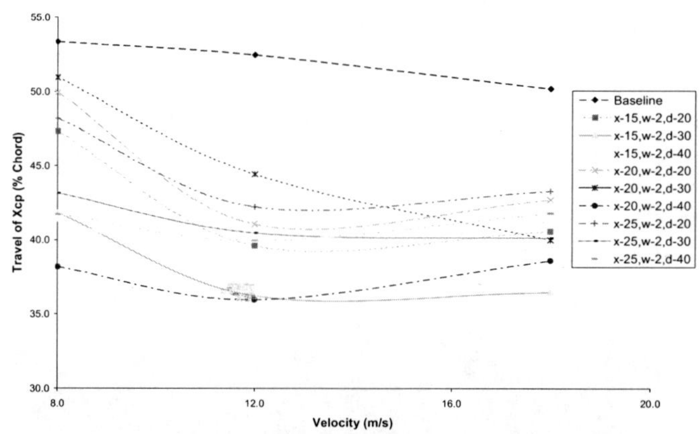

Figure 8: Average Travel of Xcp across AOA as % of Chord

From Figure 8, two configurations appear as reducing the movement of the center-of-pressure significantly from the baseline. The x-20, w-2, d-40 and the x-15, w-2, d-30 slot geometries are located closest to the bottom of the figure, indicating the least movement as a percentage of the airfoil chord. The average travel of the center-of-pressure was calculated for each slot configuration across the three velocities and four angels of attack giving an overall average for the Xcp shift. The x-20, w-2, d-40 and the x-15, w-2, d-30 slot configurations were found to restrict the travel up to 40% more on average than the baseline airfoil.

The initial simulations include a negative three degree angle of attack which was included for instances in which the nose pitches down or the relative wind direction changes. An example of the results produced by the negative angle of attack can be seen in Figure 9. This figure shows the position of the center-of-pressure on the x-axis and location of Xcp on the y-axis with zero indicating the leading edge of the airfoil and one indicating the trailing edge.

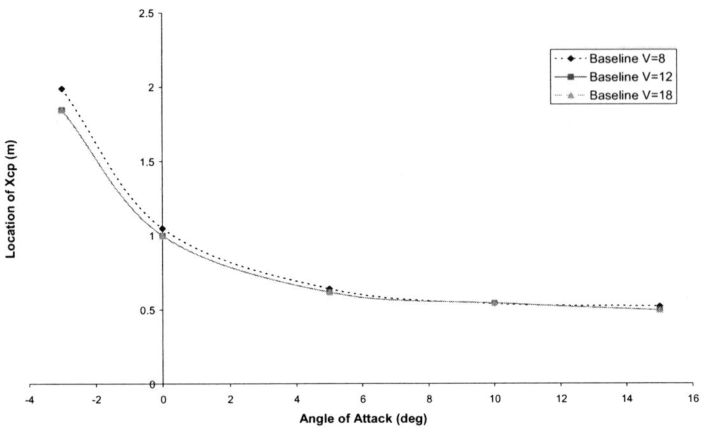

Figure 9: Baseline Travel of Xcp along Chord Length with respect to AOA

As the angle of attack decreases, the location of the center-of-pressure begins to travel to the trailing edge of the airfoil and with a negative angle of attack, travel outside the boundaries of the body. During many of the simulation cases, the center-of-pressure location existed one and a half to two times the chord length outside the airfoil for negative angles of attack. The extreme shift in the center-of-pressure location for the negative angle of attack could potentially be explained through the Venturi effect. Due to the camber and position of the airfoil relative to the ground, a nozzle-like figure was created two-dimensionally as seen in Figure 10.

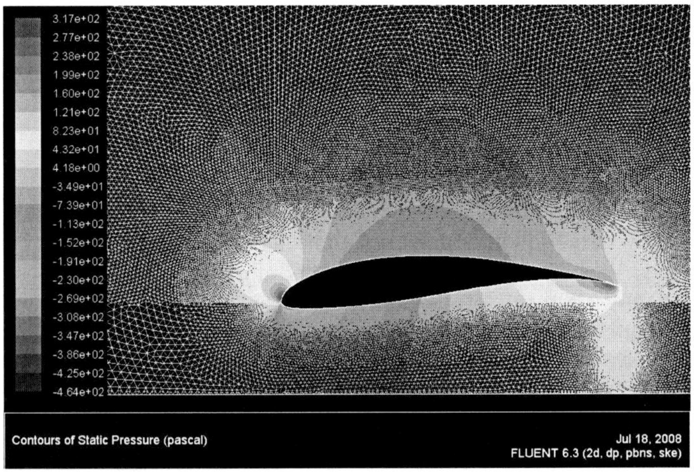

Figure 10: Static Pressure Distribution of Baseline Model at V=18 m/s

The creation of this shape resulted in a decrease in the pressure at the nose and an increase in the pressure build-up underneath the trailing edge, causing the center-of-pressure location to shift significantly beyond the tail.

The individual results for the two most beneficial slot arrangements have been placed into a series of tables which show the center-of-pressure locations at differing angles of attack and velocities. Table 2 shows the baseline data which indicates an Xcp outside of the 1 meter chord length for an angle of attack at zero. This is also a potential result of the Venturi effect. Table 2 shows the positions of the center-of-pressures for the three velocities as a function of angle of attack. The "% Travel" indicates the range that the Xcp traveled as a percentage of the chord length as both a function of the angle of attack and velocity. From Table 2 the average of the travel of the center-of-pressure across all velocities and angles of attack was determined to be 52.0%.

Table 2: Baseline Center-of-Pressure Data

AOA (deg)	Xcp (V=8)	Xcp (V=12)	Xcp (V=18)	% Travel
0.0	1.08	1.05	0.999	1.04
5.0	0.671	0.641	0.617	0.643
10.0	0.592	0.539	0.544	0.558
15.0	0.545	0.522	0.497	0.521
% Travel	53.4	52.5	50.2	52.0

Table 3 shows the x-15, w-2, d-30 slotted airfoil data, which was one of two slot configurations that appeared to limit the center-of-pressure travel more effectively than other models. The average travel of the center-of-pressure over all the angle of attacks and velocities was 38.2% which is an approximate 27% decrease in the travel from the baseline model.

Table 3: x-15, w-2, d-30 Center-of-Pressure Data

AOA (deg)	Xcp (V=8)	Xcp (V=12)	Xcp (V=18)	% Travel
0.0	0.953	0.891	0.880	0.908
5.0	0.724	0.700	0.659	0.626
10.0	0.560	0.557	0.547	0.555
15.0	0.535	0.528	0.515	0.526
% Travel	41.8	39.7	36.5	38.2

Table 4 shows the data for the x-20, w-2, d-40 slotted airfoil. This airfoil showed slight improvement in the average travel of the center-of-pressure over the x-15, w-2, d-30 airfoil with an average travel of 37.5% compared to the 38.2% mentioned previously.

Table 4: x-20, w-2, d-40 Center-of-Pressure Data

AOA (deg)	Xcp (V=8)	Xcp (V=12)	Xcp (V=18)	% Travel
0.0	0.904	0.881	0.899	0.894
5.0	0.618	0.618	0.616	0.617
10.0	0.546	0.544	0.537	0.542
15.0	0.522	0.521	0.512	0.518
% Travel	38.2	36.0	38.6	37.6

While the 18 m/s velocity was shown to have less travel with the 15% chord position, the 20% chord adequately compensated for the center-of-pressure travel at lower velocities. Therefore, the x-20, w-2, d-40 was examined further with the initial convergence criteria to estimate whether its configuration would achieve the best possible stability for the slotted style of pitch control.

Since the chord position was a midrange value but the vertical slant was a maximum value, the slant of the slot would have to be increased by 10 degrees to check whether the pitch stability improved. This would confirm whether the x-20, w-2, d-40 was an improved configuration out of these test cases. Table 5 shows the resulting data from the *d*=50 simulation.

Table 5: x-20, w-2, d-50 Center-of-Pressure Data

AOA (deg)	Xcp (V=8)	Xcp (V=12)	Xcp (V=18)	% Travel
0.0	0.837	0.856	0.885	0.860
5.0	0.601	0.605	0.600	0.602
10.0	0.539	0.539	0.530	0.536
15.0	0.516	0.515	0.504	0.512
% Travel	32.1	34.1	38.1	34.8

On average, the new slot configuration was able to restrict the center-of-pressure movement by another 6% compared to the x-20, w-2, d-40 slot. At higher velocities the center-of-pressure travel begins to stabilize, however, velocities as low as the 8 m/s still see up to 15% less shift from the 40 degree angle to the 50 degree angle. Therefore the vertical slot was increased again by 10 degrees to determine whether the center-of-pressure location would continue to travel.

Knowing that the greatest shift of the center-of-pressure will occur from the largest to smallest angle of attack, the test case was restricted to 0 degrees and 15 degrees. The average travel of the center-of-pressure is approximately 33% of the chord which is a 5 % decrease from the *d*=50 case as can be seen in Table 6.

Table 6: x-20, w-2, d-60 Center-of-Pressure Data

AOA (deg)	Xcp (V=8)	Xcp (V=12)	Xcp (V=18)	% Travel
0.0	0.804	0.835	0.871	0.836
15.0	0.493	0.515	0.509	0.506
% Travel	31.1	31.9	36.2	33.1

Using the initial convergence criteria, a trend is appearing that as the vertical angle of the slot increases, the travel of the center-of-pressure begins to stabilize at lower velocities regardless of the angle of attack. Since the control of the travel continued to improve with an increase in slant, the geometry was again altered by 10 degrees. The results of the simulation are shown in Table 7.

Table 7: x-20, w-2, d-70 Center-of-Pressure Data

AOA (deg)	Xcp (V=8)	Xcp (V=12)	Xcp (V=18)	% Travel
0.0	0.808	0.835	0.855	0.833
15.0	0.517	0.525	0.514	0.519
% Travel	29.0	31.0	34.1	31.4

On average, there is a 5% decrease in the range of the center-of-pressure from the last slot configuration. Due to the close range with which the slots are achieving the stability of the center-of-pressure, the comparison will be continued with the extended convergence criteria for additional accuracy.

Analysis of Secondary Convergence Data

The optimal geometries of x-15, w-2, d-30 and x-20, w-2, d-40 were selected for additional comparison from the initial convergence criteria. The results were compared against the baseline data which was also recalculated using the extended criteria. A comparison of these results for V=18 m/s can be seen in Table 8.

Table 8: Comparison of Test Cases for V=18 m/s

AOA (deg)	Baseline	x-15,w-2,d-30	x-20,w-2,d-40
0.0	0.584	0.597	0.590
15.0	0.423	0.444	0.452
% Travel	16.0	15.3	13.8

As shown previously with the initial convergence data, the x-20, w-2, d-40 slot configuration was able to control the travel of the center-of-pressure more effectively, compared to the x-15, w-2, d-30 geometry. At a velocity of 18 m/s, the x-20, w-2, d-40 slot configuration exhibited a 14% change on the center-of-pressure travel, whereas at a velocity of 8 m/s, the travel was reduced 6.5 %.

Since the position of the optimal slot is a midrange design value and the vertical angle is a maximum value, higher vertical angles were assessed to examine the range of the pitch stability. As previously mentioned with the initial convergence criteria, there were additional slot geometries tested for optimization. For the extended criteria, the x-20, w-2, d-50 and x-20, w-2, d-60 configurations indicated an increase in travel as seen in Table 9.

Table 9: Comparison of Additional Test Cases for V=18 m/s

AOA (deg)	x-20,w-2,d-40	x-20,w-2,d-50	x-20,w-2,d-60
0.0	0.590	0.632	0.629
15.0	0.452	0.448	0.447
% Travel	13.8	18.4	18.2

These results suggest that the x-20, w-2, d-40 configuration was most capable of controlling the travel of the center-of-pressure to within approximately 14% of the chord compared to the 50% of the chord experienced with the baseline.

Verification of Results

In order to validate the results of this simulation, the baseline data from an unmodified Wortmann FX 63-137 had to be found. Due to the necessity for accuracy, the secondary convergence criteria were used for comparison.

The results presented from this simulation were comparatively analyzed against previous work performed both computationally and experimentally. In 2008, Smith computationally analyzed the aerodynamic characteristics of the Wortmann FX 63-137, primarily the lift and drag coefficients. He found the trend of the lift coefficients as a function of angle of attack for various h/c ratios. Smith validated his results by proving his method of model generation was accurate by creating simulated models of the NACA 4412 and comparing them to experimental results [9]. The experimental data for the FX 63-137 is taken from Abbot and Von Doenhoff which is freestream flow and therefore does not take into account ground effect. Figure 11 shows the comparison of the lift coefficients between the experimental, computational and current research with respect to the angle of attack.

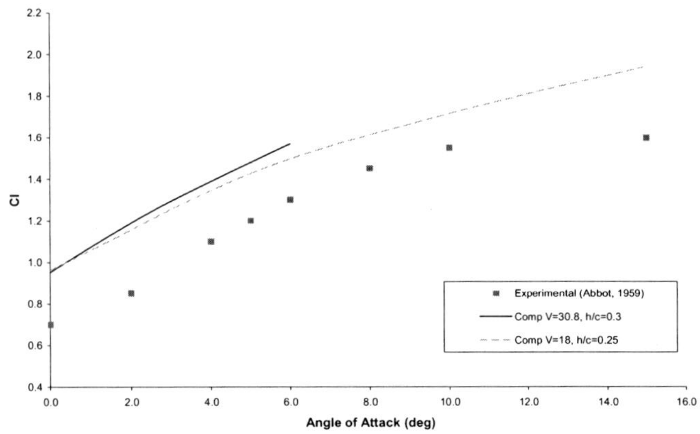

Figure 11: Comparative Analysis of Results with Previous Research

In Figure 11, it can be seen that the computational coefficients are 10% to 40% higher than the experimental values found in freestream. This is due to the pressure increase as the moving bodies approach the ground plane [9]. As seen in previous research, the lift coefficient is supplemented by its proximity to the ground, in which the optimal amount occurs at a height that is 25% of the chord [14]. The data provided by Smith is slightly higher than the values found through this research due to the increased velocity used during that simulation.

In addition to the comparison of the ground effect data to the freestream experimental data, a comparison was necessary between the baseline operating in freestream flow and the experimental data provided by Abbot and Von Doenhoff. The baseline model was altered so that the airfoil was equally distance from all walls inside the model, permitting freestream simulation. The freestream baseline model was simulated with a velocity of 2 m/s. This velocity was found using Equation 2 and a Reynolds number of approximately 155,000. The given Reynolds number corresponds to that of the original experimental data. The comparison between the experimental and computational data can be seen in Figure 12.

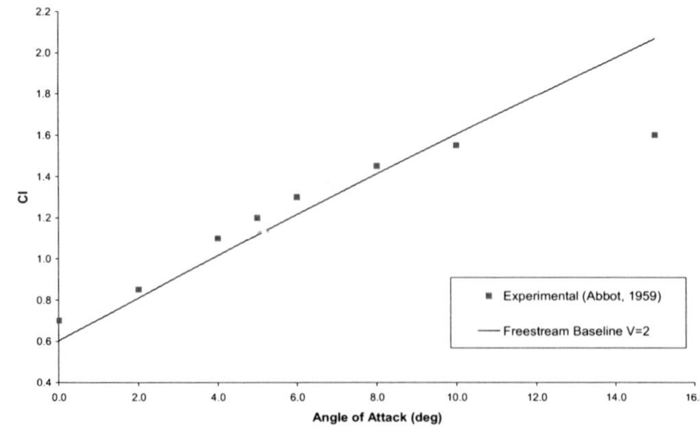

Figure 12: Comparative Analysis of Freestream Results

The freestream computational data follows similar trend lines to that of the experimental data. The discrepancy between the data is a potential result of the inherent differences that exist between the two methods of research. Computational research operates in an ideal environment with calculated effects as a result of applied forces. The data obtained from the simulation is the optimal result for that situation. Experimental data has the potential to include extraneous influences as a result of exposure to laboratory conditions. In addition, it operates in a semi-realistic atmosphere, which most closely relates to the actual operating environment.

CONCLUSIONS

Previous research showed that the center-of-pressure travel was only minimal at a steady angle of attack [9]. This research has shown positive results with respect to the effectiveness of slots, on passively controlling the movement of the center-of-pressure at varying angles of attack. Of the three positions along the chord line, the 20% position was most advantageous in stabilizing movement. Along with that characteristic, the 40 degree vertical slot was the most capable of the selected geometries in limiting the travel of the center-of-pressure and hence controlling pitch.

The ability of slot placement to passively control the center-of-pressure and hence the pitch stability has a wide range of applications outside that of a ground effect glider. Pitch instability is an inherent problem in many high speed technologies, including aircraft, missiles, and rockets. Further research in the area will allow for this slot configuration to be applied to 3-dimensional models to verify its efficiency.

REFERENCES

[1] Abbot, I. a. (1959). *Theory of Wing Sections Including a Summary of Airfoil data.* Dover, N.Y.

[2] Appich, W. a. (1979). Aerodynamic Effects of Body Slots on a Guided Projectile with Cruciform Surfaces. *Atmospheric Flight Mechanics Conference.* Boulder, CO: AIAA-1979-1658.

[3] Barber, T. E. (November 1999). A Technical Note on the Appropriate CFD Boundary Conditions for the Prediction of Ground Effect Aerodynamics. *The Aeronautical Journal* .

[4] Borst, H. (1979). Analysis of Vehicles with Wings Operating in Ground Effect. *Advanced Marine Vehicles Conference.* Baltimore, MD: AIAA.

[5] Celik, I. (2005). Procedure for Estimation and Reporting of Discretization Error in CFD Applications. *ASME Journal of Fluids Engineering.*

[6] Chun, H. a. (2003). Turbulence Flow Simulation for Wings in Ground Effect with Two Ground Conditions: Fixed and Moving Ground. *International Journal of Maritime Engineering* .

[7] De Divitiis, N. (February 2005). Performance and Stability of a Winged Vehicle in Ground Effect. *Journal of Aircraft* , Vol. 42, No.1.

[8] Mashalik, A. (Juneo 2001). The Main Research Problems Solved in Designing Russian Ekranoplans and which are Necessary to be Solved for Wide Practical Realization of Ekranoplans. *EAGES 2001 International Ground Effect Symposium.* Toulouse, France.

[9] Smith, J. (2008). Validation of an Airfoil in the Ground Effect Regime Using 2-D CFD Analysis. *AIAA Aerodynamic Measurment Technology and Ground Testing Conference.* Seattle, WA: AIAA.

[10] Williams, K. *Concept of AirRay.*

[11] Raymond, A. (1921). *Ground Influence on Aerofoils.* NACA Technical Note - 67.

[12] Ollila, R. (1979). Historical Review of WIG Vehicles. *Advanced Marine Vehicles Conference.* Baltimore MD: AIAA-1979-2033.

[13] Ansys, Inc. *Fluent 6.2 User's Gruide Version 6.3.* 2005.

[14] Rozhdestvensky, K. (2000). *Aerodynamics of a Lifting System in Extreme Ground Effect.* Berlin: Springer.

[15] Rozhdestvensky, K. (2006, May). Wing-in-Ground-Effect Vehicles. *Progress in Aerospace Sciences , 2 (3)* .

Proceedings of the ASME 2010 International Mechanical Engineering Congress & Exposition
IMECE2010
November 12-18, 2010, Vancouver, British Columbia, Canada

IMECE2010-40463

INTEGRATION OF GPS AND ACCELEROMETER UNCERTAINTIES TO IMPROVE THE ESTIMATION OF THE POSE OF AUTONOMOUS VEHICLES

Wilson O. Achicanoy M.
Department of Mechanical Engineering
Universidad de los Andes
Bogotá
Colombia
Email: wo.achicanoy115@uniandes.edu.co

Carlos F. Rodríguez H.
Department of Mechanical Engineering
Universidad de los Andes
Bogotá
Colombia
Email: crodrigu@uniandes.edu.co

ABSTRACT

Uncertainty fusion techniques based on Kalman filterin are commonly used to provide a better estimation of the state of a system. A comparison between three different methods to combine the sensor information in order to improve the estimation of the pose of an autonomous vehicle is presented. Two sensors and their uncertainty models are used to measure the observables states of a process: a Global Positioning System (GPS) and an accelerometer. Given that GPS has low sampling rate and the uncertainty of the position, calculated by double integration from the accelerometer signal, increases with time, fi st a resetting of the estimator based on accelerometer by the GPS measurement is done. Next, a second method makes the fusion of both sensor uncertainties to calculate the estimation. Finally, a double estimation is done, one for each sensor, and a estimated state is calculated joining the individual estimations. These methods are explained by a case study of a guided bomb.

INTRODUCTION

Methods for sensor fusion and optimal state estimation have been proved last years [1, 2]. Research work has been principally directed to improve Kalman estimators at different scenarios, including navigation; Kalman variations as the Extended Kalman Filter (*EKF*) and the Unscented Kalman Filter (*UKF*) have contributed to improvement of navigation systems, as is found in [3, 4].

Navigation has been enhanced by sensor technology, and estimators have reached more accuracy using more than one sensor to measure the observable states of a process. But, uncertainty is always present in measurements and state estimation is conditioned to measurement covariance accuracy, it means that uncertainty changes have impact on the estimation performance. On the other hand, multi-sensor configuration produce cross-correlation between sensors and sensor fusion robustness depends on process and measurement covariance. In [5] is proved that process covariance changes the results of the fusion methods: The less process covariance, the less cross-covariance is generated.

In this work sensor fusion techniques based on Kalman estimation are proved and compared. The mean assumption is that the covariances of the sensor model can be reset to initial conditions along the trajectory. GPS and accelerometer covariances can be modeled by functions that depend on variables as the state and the control input, therefore Kalman estimator can use these variables to improve the covariance at any given times. Based on two sensor fusion methods (Measurement Fusion and Track to Track [1]) and *EKF* estimator, three methods are presented and proved. To show these methods a case study of a guided bomb with GPS and accelerometer is presented.

The paper is organized as follows: First, the description of problem is done. Next, the case study is introduced (including the system model, the measurement model, the *EKF* estimator, and the simulation conditions). After that, the three alternatives of sensor fusion are simulated. Finally, the conclusion and future work are presented.

PROBLEM STATEMENT

In recent years, optimal state estimation [2], based on Kalman filte [6] and sensor fusion methods [1], has been developed. In general, Kalman filte performs state estimation based on the system dynamics and a set of measurements. A multisensor measurement system use many different sensors to acquire the observable states of a process and can be modeled by the equation:

$$\begin{aligned} \mathbf{y}_i &= h_i(\mathbf{x}, \mathbf{u}, \nu_i, t) \\ i &= 1, \ldots, N \end{aligned} \tag{1}$$

where $\mathbf{y}_i \in R^r$ is the ith measurement vector ($r \leq n$), $\mathbf{x} \in R^n$ is the state vector, $\mathbf{u} \in R^m$ is the input vector (control), $\nu_i \in R^n$ is the ith measurement noise vector, t is the time, N is the number of sensors, and h_i is a linear or non-linear function. It is assumed that each sensor is assigned to unique independent measurement equation.

For Kalman estimation is common to suppose ν_i as a Gaussian zero-mean white process with constant covariance matrix $\mathbf{R}_i$. Given that several factors may affect the measurements, $\mathbf{R}_i$ is not constant along the process. Consequently, measurement uncertainty changes and state estimation too. For instance, vehicles with inertial navigation systems undergo measurement covariance changes, due to internal sensor properties as bias change, limited bandwidth, non-orthogonal axis, etc. Additionally, external factors as ambient temperature variations, or inappropriate attachment of the sensor to the vehicle, may produce covariance alterations. Measurement uncertainty is difficul to be represented, but variation of measurement covariance along a trajectory can be approximated by means of tests or laboratory experiments.

CASE STUDY

To estimate the pose of a guided bomb, the description of the problem can be separated in three principal sections: the system model, the measurement system, and the estimator. In this work, a simple system model of a guided bomb assumes that: (i) a guided bomb is a rigid body with the mass concentrated at a point (center of mass - *cm*), (ii) the motion of the *cm* is constrained in a vertical plane XZ, (iii) the aerodynamic and gravitational forces acting on the bomb are applied at the *cm*, and (iv) a control of attitude (path angle) is possible. Additionally, the measurement system is based on GPS and accelerometer, and an *EKF* filte is used as estimator. In this section the conditions for simulation are described too.

System Model

The guided bomb's equations of motion in terms of an inertial coordinate frame (fla earth due to the short range of conventional non-thrusted bombs) can be written as [7, 8]:

$$\begin{aligned} \frac{dx}{dt} &= V\cos\gamma \\ \frac{dz}{dt} &= V\sin\gamma \\ \frac{dV}{dt} &= -\frac{1}{m}\left(D_0 + u^2 D_i\right) - g\sin\gamma \\ \frac{d\gamma}{dt} &= \frac{g}{V}(u - \cos\gamma) \end{aligned} \tag{2}$$

where x is the longitudinal range, z is the altitude, V is the velocity, γ is the path angle (between vector V and axis x), m is the constant mass, and g is the gravitational acceleration. Additionally, the drag force D has been modeled as a parabolic form $D = qSC_{D0} + (KL^2)/(qS)$; where $q = (1/2)\rho V^2$ is the dynamic pressure, ρ is the atmospheric density, S is the reference area (mean value of the cross sectional area of the guided bomb), C_{D0} is the zero-lift drag coefficient $K = dC_D/dC_L^2$ is the change of drag due to lift, C_D is the drag coefficient C_L is the lift coeffi cient, and L is the lift force.

For control design, the lift force L is subjected to the constraint $L \leq (mg)\,u_{max}$, where u_{max} is the limit of the load factor that weights the required lift to control the guided bomb and depends on the structural limits, the control surface actuators, and the stability of the autopilot [7]. Assuming this constraint, the load factor u is given by $u = L/mg = (qSC_L)/(mg)$. Then, the zero-lift drag D_0 and the lift-induced drag D_i in Eqn. (2) are, respectively:

$$D_0 = qSC_{D0} \tag{3}$$

$$D_i = \frac{K(mg)^2}{qS} \tag{4}$$

Measurement Model

Measurement equations for a GPS and an accelerometer are developed. Both, range (coordinate x) and altitude (coordinate z) of the guided bomb in the vertical plane XZ are measured by the GPS and computed from the two-axis accelerometer. The coordinates of the bomb are directly obtained from the GPS (assuming that a coordinate transformation from geographic coordinates $(longitude, latitude, altitude)$ to local Cartesian coordinates (x, z) has been previously done) whereas double integration is applied on accelerometer measurements. Zero-bias error is initially assumed.

The measurement process $h_i(\mathbf{x}, \mathbf{u}, \nu_i, t)$ is considered as linear for each sensor, then, the accelerometer and the GPS measurements, y_A and y_{GPS}, respectively, are:

$$\mathbf{y}_A = C\mathbf{x} + v_A$$
$$\mathbf{y}_{GPS} = C\mathbf{x} + v_{GPS} \quad (5)$$

where $\mathbf{x}$ is the real state, $C = \begin{bmatrix} 1 & 0 & 0 & 0 \\ 0 & 1 & 0 & 0 \end{bmatrix}$ is the measurement matrix, and v_A and v_{GPS} are Gaussian zero-mean white noise. The covariance matrices for each noise are:

$$R_A = \begin{bmatrix} \sigma_{Ax} & 0 \\ 0 & \sigma_{Az} \end{bmatrix}$$
$$R_{GPS} = \begin{bmatrix} \sigma_{GPSx} & 0 \\ 0 & \sigma_{GPSz} \end{bmatrix} \quad (6)$$

Taking into account calibration process of two commercial sensors (accelerometer ADXL330 [9] and GPS VBOX [10]), two measurement covariance models have been proposed with: (i) a parabolic variance σ_A for the accelerometer axis that depends on distance, and (ii) a constant variance σ_{GPS} for the GPS axis. These variances can be written as:

$$\sigma_{Ap} = c_0 + c_1 p + c_2 p^2$$
$$\sigma_{GPSp} = c \quad (7)$$

where p is the distance x or z for each axis, c_i, $i = \{0, 1, 2\}$, is a constant coefficien calculated in accelerometer calibration tests, and c is a constant determined by the GPS accuracy. Additionally, independence between axes and equal axis variance have been considered, i.e. $\sigma_{Axz} = \sigma_{Azx} = \sigma_{GPSxz} = \sigma_{GPSzx} = 0$, $\sigma_{Ax} = \sigma_{Az}$, and $\sigma_{GPSx} = \sigma_{GPSz}$. Equations (6) and (7) represent non-correlation between the two sensors.

EKF Estimator

EKF [2] is a variation of the linear Kalman filte [6] for a non-linear system:

$$\dot{\mathbf{x}} = f(\mathbf{x}, \mathbf{u}, \omega, t) \quad (8)$$

where ω is a Gaussian zero-mean white process with covariance matrix $\mathbf{Q}$. *EKF* uses a firs order Taylor series to approximate the non-linear system to a linear one, evaluated at the current estimated state $\hat{\mathbf{x}}$. Kalman filte algorithm can be divided in two stages [2]: (i) Prediction and (ii) Correction. The former calculates the evolution of the estimated state $\hat{\mathbf{x}}$ and the estimation error covariance $\mathbf{P}$ by means of the non-linear process and the linearization matrices calculated from the Taylor approximation. The correction stage updates the above parameters with the Kalman gain $\mathbf{K}$, the measurements, and the covariances of the system and the measurements.

Simulation Conditions

Given that the air density depends on the altitude z, a general model of the ambient atmosphere density could be included into the model of the guided bomb. ρ from $z = 0$ m up to $z = 11000$ m is assumed as [7]:

$$\rho = \frac{P_a}{RT_a} \quad (9)$$

where $P_a = P_0(1 - 2.255692257 \times 10^{-5} z)^{5.2561}$ is the ambient atmospheric pressure (P_0 is the pressure at sea-level), $T_a = T_0 - 0.006499708z$ is the ambient atmospheric temperature (T_0 is the temperature at sea-level), and R is the air gas constant.

A ballistic trajectory for a guided bomb is simulated with initial conditions given by the initial state vector $\mathbf{x}_0 = [0, 6000, 300, -10\frac{\pi}{180}]^T$ and the initial estimation error covariance $\mathbf{P}_0 = I_4$ (identity matrix of size four). The system covariance $\mathbf{Q}$ is assumed equal to zero through the trajectory and the input $u = 0.5\sin[2\pi(0.001)t]$ is an arbitrary function that guarantees fligh stability. Additionally, all noise process for each *EKF* variation (to be compared each other) are simulated using the same random numbers. Simulations are done in Matlab® with the parameters listed in Tab. 1.

SENSOR FUSION METHODS

According to Eqn. (6), GPS measurements are more reliable, in time intervals where disturbances do not affect the signal receiver, than accelerometer measurements. On the other hand, in a time interval, there are more accelerometer measurements than GPS measurements, due to the fast accelerometer sampling frequency. Then, an appropriate design of the estimator has to combine these two advantages.

In this work, and based on techniques as *EKF* estimator, Measurement Fusion, and Track to Track [1, 11, 12], the following solution is proposed: The covariance matrix $\mathbf{R}_i(\mathbf{x})$ is reset or updated to a minimum value at GPS sampling times, by a best measurement from the GPS or with the estimated state $\hat{\mathbf{x}}$ given by a sensor fusion method.

INDEPENDENT *EKF* ESTIMATORS

To establish a base line solution, estimators are independently computed: one estimator based on accelerometer EKF_A

TABLE 1. SIMULATION PARAMETERS.

Parameter	Quantity	Value	Units
g	Gravitational acceleration	9.8	m/s^2
m	Mass	218	Kg
D	Guided bomb's diameter	0.275	m
S	Guided bomb's reference area	$2\pi D^2/4$	m^2
C_{D0}	Zero-lift drag coefficien	0.02	
K	Change of drag with the lift	0.5	
t_f	Total fligh time	30	s
d_t	Integration time step	0.01	s
f_A	Accelerometer sampling rate	10	Hz
F	GPS/Accelerometer sampling factor	$1/5$	
f_{GPS}	GPS sampling rate	Ff_A	Hz
$[C_0, C_1, C_2]$	Accelerometer variance coefficient	$[10, 5, 2]$	m
σ_{GPSp}	GPS variance in x or y coordinate	5	m
P_0	Atmospheric pressure at sea-level	101314.6	N/m^2
T_0	Atmospheric temperature at sea-level	288.1667	K
R	Gas constant (air)	286.9924	$(N.m)/(Kg.K)$

and other estimator based on GPS EKF_{GPS}. Also, two disturbances for the GPS are considered in a fi ed time interval: The absence of measurements due to receiver failures and an exaggerated change of its precision (this analysis does not take into account effects of the disturbance on the accelerometer).

Fig. 1 shows the ballistic trajectory error ε, i.e. the Euclidean distance between the real trajectory and the estimated trajectory. ε_1 is the error for EKF_A and ε_2 is the error for EKF_{GPS}. Both estimators are working without disturbance. Since accelerometer variance is an increasing function that depends on distance and GPS variance is constant, EKF_A is less accurate than EKF_A. ε_3 is the error when the disturbance deletes GPS data from $t = 5.0\ s$ to $t = 15.0\ s$ (absence of measurements) and ε_4 is the error when the disturbance causes exaggerated GPS variance (it is assumed that $\sigma_{GPSx} = \sigma_{GPSz}$ changes from $5.0\ m$ to $50.0\ m$). In last scenarios, disturbance leads to wrong estimated states.

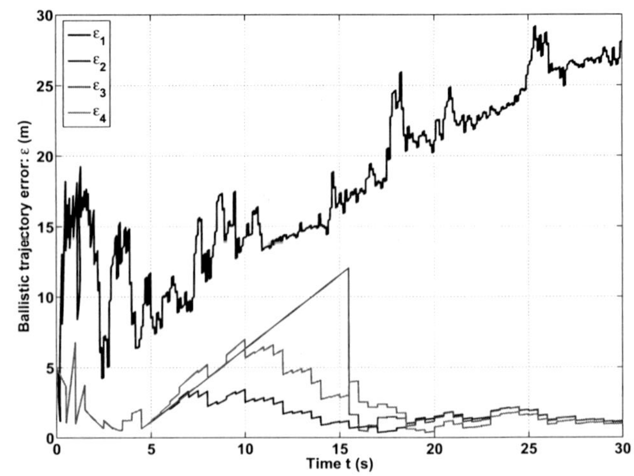

FIGURE 1. Ballistic trajectory error ε for EKF estimators. ε_1 and ε_2 are the errors (without disturbance) for EKF_A and EKF_{GPS}, respectively. ε_3 and ε_4 are the errors for EKF_{GPS} with disturbance (absence of measurements and exaggerated variance change, respectively). The disturbance is applied at time interval $[5.0, 15.0]\ s$.

Updated EKF_A

This solution proposes the resetting of the EKF_A by the GPS measurements. Update means that the GPS measurements are used to correct the estimated state of the EKF_A and the accelerometer variance is reset with the estimated state $\hat{\mathbf{x}}$. This is done at GPS sampling time. Fig. 2(a) shows the error ε without disturbance. ε_1 and ε_2 are the errors for EKF_A and EKF_{GPS}, respectively, and ε_3 is the error for the Updated EKF_A. $\varepsilon_3 < \varepsilon_1$ for almost all the time. But it does not guarantee $\varepsilon_3 < \varepsilon_2$, because EKF_A algorithm does not process information of the GPS variance and the Kalman gain $\mathbf{K}$ does not appropriately weight the GPS measurements. Updated EKF_A is better than the EKF_A, but is not better than the EKF_{GPS}.

Figure 2(b) shows the errors when absence of GPS measurements at the time interval $[15.0, 20.0]\ s$. ε_1 is the error for the EKF_{GPS} without disturbance, ε_2 is the error for the EKF_{GPS} with disturbance, and ε_3 is the error for the Updated EKF_A with disturbance too. Actually, the Update EKF_A is not robust against disturbance. Although the error is reduced through the disturbance time, it increases due to bias in the accelerometer variance. After the disturbance the resetting can not remove this bias and ε_3 is linear and smooth.

Figure 2(c) shows the errors when the disturbance produces exaggerated measurements in the same time interval. ε_1 is the error for the EKF_{GPS} without disturbance, ε_2 is the error for the EKF_{GPS} with disturbance, and ε_3 is the error for the Updated EKF_A with disturbance too. It is assumed that GPS variance changes from $\sigma_{GPSx} = \sigma_{GPSz} = 5.0\ m$ to $\sigma_{GPSx} = \sigma_{GPSz} = 50.0\ m$. Despite GPS measurements are widely spread, Updated EKF_A has some robustness and can recover a good estimation after the disturbance time. This results indicates that exaggerated measurements are better than their absence, and is suitable to keep update process even with the worst GPS measurements.

EKF_{UMF}

This process of uncertainty fusion, based on Measurement Fusion [1], builds an augmented vector of accelerometer and GPS measurements $\mathbf{y}_a = \begin{bmatrix}\mathbf{y}_A\ \mathbf{y}_{GPS}\end{bmatrix}^T$ and an augmented covariance matrix $\mathbf{R}_a = \begin{bmatrix}\mathbf{R}_A & 0 \\ 0 & \mathbf{R}_{GPS}\end{bmatrix}$ that are used to update the predicted state $\hat{\mathbf{x}}$ (into the Correction stage of the EKF algorithm).

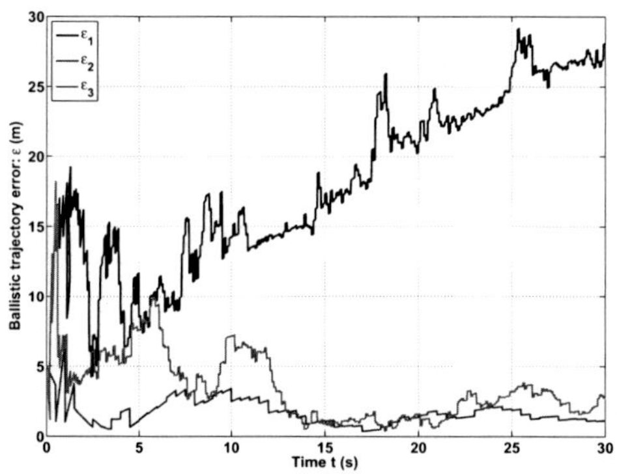

(a) ε_1 is the error for the EKF_A, ε_2 is the error for the EKF_{GPS}, and ε_3 is the error for the Updated EKF_A.

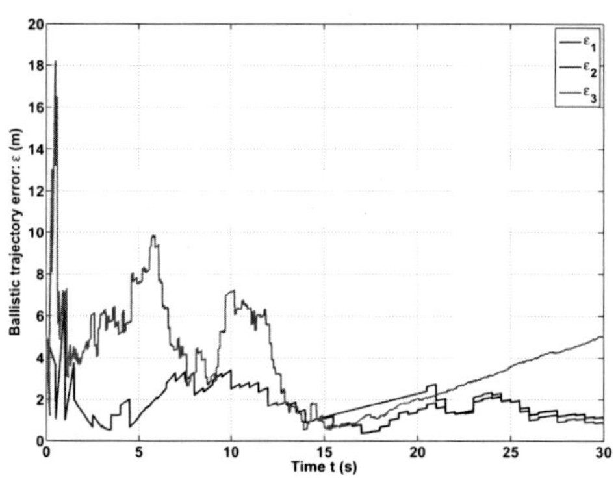

(b) ε_1 is the error for the EKF_{GPS} without disturbance, ε_2 is the error for the EKF_{GPS} with disturbance, and ε_3 is the error for the Updated EKF_A with disturbance.

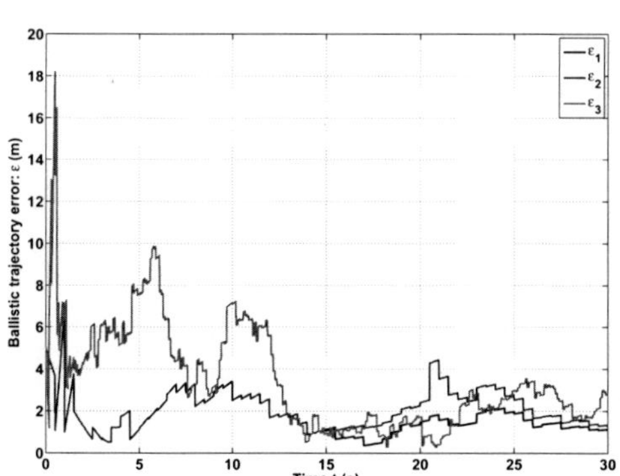

(c) ε_1 is the error for the EKF_{GPS} without disturbance, ε_2 is the error for the EKF_{GPS} with disturbance, and ε_3 is the error for the Updated EKF_A with disturbance.

FIGURE 2. Ballistic trajectory error ε for the Updated EKF_A. (a) Without disturbance. (b) Absence of GPS measurements. (c) Exaggerated GPS variance change. The disturbance occurs at the time interval $[15.0, 20.0]$ s.

To achieve accuracy, the accelerometer is updated with the estimated state $\hat{\mathbf{x}}$ generated at GPS sampling time, thus only one EKF is run.

Fig. 3(a) shows the errors ε without disturbance. ε_1 is the error for the EKF_A, ε_2 is the error for the EKF_{GPS}, and ε_3 is the error for the EKF_{UMF}. $\varepsilon_3 < \varepsilon_1$ for almost all the time and $\varepsilon_3 < \varepsilon_2$ in some time intervals. EKF_{UMF} attains good estimation, specially at early time, where Kalman estimation is still searching for an appropriate estimation error covariance $\mathbf{P}$ and appropriate Kalman gain $\mathbf{K}$. At the end of the simulation ε_3 converges to EKF_{GPS} error ε.

Figure 3(b) shows the errors when absence of GPS measurements happen at time interval $[15.0, 20.0]$ s. ε_1 is the error for the EKF_{GPS} without disturbance, ε_2 is the error for the EKF_{GPS} with disturbance, and ε_3 is the error for the EKF_{UMF} with disturbance too. EKF_{UMF} is robust and maintains a low error ε.

Similarly, Fig. 3(c) shows the errors when the disturbance produces exaggerated measurements in the same disturbance time. ε_1 is the error for the EKF_{GPS} without disturbance, ε_2 is the error for the EKF_{GPS} with disturbance, and ε_3 is the error for the EKF_{UMF} with disturbance too. EKF_{UMF} is practically not affected by the disturbance and it is highly robust. This fusion works appropriately, because EKF estimator takes into account both measurements and covariances of each sensor. Also, the accelerometer covariance resetting improves the estimation.

EKF_{UTT}

Uncertainty fusion based on Track to Track [1] runs a EKF estimator for each sensor, but makes Prediction with the same estimated state and Correction with Least Square LQ estimation. This means that the estimated states of both EKF estimators are joined in one suboptimal estimator. The Millman's formula [12] is used to estimate a unique estimated state $\hat{\mathbf{x}}$ and the accelerometer covariance is reset with the suboptimal estimated state at GPS sampling time. Two EKF estimators are run.

Fig. 4(a) shows the errors ε without disturbance. ε_1 is the error for the EKF_A, ε_2 is the error for the EKF_{GPS}, and ε_3 is the error for the EKF_{UTT}. Since the fina state is estimated with LQ (weighted mean), ε_3 is between ε_1 and ε_2 for almost all the time. The error ε_3 remains around maximum GPS covariance value since accelerometer covariance is an increasing function and GPS covariance is constant.

Figure 4(b) shows the errors when absence of GPS measurements happen at time interval $[15.0, 20.0]$ s. ε_1 is the error for the EKF_{GPS} without disturbance, ε_2 is the error for the EKF_{GPS} with disturbance, and ε_3 is the error for the EKF_{UTT} with disturbance too. EKF_{UTT} is robust and maintains a constant value through the disturbance time. The estimation is not better than the EKF_{GPS} estimation.

Fig. 4(c) shows the errors when the disturbance produces exaggerated measurements in the same disturbance time. ε_1 is the

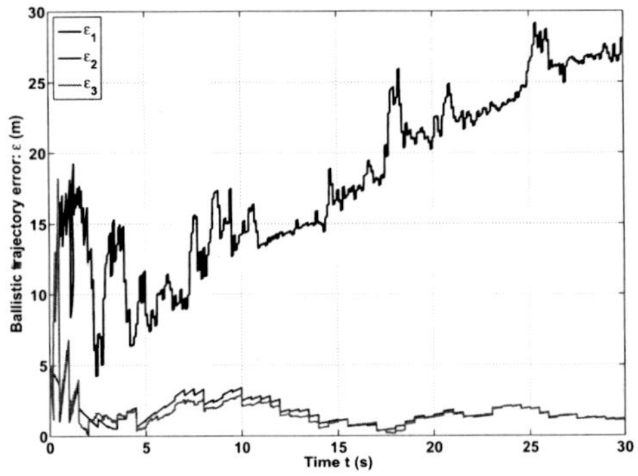

(a) ε_1 is the error for the EKF_A, ε_2 is the error for the EKF_{GPS}, and ε_3 is the error for the EKF_{UMF}.

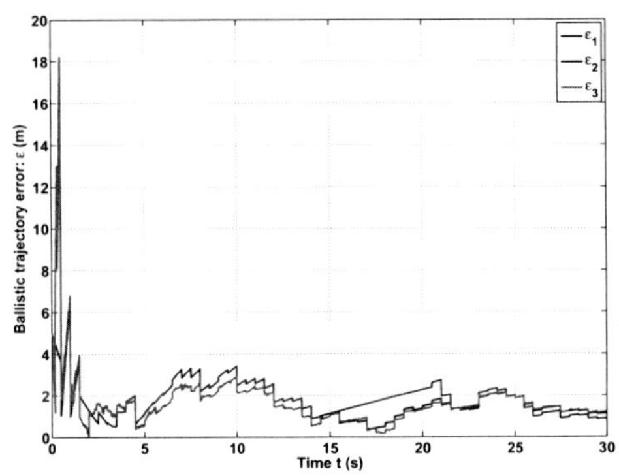

(b) ε_1 is the error for the EKF_{GPS} without disturbance, ε_2 is the error for the EKF_{GPS} with disturbance, and ε_3 is the error for the EKF_{UMF} with disturbance.

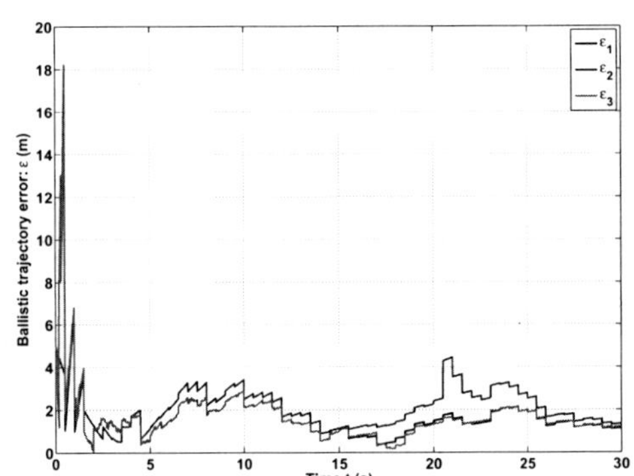

(c) ε_1 is the error for the EKF_{GPS} without disturbance, ε_2 is the error for the EKF_{GPS} with disturbance, and ε_3 is the error for the EKF_{UMF} with disturbance.

FIGURE 3. Ballistic trajectory error ε for the EKF_{UMF}. (a) Without disturbance. (b) Absence of GPS measurements. (c) Exaggerated GPS variance change. The disturbance occurs at the time interval $[15.0, 20.0]$ s.

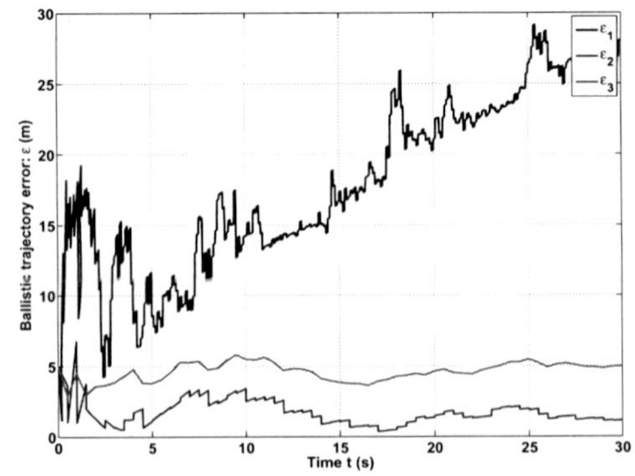

(a) ε_1 is the error for the EKF_A, ε_2 is the error for the EKF_{GPS}, and ε_3 is the error for the EKF_{UTT}.

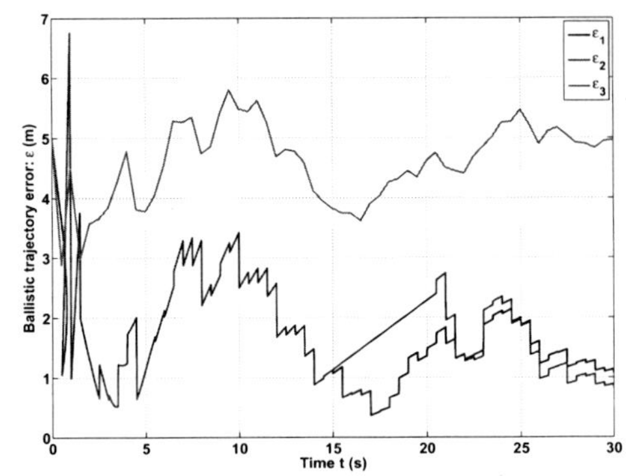

(b) ε_1 is the error for the EKF_{GPS} without disturbance, ε_2 is the error for the EKF_{GPS} with disturbance, and ε_3 is the error for the EKF_{UTT} with disturbance.

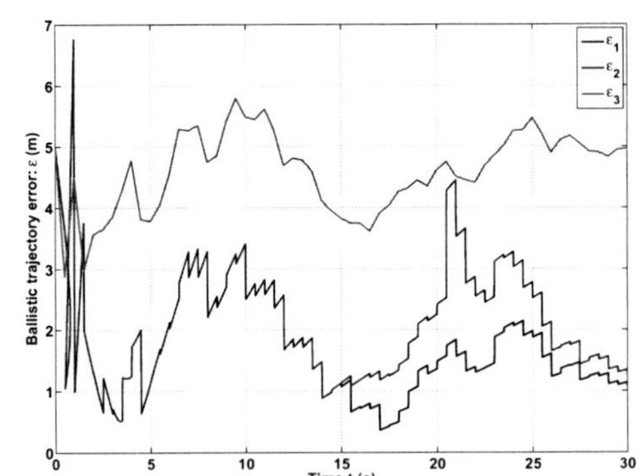

(c) ε_1 is the error for the EKF_{GPS} without disturbance, ε_2 is the error for the EKF_{GPS} with disturbance, and ε_3 is the error for the EKF_{UTT} with disturbance.

FIGURE 4. Ballistic trajectory error ε for the EKF_{UTT}. (a) Without disturbance. (b) Absence GPS measurements. (c) Exaggerated GPS variance change. The disturbance occur at the time interval $[15.0, 20.0]$ s.

error for the EKF_{GPS} without disturbance, ε_2 is the error for the EKF_{GPS} with disturbance, and ε_3 is the error for the EKF_{UMF} with disturbance too. As is expected by Track to Track fusion method, fina estimation is robust against disturbance. The estimation is not better than EKF_{GPS} estimation.

CONCLUSION AND FUTURE WORK

Sensor fusion methods have been proved by simulation. The results show that EKF_{UMF} has the minimum error ε. Additionally, the updated accelerometer covariance, that has been reset by the estimated state, improves the results and the robustness of this estimator.

EKF_{UTT} shows the best robustness and updated accelerometer covariance has not meaningful effects on the error ε reduction. If functions related to sensor's uncertainty has minimums and maximums along the trajectory, as a positive sinusoid, EKF_{UTT} could improve the estimation, because the estimator will follow the minimum sensor covariance (robustness against sensor failures).

To achieve optimal state estimation, with multi-sensor fusion, uncertainties of accelerometer and GPS measurements has to be included in the EKF estimator. On the other hand, it is important to keep all the measurements, even if they are critically affected by a disturbance. In navigation, lack of GPS measurements are common along the vehicle trajectory, and a method to interpolate the estimation in those time intervals has to be implemented. Then, Update EKF_A can be used as a parallel estimator to fil the process.

Future work is addressed to the development of an optimal control that uses these EKF estimators. Analysis can be done with other variance functions and cross-covariances, generated by correlation between sensors, can be included too. Covariance constrains added to the solution of the optimal control problem must be considered for future simulations.

ACKNOWLEDGMENT

This work is supported by Industria Militar de Colombia - INDUMIL and the Universidad de los Andes.

References

[1] Mitchell, H. B., 2007. *Multi-sensor data fusion: an introduction*. Springer Verlag.

[2] Simon, D., 2006. *Optimal State Estimation: Kalman, H Infinit , and Nonlinear Approaches*. Wiley & Sons.

[3] St-Pierre, M., and Gingras, D., 2004. "Comparison between the unscented kalman filte and the extended kalman filte for the position estimation module of an integrated navigation information system". In Intelligent Vehicles Symposium, 2004 IEEE, pp. 831 – 835.

[4] Tang, K., Wu, M. M., and Hu, X., 2007. "Multiple model kalman filterin for mems-imu/gps integrated navigation". In Industrial Electronics and Applications, 2007. ICIEA 2007. 2nd IEEE Conference on, pp. 2062 –2066.

[5] Zhao, J., Xu, T., Ren, L., and Cui, Z., 2004. "Effect of the common process noise on performance of two-sensor fused-track". In Radar Conference, 2004. Proceedings of the IEEE, pp. 225 – 229.

[6] Kalman, R. E. "A new approach to linear filterin and prediction problems". *Transactions of the ASME – Journal of Basic Engineering,* ***82***.

[7] Siouris, G. M., 2006. *Missile Guidance and Control Systems*. Springer, NY.

[8] Geng, L., and Zheng, Z., 2007. "Application of maximum principle with fina state constraints in gliding control for guided bombs". In Automation and Logistics, 2007 IEEE International Conference on, pp. 205 – 209.

[9] Devices, A., 2006. *ADXL330 Datasheet*. See URL `http://www.analog.com`.

[10] Racelogic, 2009. *RLVB3iR2G2 Datasheet*. See URL `http://www.racelogic.co.uk`.

[11] Zhou, W., Zhao, C., and Guo, J., 2009. "The study of improving kalman filter family for nonlinear slam". *Journal of Intelligent and Robotic Systems,* ***56***(5), dec, pp. 543 – 564.

[12] Shin, V., Lee, Y., and Choi, T., 2006. "Generalized millman's formula and its application for estimation problems". *Signal Processing,* ***86***(2), pp. 257 – 266.

AUTHOR INDEX

Proceedings of the ASME 2010 International Mechanical Engineering Congress and Exposition Volume 1 – Advances in Aerospace Technology